MODERN SUPERABSORBENT POLYMER TECHNOLOGY

MODERN SUPERABSORBENT POLYMER TECHNOLOGY

Edited by

Fredric L. Buchholz

Andrew T. Graham

WILEY-VCH

New York / Chichester / Weinheim / Brisbane / Singapore / Toronto

This book is printed on acid-free paper. ♾

Published simultaneously in Canada.

Library of Congress Cataloging in Publication Data:

Modern superabsorbent polymer technology / edited by Fredric L.
Buchholz, Andrew T. Graham.
p. cm.
Includes index.
ISBN 0-471-19411-5 (cloth : alk. paper)
1. Acrylic resins. 2. Sorbents. I. Buchholz, Fredric L., 1952– .
II. Graham, Andrew T., 1945– .
TP1180.A35M63 1997
668.4′232—dc21

Printed in the United States of America

10 9 8 7 6 5 4 3 2 1

CONTENTS

PREFACE

The essential idea of using a superabsorbent polymer as a liquid storage medium in personal care products is a relatively old idea, dating back to 1966. Successful commercialization of superabsorbent polymers, however, occurred in about 1980. As a result of commercialization of superabsorbent polymer in disposable infant diapers, the technical literature describing the synthesis and properties of superabsorbent materials has exploded in the past 15 years. The literature includes dozens of patents in addition to journal articles and conference proceedings. It has been a daunting task to stay abreast of the developments in this field and to sort through the myriad of synthetic chemistry and analytical techniques to better understand the new technology as it develops.

In view of this problem, it seems appropriate to find a common context of technology and science in which to place the inventions and discoveries so that we may achieve a better understanding of what superabsorbent polymers are, how they are made, and why they are relevant to modern life. In this volume, we hope to provide such a context. We provide a rational thread based in polymer science that will stitch together the various technologies into a unified whole. This rational thread is that the material properties of superabsorbent polymers define their usefulness in personal care items and in the other applications and that these properties result directly from the synthetic techniques used to make the polymers. We hope to show how these basic ideas may be linked into a useful model of superabsorbent polymers.

In this volume, we focus primarily on descriptive chemistry and physics and on industrial practices to show the practical side of these materials. To a large extent, we leave derivations of important mathematical relationships as exercises for students and other interested parties, but use the results to show the interrelationships between the structures, properties, and uses of superabsorbent polymers.

In Chapter 1, the concepts of absorbency and superabsorbency are defined, compared, and contrasted. Even though the mathematics used to describe absorbency and superabsorbency are very similar, the underlying mechanisms of the processes are entirely different. An understanding of these mechanisms and how they coexist in modern absorbent products is vital to new product development.

In Chapter 2, the synthetic chemistry of superabsorbent polymer is described in detail. Superabsorbent polymers are crosslinked polymer networks, but real polymer networks are vastly different from the idealized "three-dimensional fisherman's net" that is easy to imagine. An understanding of how the conditions of synthesis affect the structure helps us create improved superabsorbent polymer. Preparative examples are given, and the effects of impurities are described. The polymerization kinetics are discussed in detail, as they apply specifically to superabsorbents. Some new methods to measure and study the kinetics of the polymerization are described and discussed.

In Chapter 3, the process chemistry is described. It may appear that there are as many processes for making superabsorbent polymers as there are manufacturers. However, there are many similarities between the technologies because the principal application is in personal care, and the dominant chemistry is the free-radical polymerization of acrylic acid. The similarities and differences between the major classes of processes, solution and suspension polymerization, and those between the individual manufacturers are detailed. The processes are broken into the component unit operations, and the effects of these operations on the polymer properties are detailed.

In Chapter 4, the analytical chemistry of superabsorbent polymers and the raw materials are surveyed. Numerous techniques for evaluation of the properties and behavior of superabsorbent polymer have been developed. We describe different methods and show how the methods relate to the properties and polymer science of superabsorbent polymer. Methods include spectroscopic methods used to probe the molecular structure of the polymers and methods used to qualitatively and quantitatively measure the absorbency and mechanical properties. A number of these methods are critically reviewed and compared according to their foundations in polymer science.

In Chapter 5, the structure and property relationships of crosslinked superabsorbent polymer are described. The relationship between the structure of the polymer network, the conditions of synthesis, and the properties, such as the swelling capacity and the elastic modulus of the gel, are developed, with a focus on the principal role of superabsorbent polymer as an absorbent in disposable goods. A model of the absorbency under load is developed, and the dependence of this measurement on the numerous parameters of the polymer system is described.

In Chapter 6, new superabsorbent polymers and forms are surveyed. As the role of superabsorbent polymer in absorbent cores expands, new materials are being developed to meet the expanding needs. Absorbent fibers and foams are described and contrasted with the granular polymers that are widely available at the present time. The potential drawbacks and benefits of biodegradable superabsorbent polymers are also discussed.

In Chapter 7, applications of superabsorbent polymers are described. Although the bulk of the research and development effort has been placed on disposable absorbent products, numerous other applications have been developed, and some of them have achieved commercial success. A detailed look is taken at the use of superabsorbent polymers in personal care products, and some of the interesting, emerging applications are described, including horticultural uses, controlled release, and construction materials.

In the end, we hope you understand superabsorbent polymers and their applications better. As a result of creating this book, we know that we do.

FREDRIC L. BUCHHOLZ
ANDREW T. GRAHAM

Midland, Michigan

ACKNOWLEDGMENTS

Few works of this kind are individual efforts, and this book about superabsorbent polymers is certainly no exception. Of course we wish to thank the principal contributors of the chapters. Tom Staples has been involved with superabsorbent polymer research at The Dow Chemical Co. since the early 1980s and was instrumental in developing our first manufacturing process. David Henton, in contrast, spent much less time in the superabsorbent polymer R&D group, but left us with some valuable new techniques for analysis. Larry Wilson divides his time between synthesis and physical chemistry of superabsorbent polymer, which is an effective combination of interests. Sergio Cutié, Pat Smith, and Bob Reim spend countless hours each year solving our most difficult problems in analytical chemistry of superabsorbent polymer. David Allan helps us take our thinking and products in new and different directions. This book would not have been written without their efforts.

Many of our other colleagues have contributed important technical data and concepts to our R&D effort and to this book. Thomas Johnson, Cynthia Powell, and Donna Shaffer provided data that appear in Chapter 2. Thomas Ciszewski, Mike Fialkowski, and Ken Salo were very helpful in creating Chapter 3. Many colleagues in the Analytical Sciences department at Dow have helped collect data and develop techniques that underpin Chapter 4. Notable among these colleagues are Selso Martinez, Andrew Pasztor, Marianne McKelvy, David Frurip, Linda Peerey, John Russel, and Benjamin Esquivel. Sandra Pesce collected all the data for the figures in Chapter 5. The following individuals have been helpful in preparing Chapter 6: Josef Burgert, Herbert Gartner, David Henton, Steve Mork, Cindy Powell, Robert Sandon, James Steinbrunner, Robert Terbrueggen, Robert Thomas, and Joseph Weir. Fran Voci helped with searches of the literature, and Mr. Johnson also provided some data that appear in Chapter 7. William Heeschen, David Williams, and Cary Black provided the photomicrographs that appear throughout the book. This book would have been empty without their collective efforts and those of all individuals who have helped us in many ways over the past dozen years.

Books written from an industrial perspective are never written without at least the consent of management. Our efforts have been blessed with enthusiastic support by Jerry DeVrieze, Peter Dulcamara, Jr., George Biltz, Ian Davenport, and Jerry Wasserman, who believe, as we do, that communication in the chemical industry is more than just technical reports and patent disclosures. This book would have remained only a dream without their encouragement.

Finishing a book always requires more time than expected at the outset. This book required many hours taken from time with family, and we wish to acknowledge the patience and support of our families. Pamela Buchholz and Suzie Graham convinced us that the job should be done well, even if family commitments were postponed, or sometimes even dropped. They were with us through the frustrations and joys of book writing and editing. This book would have been a less fulfilling experience without their support and encouragement.

A special acknowledgment, and our thanks, must go to Dr. Edmund H. Immergut. He brought the idea of writing this book to us, believed in the value of the effort, and helped secure the publisher.

CONTRIBUTORS

David S. Allan, Superabsorbent Products R&D, Specialty Chemicals Department, The Dow Chemical Company

Fredric L. Buchholz, Superabsorbent Products R&D, Specialty Chemicals Department, The Dow Chemical Company

Sergio S. Cutié, Analytical Sciences Laboratory, The Dow Chemical Company

Andrew T. Graham, Superabsorbent Products R&D, Specialty Chemicals Department, The Dow Chemical Company

David E. Henton, Plastics Department, The Dow Chemical Company

Robert E. Reim, Analytical Sciences Laboratory, The Dow Chemical Company

Patrick B. Smith, Analytical Sciences Laboratory, The Dow Chemical Company

Thomas L. Staples, Superabsorbent Products R&D, Specialty Chemicals Department, The Dow Chemical Company

Larry R. Wilson, Superabsorbent Products R&D, Specialty Chemicals Department, The Dow Chemical Company

MODERN SUPERABSORBENT POLYMER TECHNOLOGY

1

ABSORBENCY AND SUPERABSORBENCY

Fredric L. Buchholz

Superabsorbent polymers are fascinating substances. The most commonly available superabsorbent polymers are hard, dry, granular powders that look much like clean, white sand or granular table sugar. Other shapes, such as microspherical powders and fine fibers, are also known. Any similarity to sand or sugar is immediately forgotten when the superabsorbent polymer is put into contact with water or water solutions. At first, the slurry of water and the particles of the hard, dry granules seems unremarkable, but gradually the superabsorbent polymer absorbs the water, turning into a soft, rubbery gel. Pads of fluffed cellulose pulp fibers will absorb about 12 g of water or other aqueous solutions per gram of dry fiber,[1] whereas superabsorbent polymers will absorb up to 1,000 g of water per gram of polymer and up to about 100 g of dilute salt solution per gram of polymer. Unlike the behavior of a soaked pad of cotton or a sponge, the superabsorbent polymer gel will not release the water when squeezed with the fingers.

The advent of superabsorbent polymers is a classic case of the substitution of materials. The principal use of superabsorbent polymers is as an absorbent for water and aqueous solutions, primarily in diapers, adult incontinence products, and feminine hygiene applications. In these applications superabsorbent materials are inexorably replacing traditional absorbent materials such as cloth, cotton, paper wadding, and batts of cellulose fiber. The substitution of superabsorbent polymers for fibrous wood pulp in disposable sanitary products has taken place gradually over a time span of about 20 years and is probably not yet complete. To explain better this process of substitution, and superabsorbents generally, the essential characteristics of the tradition-

Modern Superabsorbent Polymer Technology. Edited by Fredric L. Buchholz and Andrew T. Graham. ISBN 0-471-19411-5.

al absorbents are described first, followed by a description of the behavior of superabsorbents so that a comparison can be made.

1.1. ABSORBENTS

1.1.1. Traditional Absorbent Materials

Traditional absorbent materials may be divided into two general classes: fiber masses and foams. Fiber masses may be subclassified into ordered structures such as woven cloth and into random structures such as cotton wadding, cellulose fiber batts, and paper. A few examples are shown in Figure 1.1. The substances that make up the absorbent structures vary significantly and include naturally occurring, plant-based cellulosics such as cotton and wood pulp, natural sponges, synthetically produced polyacrylics that may be spun into fibers, and polyurethanes that can be crafted into a foam. But these materials have several features in common that lead to their absorbency.

One general feature common to all traditional absorbent materials is the presence of open space or pores within the structure of the material. The apparent size of pores in absorbent materials varies enormously, but is certainly larger than the dimensions of most molecules. The mean size of the spaces ranges from 0.5 μm in some papers[2] to several hundred micrometers in bulk fiber masses.[3] In addition to the simple presence of open space, the absolute amount of open space and the size of the openings are important for high absorbency. The porosity or void volume fraction of the material is a measure of the absolute open volume available and is closely related to the bulk density of the material. The bulk density is the volume-weighted density of the material and depends on the relative amount and absolute density of the solid substance making up the absorbent material. The bulk density of materials may be controlled by the processing methods employed to make the absorbent material, such as compression. The water absorbency and bulk density of some common absorbents are shown in Table 1.1. These examples represent the variety of methods used to make absorbents. For example, fibers may be solvent-laid as in paper-making, air-laid as in diaper manufacture (wood pulp fluff), or woven as in cloth manufacture. Synthetic substances may also be foamed to vary the properties of the absorbent material and the economics of the manufacturing process.

Table 1.1. The water absorbency of some common absorbent materials

Absorbent Material	Bulk Density (g/cm^3)	Water Absorbency (g/g)
Whatman No. 3 filter paper	3.424	1.8
Facial tissue	0.166	4.0
Cellulose acetate sponge	0.048	7.8
Polyurethane sponge	0.040	10.5
Wood pulp fluff	0.030	12.0
Cotton ball	0.018	18.9

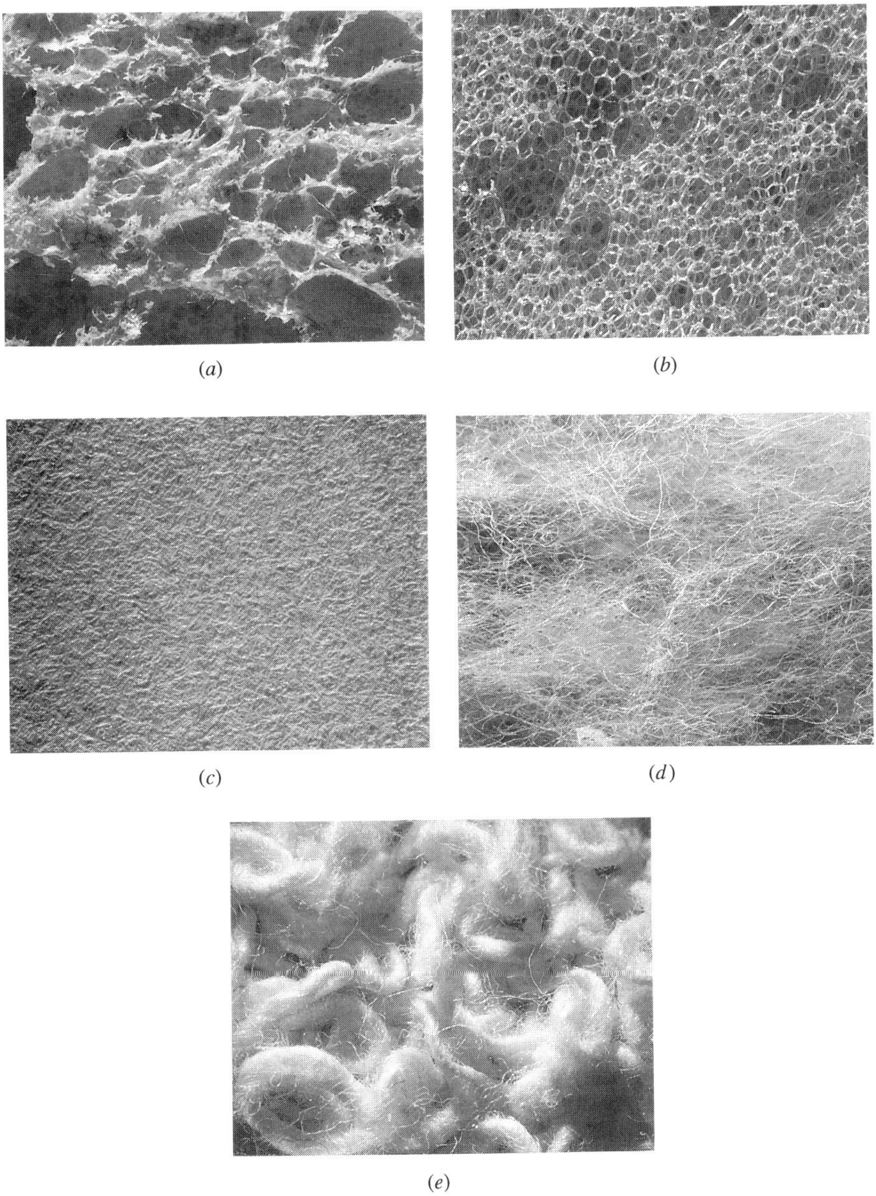

Figure 1.1. Photomicrographs of absorbent materials taken at ×10 magnification: **(a)** cellulose acetate sponge, **(b)** polyurethane sponge, **(c)** filter paper, **(d)** cotton ball, and **(e)** terry cloth.

1.1.2. Physics of Absorbent Materials

The open space in absorbents can act as a reservoir for fluid storage only if a means for transporting the fluid into the open space is available. When the open space forms a connected path throughout the material, it becomes possible to transport liquid into and through the material. The transport of liquid into and through a porous medium requires, additionally, that a force or pressure of some kind be applied to the liquid.

The principal factors affecting the flow through absorbent materials may be related by means of Darcy's law, which is a general, empirical relationship describing the flow through a porous medium. For steady, slow flow in one direction, Darcy's law is given by

$$\frac{dv}{dt} = -\frac{\kappa_D A \Delta P}{\mu L} \tag{1.1}$$

where the volumetric flow rate dv/dt is given in terms of the cross-sectional area A of the sample, the length L of the sample, the viscosity μ of the fluid, the specific permeability κ_D of the sample, and the pressure difference ΔP driving the flow process. The specific permeability is a measure of the effective connectedness of the porous structure, be it a mat of fiber or a slab of foam, and may be specified in terms of the void fraction and extent of connectedness of the absorbent material. Numerous models of absorbent and porous media have been constructed that connect the specific permeability to the structure.[4] The structural factors are usually limited to an effective diameter of the pore, based on the diameters of the fibers or particles that make up the absorbent medium, and the porosity of the medium.

Many different forces may act on the liquid to move it into and through the open spaces, including gravity, hydrostatic pressure, and capillary pressure. The latter is of great importance in absorbent materials because it is responsible for the wicking action of the absorbent medium that acts to initially transport the liquid into the dry absorbent material. Consider for a moment a simple, cylindrical capillary tube, containing a liquid that perfectly wets the walls of the capillary and air. The contact angle of the liquid with the wall is zero, and the meniscus is spherical, for which the radius of curvature is equal to the radius of the capillary tube. In this case, the capillary pressure p_c is given by

$$p_c = \rho g h = \frac{2\sigma}{R} \tag{1.2}$$

where ρ is the density of the liquid, h is the height of rise of the liquid, g is the acceleration of gravity, σ is the surface tension of the liquid, and R is the radius of the capillary. Therefore, the height or distance of rise in a capillary tube is greatest in tubes of small radius. However, in applications of absorbent materials, the rate of absorption is also critically important.

The rate of wicking of liquid in capillaries is stated by the Lucas–Washburn equation:

$$\frac{dh}{dt} = \frac{R^2}{8\mu h}\left[\frac{2\sigma\cos\theta}{r} - \rho g h\right] \tag{1.3}$$

In Eq. 1.3, dh/dt is the change in height h of liquid of viscosity μ versus time t (wicking rate) in a vertical tube of radius R under the influence of the capillary pressure and gravity. The capillary pressure is represented by the first term inside the brackets and is given by the surface tension σ of the liquid, the radius of curvature r of the meniscus of the air–liquid interface in the pore, and the cosine of the contact angle θ of the liquid with the substance forming the walls of the pore. The effect of gravity on the wicking rate in the pore (capillary) is given by the second term in the brackets and consists of the density ρ of the liquid, the gravitational acceleration g, and the height of the liquid column. Liquid will rise in the capillary until the interfacial tension is balanced by the weight of liquid in the column. The contact angle is a measure of the wettability of the solid surface by the liquid and ranges from 0° (perfect wetting) to 90° (nonwetting). Wicking is driven, therefore, by the spreading of a liquid of high interfacial tension (with air) along the surface of the solid absorbent material. The maximum height of wicking in a capillary is achieved with liquids that perfectly wet the solid surface of the walls of the pore or capillary and with small capillary radius (as in Eq. 1.2). Maximum rate of wicking requires a large capillary radius and a liquid with low viscosity. To complicate matters, most absorbent materials have extremely irregular pores, with many different radii, and with the radius changing size along the flow path. This irregular structure contributes to poor wicking in absorbents. Liquid also may be transported by other mechanisms, and these are also important in practical absorbents.

Another force that can drive fluid flow is an externally supplied hydrostatic pressure difference between the faces or ends of the absorbent material. In contrast to wicking, liquid flow under an imposed hydrostatic pressure can be maintained in a saturated pore or capillary if a pressure difference is maintained across the pore or across the ends of a capillary. Whereas the capillary pressure "pulls" fluid into the pore, an applied hydrostatic pressure "pushes" liquid into the pore. When the entire capillary is wet, the meniscus at the wet–dry interface disappears and the capillary pressure vanishes, but any imposed hydrostatic pressure remains.

The structure of the absorbent also governs its mechanical rigidity or modulus, which in turn determines the propensity of the wetted structure to be dewatered by application of external forces such as mechanical compression. Mechanical dewatering is usually quite easy for most traditional absorbents, and this feature is a principal deficiency in applications where liquid retention is important, for example, in diapers or paper towels. The compressibility of the materials is determined by the modulus of the porous structure. The Young's modulus E_f of the porous structure is determined by the modulus E_0 of the solid, structural substance and its volume fraction ϕ in the absorbent material. For example, the apparent modulus of foams may be calculated as follows[5–7]:

$$\frac{E_f}{E_0} = \phi(2 + 7\phi + 3\phi^2)/12 \tag{1.4}$$

Even with sufficient structural rigidity to prevent compressional collapse, traditional absorbents may be dewatered by application of additional gravitational force. In a centrifuge, for example, dewatering may take place by overcoming the capillary pressure of the material. Resistance toward extra gravitational force requires very small capillaries, which have a large capillary pressure. However, for a fully saturated absorbent, any application of an accelerational force will cause some dewatering (see Eq. 1.1).

In summary, the absorbency of traditional materials is a function of the relative volume of open space in the sample (the porosity), the extent of connectedness of that volume, the size of the individual pores, and the mechanical strength of the porous structure, especially when wet. These factors are controlled by the choice of the structural material or fiber type, by the radius and length of the fibers and the pores, and by the method used to manufacture the absorbent structure.

1.2. SUPERABSORBENTS

1.2.1. Physics of Superabsorbents

Superabsorbent polymers are crosslinked networks of flexible polymer chains. The most efficient water absorbers are polymer networks that carry dissociated, ionic functional groups. A schematic drawing of such a network is shown in Figure 1.2. Except for the molecular-sized chains that make up the network, this picture of a network is remarkably similar looking to the mass of cotton fibers shown in Figure 1.1d. However, whereas traditional absorbents, like cotton, take up liquids by a flow

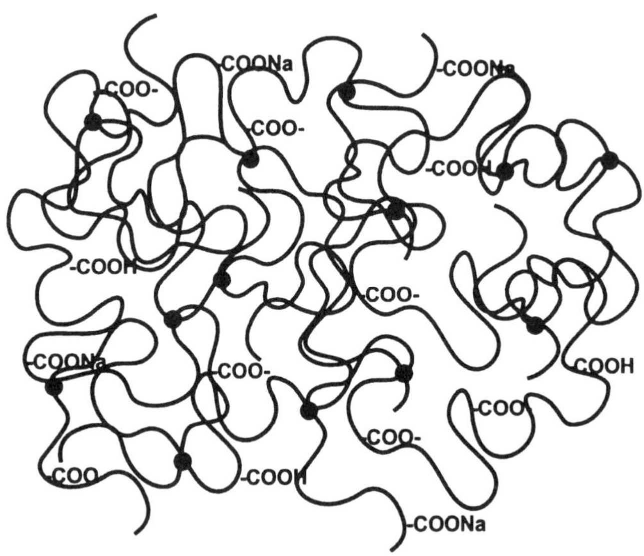

Figure 1.2. Schematic representation of a crosslinked polymer network. The dots represent the crosslinks between the coiled polymer chains, which have ionic carboxylate functional groups spaced along their length.

or convective mechanism, superabsorbents function by means of a diffusive mechanism. Diffusion is a molecular phenomenon wherein molecules move in a random, nonorganized manner as a result of their thermal energy. Increasing the temperature makes the molecules move faster, and the diffusive processes occur therefore at a faster rate. The molecules are in constant motion and will move from their average position in space toward regions of lower activity by purely statistical means.

Water moves into a particle of superabsorbent polymer because the activity of water is initially lower in the interior of the particle. As water diffuses into the particle, the particle volume increases and the polymer chains that form the superabsorbent polymer network must also move, generally in a direction opposite to that of the water molecules, in order to accommodate the volume of the additional molecules of water. The change in volume is apparent in Figure 1.3, which shows a particle of superabsorbent polymer before and after swelling in a dilute salt solution. Because polymer molecules are much larger than water molecules, they diffuse much more slowly. Not only are the polymer molecules larger, but they are also connected one to the other by the crosslinks, so that the polymer does not eventually dissolve in the water and become useless as a superabsorbent. Therefore, to move in a certain direction, the polymer chains must move cooperatively, requiring more time than if each polymer chain, or each monomer segment, moved individually. The slow polymer motion is a principal rate-limiting phenomenon in polymer swelling. This slow motion and the factors that affect it are discussed more fully in Chapter 5.

The polymer chains each have a preferred size and orientation in space that depends on their environment and their own thermal motion, which keeps them in constant, random motion about an average location. When water diffuses into the superabsorbent polymer, the favored, random configuration is disturbed, because the chain elongates as a necessary consequence of the change in volume of the particle. At first the disturbance is not very large, but as more water diffuses into the particle, the magnitude of the disturbance increases. Eventually, the force resulting from the elongation of the polymer chains balances the force that results from the

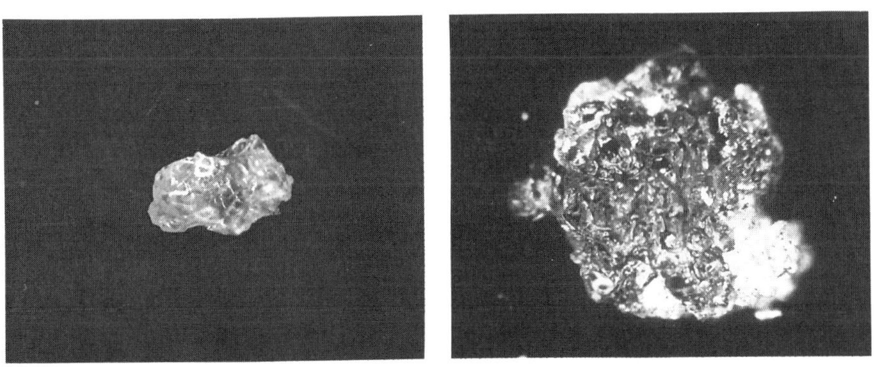

(*a*) (*b*)

Figure 1.3. Water absorption by superabsorbent polymer: **(a)** dry granule and **(b)** swollen gel particle. Magnifications ×25.

difference of water concentration between the inside and the outside of the particle, and water stops moving into the particle. The position of the resulting equilibrium depends on the interconnected structure of the polymer in the particle and on the nature of the liquid that diffuses into the particle.

The process of diffusion is described by Fick's law (Eq. 1.5), which relates the diffusional flow rate f to the concentration gradient dc/dx of the diffusing substance in the sample

$$\frac{f}{A} = -D\frac{dc}{dx} \tag{1.5}$$

where f is the flow rate in the x direction, A is the cross-sectional area of the sample, and D is the diffusion coefficient (proportionality constant).

It is important to recognize that different mechanisms control the motion of fluids in absorbent and superabsorbent materials, even though the flow phenomena may be described mathematically in similar ways.[8] For absorbents, the flow rate is proportional to the pressure gradient according to Darcy's law (Eq. 1.1). The gradient may be a result of a positive, hydrostatic liquid pressure head or a negative, capillary suction pressure. Therefore, the flow depends on the leveling of water at two different heights by gravitational force, or the surface tension of a liquid pulling water along a wetting surface. For superabsorbents, the absorption is proportional to the gradient of chemical activity, according to Fick's law. The flow depends therefore on the random motion of individual molecules and on their size and thermal energy. This is not to say that physical phenomena such as liquid leveling and surface tension of the liquid are not important in the functioning of superabsorbents. In any practical application of superabsorbent polymers, liquid must flow to the surface of the superabsorbent polymer before the diffusive process can begin, and the ability of the liquid to spread on the surface of the superabsorbent polymer, and flow between the particles, will affect the process. A simplified approach to convective flow in particulate masses of deformable gels is described in Chapter 5. Because the absorption process in superabsorbents depends on the size and thermal energy of the molecules, the resulting changes in the state of the molecules after absorption are of principal interest.

When superabsorbent polymers absorb liquids, an elastic gel forms. The gel is a soft, deformable solid composed of water and the expanded polymer chains. The expansion of the polymer chains is limited by the crosslinks. The behavior of the gel at equilibrium may be described by an equation of state, which is analogous to the equations of states of gases, for example, the ideal gas law

$$PV = nRT \tag{1.6}$$

where P is the pressure in the gas, V is its volume, n is the number of gas molecules, R is the gas constant, and T is the temperature of the gas. Similarly, the state of the gel may be given in terms of the pressure, which is manifested as the rigidity (modulus) of the gel and which derives from the elasticity of the polymer chains in the gel, the volume, which depends on the amount of liquid absorbed, the number of polymer chains that forms the gel, and the temperature of the gel. The interrelationships of these factors are described more fully in Chapter 5. The principal features of interest in the technology of superabsorbents are the volume of the gel and how to optimize the volume in light of the other properties and the methods used to synthesize the polymer. These factors are described more fully in Chapters 2 and 3. The fundamental relation, however, is that the volume is inversely proportional to the elastic modulus of the gel, just as the volume of an ideal gas is inversely proportional to the pressure of the gas. This relationship is demonstrated in Figure 1.4, which shows the relationship of the equilibrium swollen volume of a superabsorbent polymer gel to its shear modulus *G*. Both quantities are controlled by the extent of crosslinking in the sample.

1.2.2. Superabsorbent Polymer Types

1.2.2.1. Polyelectrolyte Vs. Nonpolyelectrolyte Hydrogels. Absorbent polymers may be divided into two classes: nonionic and ionic polymers. The ionic polymers may be further classified into anionic and cationic polymers. Nonionic polymers absorb water and aqueous fluids by means of the energetic and entropic interactions made possible by mixing the aqueous fluids with the hydrophilic groups that are present along the polymer chain. The hydrophilic groups are solvated by the water molecules through hydrogen bonds. In addition, the entropy of the system may be increased during mixing by a decrease in the partially ordered struc-

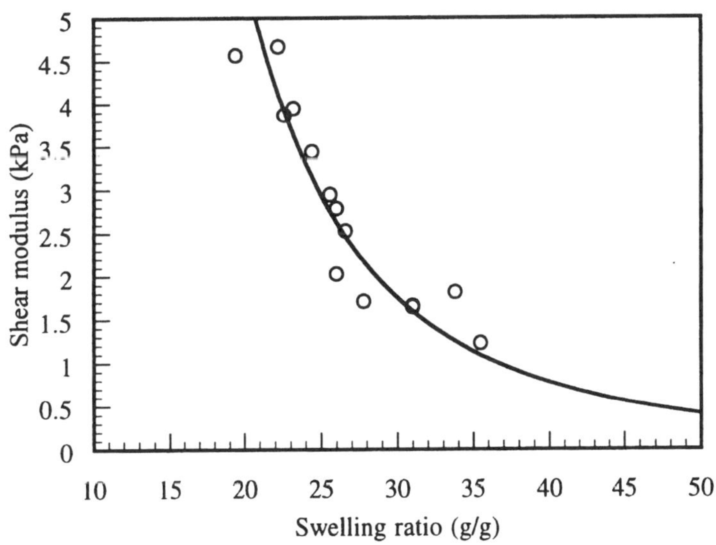

Figure 1.4. The correlation of shear modulus to the swelling ratio for superabsorbent polymers swollen to equilibrium in 0.9% NaCl solution. The samples vary in crosslink density. The shear modulus scales as the −2.83 power of the swelling ratio in this figure.

ture of pure, liquid water. These are the same phenomena that lead to the dissolution of water-soluble polymers. The principal distinction of absorbent polymers over water-soluble polymers is the presence of crosslinks within the molecular structure of the absorbent polymers. The crosslinks connect the various polymer chains into a huge, insoluble molecule that nevertheless can change its shape and volume as it becomes solvated by water. The result of the absorption of solvent is a swollen, soft gel.

Superabsorbent polymers are in distinct contrast to the nonionic absorbent polymers. The energetic and entropic mechanisms present with nonionic polymers are still operative, but are largely overshadowed in magnitude by the presence of the ionic charges that are spaced along the polymer chains of superabsorbent polymers, as shown above in Figure 1.2. The presence of the polymeric ions results in increased absorption because the ions are more strongly solvated than nonionic functional groups, as a result of strong ion–dipole interactions. The ions that are fixed to the polymer chains (negative carboxylate groups or sulfonate groups, or positively charged quaternary ammonium groups, for example) also repel each other (coulombic interaction) and tend to stiffen the polymer chain. As a result, the stiffened chain occupies a greater solvated volume. Because electrical neutrality is required for the polymer, each of the charges along the polymer chain must be matched with an ion of opposite charge. These counterions also are solvated strongly. As a result of their large number in the polymer system, a large driving force for swelling (osmotic pressure) is present. From a thermodynamic perspective, there is no difference in the effects of positively charged polymer chains compared with negatively charged chains, but there is a difference in the mass efficiency of the absorption process, as described below.

1.2.2.2. Covalent Vs. Ionic Vs. Physical Gels. The nature of the crosslinks plays a significant role in the properties of superabsorbent polymers. There are three principal bonding types that are used to bind the polymer chains together: covalent, ionic, and hydrogen bonds. Two basic methods are used to introduce covalent crosslinks. Covalent crosslinks are formed when the major monomer (e.g., acrylic acid) is copolymerized with a di-, tri-, or tetra-vinyl monomer such as *N,N'*-methylenebis(acrylamide), 1,1,1-trimethylolpropanetriacrylate, or tetraallyloxyethane, in a free-radical initiated addition polymerization.[9] The structures of some of these crosslinkers are shown in Chapter 2 (see Fig. 2.2). Covalent crosslinks are also introduced by reacting the preformed polymer chains with a di- or tri-functional reagent that reacts with (for example) the carboxylic acid groups by means of a condensation or addition reaction. This type of crosslinking is analogous to the vulcanization of rubber. An example of a condensation reaction used with superabsorbent polymer is a simple esterification of the carboxylic acid groups with ethylene glycol or glycerol. Such an esterification of poly(acrylic acid) and poly(vinyl alcohol) was used to make one of the early examples of a chemomechanical, polyelectrolyte gel[10] that may be induced to change its state by the application of an electrical potential across the sample. An example of an addition reaction is the addition of the carboxylic acid function to the oxirane rings of ethyleneglycol diglycidyl ether,[11] as shown in Chapter 2 (see section 2.7.3). One benefit of covalent crosslinks is that they are permanent and stable with time, so that the properties are maintained during use. By proper choice of the crosslinker and the synthetic conditions, some control over the placement of crosslinks is achieved.

Ionic crosslinks are formed by reacting a polyvalent ion of opposite charge with the charged polymer chains. The crosslink forms as a result of charge association of the unlike charges. Examples are crosslinking of carboxylic acid with aluminum [12] or chromium ion or with an organic polycation such as a diquaternary ammonium compound. Because the bond is formed by ion association (charge neutralization), the chemical structure of the crosslinker is less important in determining the placement of the crosslinks compared with covalent crosslinks. If ionic components are present in the liquid to be absorbed, ion exchange may occur with the ionic crosslinks, which may alter the nature of the crosslinks and the behavior of the polymer in ways that may be unforeseen. Also because the interionic reaction is very fast, the incorporation of the crosslink and the resulting structure of the crosslinked polymer can be difficult to control.

A third type of crosslink is the physical crosslink, which is usually formed by means of hydrogen bonding of segments of one chain with the segments of another chain. A classic example is the gelling of gelatin solutions, wherein the regular structure of the polypeptide chains allows strong hydrogen bonding between chains and the formation of associated regions of polymer segments that act as crosslinks. A principal drawback of such crosslinks is that they can be destroyed by merely heating the polymer. The crosslink region typically extends over many segments of the polymer chains, leading to poor mass efficiency of crosslinker.[13]

1.2.2.3. Survey of Polymers. A number of nonionic water-soluble polymer types are known, and these may be used as absorbent polymers when crosslinked in a suitable fashion. As before, a distinction is made between nonionic and ionic polymers. Outside of the large contribution of ions to swelling, several other characteristics of the polymer system are important for making absorbent polymers. For example, the molecular weight of the polymer chains should be large for optimum absorbency (see Chapter 5). But the maximum size of polymer chains that it is possible to make depends on the structure and reactivity of the monomers and the synthetic conditions (see Chapter 2). Also, the absorbency depends on the mass of the monomer unit and its solvating character. A high-molecular-weight, hydrophobic monomer will not yield an absorbent polymer optimized for water absorption on a weight basis. In addition, the relative cost of the monomers affects the cost-effectiveness of absorption, as does the relative complexity of the processing steps necessary to synthesize the polymer. The following brief survey of absorbent polymer types is given with these characteristics in mind.

Polyacrylamide is a well-known water-soluble polymer and has been crosslinked by numerous methods to form an absorbent polymer.[14] Unless some of the amide functions are hydrolyzed to anionic carboxylates, the polymer is nonionic. Acrylamide is one of the very few monomers that can be easily polymerized to extremely high molecular weight (~10 million g/mol), which is an advantage for making absorbent polymers. A principal drawback is the practical difficulty of reducing the quantity of residual acrylamide in the final product to extremely low levels. This is desirable because of the well-documented toxicology of acrylamide monomer. The monomer is relatively inexpensive.

Poly(ethylene oxide) has rarely been used as an absorbent polymer, even though the monomer is relatively low in cost. Processing of the gaseous, highly reactive ethylene oxide adds to the cost of the polymer, and preparing stable, very-high-molecular-weight polymers is difficult. The molecular weight is limited to about 600,000 g/mol. Crosslinking may occur either by endlinking the hydroxyl end groups with an appropriate reagent (e.g., polyisocyanates) or by means of radiation treatment. Efficient endlinking is quite difficult for very-high-molecular-weight prepolymers as a consequence of the low concentration of end groups. Radiation crosslinking typically involves some chain scission events, which lower the molecular weight and the resulting effectiveness of the absorbent polymer. In addition, poly(ethylene oxide) is highly crystalline (T_m = 66°C), and water absorption is significantly suppressed as a result.

Poly(vinyl alcohol) gels have been extensively studied as absorbent polymers. They are prepared by reacting vinyl acetate to form poly(vinyl acetate), transesterifying the poly(vinyl acetate) to poly(vinyl alcohol), and finally reacting with a crosslinker such as a polycarboxylic acid (forming ester crosslinks), sodium borate (forming borate complexes), or glyoxal. Poly(vinyl alcohol) is limited to practical molecular weights of about 700,000 g/mol as a result of significant chain transfer reactions to the monomer and can be highly crystalline (depending on the mole fraction of hydroxyl groups), both features that limit the swelling. The monomer is moderately expensive when considered on a "vinyl alcohol" basis.

The primary example of a commercially available polypeptide is gelatin, which is a common food additive. This is a relatively expensive polymer, which must subsequently be crosslinked. The extensive hydrogen-bonded structure, which leads to the well-known physical gelation of a cooled gelatin solution, also results in less mass-efficient water absorption.

The principal superabsorbent polymer used commercially is crosslinked, partially neutralized poly(acrylic acid).[15] Some graft copolymers are also employed, such as partially neutralized starch-g-poly(acrylic acid) and partially neutralized poly(vinyl alcohol)-g-poly(acrylic acid). The polymer is made from acrylic acid and sodium hydroxide or sodium carbonate (and the grafting agent if desired), which are relatively inexpensive reagents. The polymer is most often made by co-polymerizing the partially neutralized acrylic acid in aqueous solution with a di- or tri-vinyl crosslinking monomer. There are many acceptable crosslinkers available commercially (see Chapter 2), and one commonly used crosslinker is 1,1,1-trimethylolpropane triacrylate. Acrylic acid forms relatively high-molecular-weight polymers, with molecular weights typically in the range of 1–3 million g/mol. Acrylic acid is not usually co-polymerized with monomers other than the crosslinkers, because most other monomers are more expensive than acrylic acid and usually provide no performance advantage to offset the higher cost.

Another well-known superabsorbent polymer is the "super slurper," which was developed at laboratories of the United States Department of Agriculture in the 1970s. It consisted of hydrolyzed starch-g-poly(acrylonitrile) and was offered as an alternative product (and process) to the wholly synthetic superabsorbents developed earlier.[16,17] The polymer was made by graft polymerizing acrylonitrile to starch and subsequently hydrolyzing the pendent nitrile groups to carboxyl and carboxamide groups with sodium hydroxide. Crosslinks were provided by hydrogen-bonded, physical crosslinking within the partially swollen starch granules. The products absorbed water to yield rather soft, low modulus gels with a relatively large amount of soluble polymer compared with covalently crosslinked superabsorbent polymers. Although the raw materials were inexpensive, the preparation of the product required many process steps and great care in the handling of acrylonitrile because of its physical and toxicological properties. The technology evolved into the graft polymerization of acrylic acid with starch that is used commercially.[18]

Sodium carboxymethylcellulose (CMC) and alginic acid are prominent members of a class of ionic polymers that are based on naturally occurring polysaccharides. Other ionic polysaccharides include the carrageenans (from seaweeds), pectins (from plant extracts), and xanthan (from microbial fermentation process). These polymers also are superabsorbent when crosslinked. CMC is a semisynthetic polymer made by reacting chloroacetic acid with sodium cellulose in a slurry with isopropanol and water. Sodium cellulose is made by swelling highly crystalline cellulose with an aqueous solution of sodium hydroxide to decrystallize the cellulose and allow homogeneous penetration of the chloroacetic acid. Each cellulose repeat unit is substituted with about 0.8 units of chloroacetic acid, providing an ionic polymer with an ionic repeat unit of molecular weight of about 226 g/Eq. The product is moderately expensive when the polymer is purified from the sodium chloride and sodium glycolate that are co-produced in the synthesis. The rigidity of the cyclic, cellulose backbone polymer provides for good superabsorbency when the soluble CMC is crosslinked. Crosslinked CMC has several disadvantages, however. The molecular weight of cellulose that is isolated in a pulping process from tree fiber is low (ca. 350,000 g/mol) compared with some of the synthetic polymers, but, at additional cost, cotton linters can provide cellulose with higher molecular weight (ca. 1.2 million g/mol) necessary for improved absorbency.[19] Preparing a superabsorbent from CMC also requires crosslinking of a viscous polymer solution, which adds to the cost of this material. In addition, crosslinking CMC is inefficient with typical crosslinkers like divinyl sulfone or glyoxal, because the most reactive hydroxyl groups of cellulose have been substituted with carboxyl groups.[20]

Sodium alginate is a polysaccharide produced by brown seaweeds such as the giant kelp and isolated by extraction and precipitation of the polysaccharide from the seaweed. The molecular weight of the product is about 250,000 g/mol. The polymer is formed of D-mannuronic acid and L-guluronic acid units ($C_6H_{10}O_7$) that are linked in various blocks and random arrangements. Each residue has a carboxylic acid group, which, when neutralized, provides the ionic character. As with CMC, the alginates must be additionally crosslinked to function as a superabsorbent.[21] The polymer is relatively expensive as a result of the harvesting and extraction processes used in its isolation.

A number of cationic polymers are also known and can function as superabsorbents when crosslinked, including poly(diallyldimethylammonium chloride), poly(vinyl pyridine), poly(vinylbenzyltrimethylammonium salts), cationic starches,[22] and hydrolyzed chitin (chitosan), which is derived from the exoskeletons of arthropods such as lobster. Frequently, a cationic monomer is co-polymerized with a less-expensive, nonionic monomer such as acrylamide to make the cationic polymer. In general, a combination of complex processing conditions and relatively low molecular weight leads to higher cost and lower effectiveness of the cationic polymers compared with the anionic polymers.

1.3. COMPOSITE ABSORBENT STRUCTURES

Superabsorbent polymers are typically used in admixture with other materials. For example, superabsorbent polymer granules are mixed with wood pulp fluff in diapers, as in Figure 1.5, and with soil particles when used as an agricultural soil amendment. The properties of such composite mixtures depend on the properties of the different components and on the interactions (if any) of the components. The optimization of flow and absorption of liquid in composites will be complex, because both convective flow (Darcy's law, Eq. 1.1) and diffusive absorption (Fick's law, Eq. 1.5) will be present. The principal factors affecting the flow of liquid in mixtures are the particle size and particle size distribution of each component of the mixture, the homogeneity of the mixture, the interaction of the components with the liquid, and the dynamic features of the system, i.e., how the properties of the mixture change with time. As an example of the complexity inherent in optimizing a composite, consider the particle size and particle size distribution of just the superabsorbent polymer component. Rapid diffusive absorption requires small particle radius, but rapid convective flow requires large pores that result from packing together large particles. Evenly sized pores present in a bed of particles of narrow size distribution

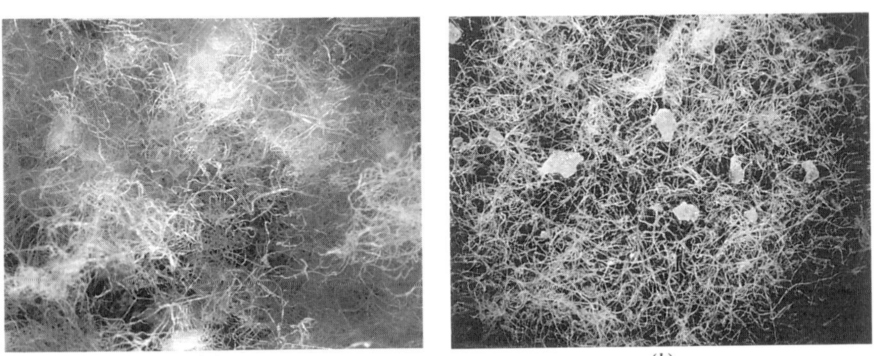

(*a*) (*b*)

Figure 1.5. Absorbent cellulose pulp fluff **(a)** and a composite of fluff and superabsorbent polymer granules **(b)**. Magnification ×10.

will improve convective flow driven by a positive liquid head, but a broad size distribution, and the resulting smaller pores, may increase the flow driven by capillary suction pressure. Addition of a second component only adds to the complexity. For example, the packing densities of fibers and granules individually are generally quite dissimilar, as seen from the bulk densities ρ_b. Granules pack much more closely ($\rho_b = 0.5$ g/cm^3) than do fibers ($\rho_b = 0.05$), and it is not clear how they will pack when mixed or how to characterize the mixture. Also the mean particle size and apparent density of cellulose fibers are different from those of superabsorbent polymers, making blending a challenging task.

Presuming that a homogeneous composite is desirable and could be made, the mechanical properties of the composite, such as the modulus, are not well understood. One expression for the modulus of a composite has been derived by Kerner [23]

$$\frac{G}{G_0} = \frac{1 + AB\phi_f}{1 - B\phi_f} \tag{1.7}$$

to express the ratio of the shear modulus G of a composite to the shear modulus of the matrix substance G_0 as a function of the volume fraction ϕ_f of the filler. The quantity A is a function of the Poisson's ratio of the composite, and the quantity B is a function of the ratio of the modulus of the filler substance to the modulus of the matrix substance. Numerous variants of this general approach have been developed.[24] In the case of mixtures of superabsorbent polymer particles with cellulose fluff, the situation is quite complex, as the system is really a three-component composite of fiber, superabsorbent particles, and air that becomes a four-component system upon absorption of liquid. The situation might be simplified by considering the fiber and air combination (fluff) as the matrix and the superabsorbent polymer as the filler, but the change in properties of the fluff matrix as one of its components (air) is replaced is not known. Of course, the composite also undergoes tremendous changes in volume during swelling of the superabsorbent polymer, and body forces may also be applied by the user during swelling. The large volume changes disrupt the original structure of the composite and cause the mechanical properties to change in a poorly understood manner. Not only is the volume of the composite changing, but also the mechanical compressibility of the composite is changing. Some recent approaches [25] to modeling this problem appear promising. Much more study is needed to advance our current understanding.

The flow of liquid into and through such a composite is also not well understood. Flow of liquid through idealized porous media that are made of uniformly sized fibers has been studied extensively.[26,27] The permeability of real fibrous systems has been well correlated to the idealized models.[28] Theories for flow through media composed of differently-sized fibers are less well developed, but progress has been made by considering the case of very coarse fibers embedded in a matrix of much finer fibers.[29] In this case the superficial velocity of fluid through the matrix increases as a result of the decreased fraction of open area in the cross section of the composite. In addition, the coarse fibers increase the path length of flow (tortuosity) of the system and create velocity gradients in the system. The sum of these effects is to decrease the permeability of the composite. The added effect of a volume increase (swelling) of one component is not included here, but studies of the effect of compression on the permeability of soft porous media may be applicable.[30,31]

The permeability of ternary mixtures of particles has been studied [32] and might be applied to the case of composites of superabsorbents and fibers. However, most flow models also are restricted to isotropic media, where the properties are the same in each spatial direction. This is unlikely to be the case in most absorbent composites because the products are made on equipment that operates continuously and has a "machine direction" along the route of manufacture. A number of compression and packing steps during manufacture make the properties in the thickness direction different from those in the length or "machine" direction. This of course adds a significant complexity to a thorough study of the system. The understanding of flow in such anisotropic media is aided by use of the "effective medium" theory.[33,34] Once again, the dynamic nature of the swelling process complicates the theories even further, and more development is necessary to adequately describe composites of fibers and superabsorbent polymers.

REFERENCES

1. G.M. Aberson, *TAPPI Spec. Tech. Asst. Publ.,* **8,** 282–307 (1972).
2. P.K. Chatterjee, ed., *Absorbency,* Elsevier Science Publishers, New York (1985).
3. A.A. Burgeni, C. Kapur, *Textile Res. J.,* **37,** 356–366 (1967).
4. F.A.L. Dullien, *Porous Media: Fluid Transport and Pore Structure,* Academic Press, New York (1979).
5. K.C. Rusch, *J. Appl. Polym. Sci.,* **13,** 2297–2311 (1969).
6. S. Swyngedau, A. Nussinovitch, M. Peleg, *Polym. Eng. Sci.,* **31,** 140–144 (1991).
7. E.A. Meinecke, R.C. Clark, *Mechanical Properties of Polymeric Foams,* Technomic Publishing Co., Westport, Connecticut (1973).
8. P.B. Weisz, *Ind. Eng. Chem. Res.,* **34,** 2692–2699 (1995).
9. F.L. Buchholz, *Chem. Br.,* **30,** 652 (1994).
10. W. Kuhn, B. Hargitay, *Z. Elektrochemie,* **55,** 490–505 (1951).
11. S. Obayashi, M. Nakamura, K. Fujiki, T. Yamamoto, *U. S. Patent 4,340,706* (1982).
12. S.H. Ganslaw, H.G. Katz, *U. S. Patent 4,090,013* (1978).
13. S.B. Ross-Murphy, In *Polymer Gels, Fundamentals and Biomedical Applications,* D. DeRossi et al., eds., Plenum Press, New York, pp. 21–40 (1991).
14. R.N. Bashaw, B.G. Harper, *U. S. Patent 3,229,769* (1966).
15. F.L. Buchholz, *Trends Polym. Sci.,* **2,** 277–281 (1994).
16. B.G. Harper, R.N. Bashaw, R.N. Atkins, *U. S. Patent 3,669,103* (1972).
17. C. Harmon, *U. S. Patent 3,670,731* (1972).
18. F. Masuda, K. Nishida, A. Nakamura, *U. S. Patent 4,045,387* (1977).
19. E.K. Just, T.G. Majewicz, In *Encyclopedia of Polymer Science and Engineering, 2nd Ed.,* H.F. Mark et al., eds., John Wiley and Sons, New York, pp. 226–269 (1985).
20. U. Anbergen, W. Oppermann, *Polymer,* **31,** 1854–1858 (1990).

21. S.T. Moe et al. *Macromolecules,* **26,** 3589–3597 (1993).
22. D. Castel, A. Richard, R. Audebert, *J. Appl. Polym. Sci.,* **39,** 11–29 (1990).
23. E.H. Kerner, *Proc. Phys. Soc.,* **69B,** 808 (1956).
24. R.D. Sudduth, *J. Appl. Polym. Sci.,* **54,** 1243–1262 (1994).
25. K.A.S. Jonsson, B.T.L. Jonsson, *AIChE J.,* **38,** 1340–1356 (1992).
26. J. Happel, *AIChE J.,* **5,** 174 (1959).
27. L. Spielman, S.L. Goren, *Environ. Sci. Tech.,* **2, 279** (1968).
28. G.W. Jackson, D.F. James, *Can. J. Chem. Eng.,* **64,** 363–374 (1986).
29. C.R. Ethier, *AIChE J.,* **37,** 1227–1236 (1991).
30. K.A.S. Jonsson, B.T.L. Jonsson, *AIChE J.,* **38,** 1349–1356 (1992).
31. M. Kataja, A. Rybin, J. Timonen, *J. Appl. Phys.,* **72,** 1271–1274 (1992).
32. N. Standish, D.N. Collins, *Powder Tech.,* **68,** 175 (1991).
33. S. Kirkpatrick, *Rev. Mod. Phys.,* **45,** 574–588 (1973).
34. P.G. Toledo, H.T. Davis, L.E. Scriven, *Chem. Eng. Sci.,* **47,** 391–405 (1992).

2

CHEMISTRY OF SUPERABSORBENT POLYACRYLATES

Thomas L. Staples, David E. Henton, Fredric L. Buchholz

Among the absorbent and superabsorbent polymer types described in Chapter 1, the crosslinked, partially neutralized poly(acrylic acids) are the principal superabsorbent polymer of commerce. Numerous other polymers have been crosslinked to form water-swellable gels, but no other polymer can provide the high charge-to-mass ratio as economically as poly(sodium acrylate). For this reason, this chapter on the chemistry of superabsorbent polymers is restricted to the formation of crosslinked networks of poly(acrylic acid) and its salts.

General descriptions of other absorbent and superabsorbent polymer types are available in the literature. Gels of some of the cellulose-based, water-swellable polymers, such as sodium carboxymethylcellulose, are described by Doelker.[1] Alginate gels are described by Segeren et al.[2] and Aslani and Kennedy[3] and in the numerous references therein. Other biopolymer gels have been described by Clark and Ross-Murphy[4] and Clark.[5] Hydrogels of poly(vinyl alcohol) are reviewed by Peppas,[6] and those of poly(ethylene oxide) are reviewed by Graham[7] and Apicella et al.[8] More extensive discussions of these and other alternatives to polyacrylate absorbents may be found in Chapter 6.

To introduce the descriptive chemistry of superabsorbent polymers, we first present two example polymerizations. Polymerization kinetics are then discussed in some detail, illustrating how process conditions affect the structure of the polymer

Modern Superabsorbent Polymer Technology. Edited by Fredric L. Buchholz and Andrew T. Graham. ISBN 0-471-19411-5.

network. Our study of the polymerization kinetics has been aided by some new experimental techniques, which are also described. Crosslinking via both co-polymerization and post-polymerization reactions is covered, and brief mention is made of kinetic modeling as it may apply to superabsorbents.

2.1. PREPARATION OF SUPERABSORBENT POLYACRYLATES

Superabsorbent polyacrylates are prepared by free-radical initiated polymerization of acrylic acid and its salts, with a crosslinker, in aqueous solution or as a suspension of drops of aqueous solution in a hydrocarbon. The polymerization chemistry is shown schematically in Figure 2.1. The two principal processes, bulk solution polymerization and suspension polymerization, share many features. The monomer and crosslinker concentrations, the initiator type and concentration, polymerization modifiers, the relative reactivities of the monomers, the basic polymerization kinetics, and the reaction temperature are all significant factors in both processes.

In either process, the monomers are dissolved in water at concentrations of 20–40 wt%, and the polymerization is initiated by free radicals in the aqueous phase. Several types of free-radical sources may be used, including thermally decomposable initiators, redox systems, photochemical initiators, and combinations of them. Redox systems used for the crosslinking copolymerizations include couples of persulfate/bisulfite, persulfate/thiosulfate, persulfate/ascorbate, and hydrogen peroxide/ascorbate. Thermal initiators include persulfates, 2,2′-azobis(2-amidinopropane)-dihydrochloride, and 2,2′-azobis(4-cyanopentanoic acid). Combinations of initiators are used especially when the polymerization takes place over a broad temperature range.

The monomers are polymerized either in the acid form (pH 2–4) or as the par-

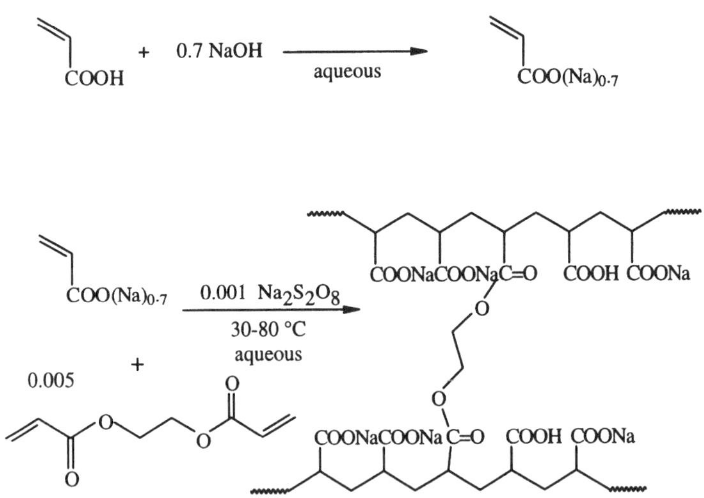

Figure 2.1. Schematic depiction of the chemistry of polymerization of superabsorbent poly(acrylates).

tially neutralized salt (pH 5–7). Inexpensive bases, such as sodium hydroxide and sodium carbonate, are used as neutralizing agents. A choice would be made based on consideration of the pH of the base solution and the resulting potential for hydrolyzing the crosslinker, the solubility limits of the base in water and on the solubility of the monomer salt in water. In suspension polymerizations, the acrylic acid must be neutralized prior to polymerization because of a substantial partition coefficient of acrylic acid in the liquid hydrocarbons used as continuous phase. The polymers made from acrylic acid (and neutralized later) or from the partially neutralized monomer are somewhat different because of the presence or absence of charged monomers and polymers during the formation of the polymer network.[9]

Small amounts of crosslinkers play a major role in modifying the properties of superabsorbent polymers. The co-polymerizable crosslinkers used in superabsorbent polymers range from di-functional compounds, such as diacrylate esters and allylmethacrylate, to tri-functional compounds, such as 1,1,1-trimethylolpropanetriacrylate and triallylamine, and to tetra-functional compounds, such as tetraallyloxyethane. The structures of these substances are shown in Figure 2.2. In addition to modifying the swelling and mechanical properties, the crosslinker affects the amount of soluble polymer formed during the polymerization as a result of its relative reactivity with acrylic acid or sodium acrylate. Efficiency of crosslinking will also depend on steric hindrance and reduced mobility at the site of pendent double bonds, the tendency of a given crosslinker to undergo intermolecular addition reactions (cyclopolymerization), and the solubility of the crosslinker in the monomer mixture. As a result of each of the above variables, a number of polymer-

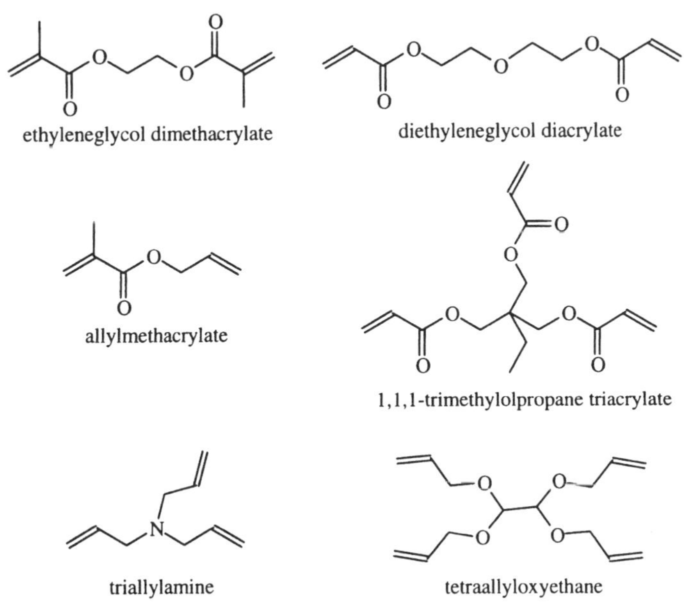

Figure 2.2. The structure of vinyl crosslinkers commonly used in the polymerization of superabsorbent poly(acrylates).

ization additives are also used to ameliorate processing difficulties on a large scale. These additives are not discussed here, because they typically have little effect on the basic chemistry. To describe the synthesis of superabsorbent polymer in a little more detail, two typical synthetic procedures are given below.

2.1.1. Example 1: Crosslinking Co-Polymerization of Acrylic Acid

The polymerization reactor is assembled from a 1-L resin reaction kettle, stoppers, polyethylene tubing, and a thermocouple, as shown in Figure 2.3. The reactor contents may be stirred by means of either a magnetic stirrer (3.2-cm PTFE-coated stir bar) or a bladed agitator on a glass or stainless steel shaft. The monomer solution is prepared in the reactor by mixing 250 g acrylic acid, 600 g distilled water, 0.20 g ethylenediaminetetraacetic acid, and 2 g ethylene diacrylate. The monomer solution is agitated and deoxygenated with nitrogen gas bubbled through the solution at a flow rate of 400 cm^3/min for 60 min.

When the requisite deoxygenation time has passed, the initiator solutions are added to the stirred monomer solution, using pipettes or syringes, in the order list-

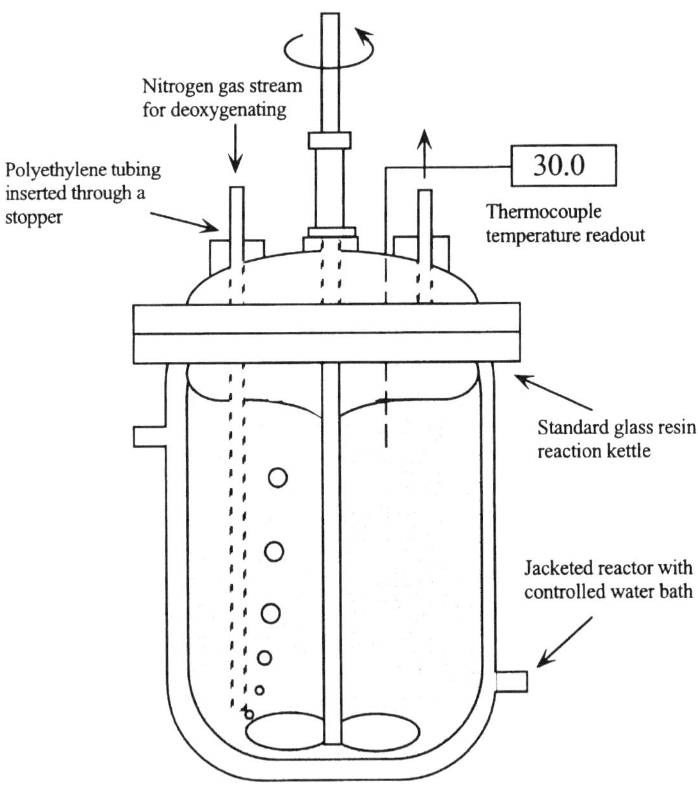

Figure 2.3. Schematic of a reactor suitable for laboratory polymerizations.

ed: 5 cm^3 of a 10% aqueous solution of sodium persulfate, 1.50 cm^3 of a 2% aqueous solution of *tert*-butyl hydroperoxide, and 6 cm^3 of a 10% aqueous solution of sodium thiosulfate pentahydrate.

An exothermic reaction begins within a few minutes if the monomer solution was adequately deoxygenated. The temperature will rise by about 70°C. After the reaction has subsided, as indicated by a drop in temperature, the reactor and its contents may be cooled by immersion in ice water. The lump of clear, rubbery gel is removed from the reactor by wetting a laboratory spatula with water, inserting it between the glass wall of the reactor and the gel, and sliding the spatula around the perimeter of the gel to free it from the glass. The lump of gel may be torn into several pieces by hand and then ground to a fine gel in a kitchen-sized meat grinder. The crumbled mass of gel is then neutralized by adding a solution of 120 g sodium carbonate dissolved in 450 g of deionized water. The carbonate solution and the crumbled gel are mixed together in a large beaker. The neutralization proceeds with a snapping and hissing action as the carbon dioxide, which is liberated during neutralization, escapes from the elastic gel. The neutralized, superabsorbent polymer gel is spread onto a tray and dried in a 100°C oven. The dried polymer may be crushed or ground and then sieved to obtain the desired particle size, which is typically in the range of 150–850 μm.

2.1.2. Example 2: Crosslinking Co-Polymerization of Partially Neutralized Acrylic Acid

The same polymerization reactor may be used as described in Example 1. A mixture of 250 g acrylic acid and 2 g 1,1,1-trimethylolpropane triacrylate (TMPTA) crosslinking agent is added to the reactor and stirred briefly to ensure that the crosslinker is dissolved. To this solution is added 1 g Versenex 80 chelating agent, which is a 40% aqueous solution of sodium diethylenetriamine pentacetate (The Dow Chemical Co.), and 1.25 g of a 5% aqueous solution of Airvol 205 poly(vinyl alcohol) (Air Products Co.), which stabilizes the crosslinker in the monomer solution. The acrylic acid in the monomer solution is then neutralized to 65 mol% by adding a sodium carbonate solution made by dissolving 120 g Na_2CO_3 in 600 g distilled water. The carbonate solution is added gradually while the CO_2 evolves from the solution.

The monomer solution is deoxygenated for 60 min, and then the following initiator solutions are added in sequence, with agitation: 5 g of a 10% aqueous solution of sodium persulfate and 0.6 g of a 10% aqueous solution of sodium erythorbate. Polymerization begins in a few minutes, starting at 25°–30°C, and the temperature rises nearly adiabatically to about 100°C (see Chapter 3, Fig. 3.2). After cooling to 70°C, the reactor contents are held at 70°C for 2–3 h to complete the reaction of the monomers. The gel is then removed from the reactor, broken into several pieces by hand and ground as in the first example, and then dried in a 100°C oven and crushed to the desired particle size.

The polymers made by these procedures are examples of basic, general-purpose superabsorbent polymers for water and aqueous solutions. Understanding the kinetics of the polymerization is essential to a complete understanding of the structures and properties that result. The kinetics are affected by the nature of the monomers and initiators used and by the specific temperatures and concentrations of the polymerization. These topics are described next in more detail. Advanced superabsorbent products have been developed, principally by means of surface crosslinking of the particles formed from the polymerization product. This chemistry is described in more detail in section 2.7.

2.2. FREE-RADICAL POLYMERIZATION OF ACRYLIC ACID AND RELATED MONOMERS

2.2.1. Monomer Properties

Important properties of acrylic acid and methacrylic acid are given in Table 2.1.[10,11] In addition to these two monomers, several multivinyl monomers are used as crosslinkers for superabsorbents, but their descriptions and selected aspects of their chemistry are discussed in the section on crosslinking.

2.2.2. Storage and Inhibition

2.2.2.1. Polymerization Inhibitors. Both acrylic and methacrylic acids are liquids that freeze somewhat below room temperature. In addition, both monomers rapidly polymerize with the evolution of large amounts of heat, as reflected in the high heats of polymerization for the relatively low monomer molecular weights shown in Table 2.1. These two factors combine to make it challenging to maintain these materials as mobile liquids while avoiding premature polymerization in large-scale, generally outdoor, storage facilities.

Not surprising, considerable care has been given over the years to developing safe methods for handling and storage of large quantities of these monomers.[12] To prevent polymerization during shipping and storage, the inhibitor of choice is generally the monomethyl ether of hydroquinone (MEHQ). About 200 ppm of MEHQ is typically added to glacial acrylic acid as an inhibitor. Because the effectiveness of

Table 2.1. Selected properties of common monomers used in superabsorbent polymers

Property	Acrylic Acid	Methacrylic Acid
CAS registry number	[79-10-7]	[79-41-4]
Molecular formula	$CH_2 = CHCOOH$	$CH_2 = C(CH_3)COOH$
Molecular weight	72.06	86.10
Melting point (°C)	13.5	16
Boiling point (°C)	141	163
Density at 20°C (g/cm^3)	1.040	1.015
ΔH_{polym} (kJ/mol)	77.4	66.1

MEHQ is significantly increased in the presence of oxygen,[13] the addition of air during storage is recommended. Many studies (cited later) have focused on the role of these two components and their synergistic interaction. Kinetic details of the effect of MEHQ on the course of polymerization are discussed later in this chapter.

2.2.2.2. Dimer Formation. While the MEHQ–air system provides effective inhibition of radical polymerization, acrylic acid also slowly dimerizes upon standing via a nonradical mechanism. One monomer molecule adds across the double bond of another to form the Michael adduct β-acryloxypropionic acid, as illustrated in Scheme 2.1.

Temperature, water content of the acid, and pH influence the rate of formation of the dimer.[14] Data from storage stability studies at room temperature show a rate of dimer formation of about 100 ppm/day in anhydrous glacial acrylic acid, as quantified by liquid chromatography.[15] Previously unpublished data (T.C. Johnson, The Dow Chemical Co.) showing dimer formation rates at a variety of conditions are given in Table 2.2. Acrylic acid was neutralized with sodium hydroxide to the indicated extent so that the final solution was 33 wt% and then stored at the temperature shown for up to 2 weeks. Two to three times during the 2-week period, a sample was taken and the dimer concentration was determined chromatographically.[15] The rates shown in Table 2.2 are the slopes of the lines of best fit.

It has been postulated that the remaining double bond of the dimer may be incorporated into the poly(acrylic acid) backbone, leading to the structure illustrated in Scheme 2.2.[14] Experiments using ^{13}C-nuclear magnetic resonance (NMR) analysis of the polymer clearly shows this incorporation and a reduced quantity of incorporated dimer after heating (see Chapter 4, section 4.2.1.2). Chromatographic determination of acrylic acid in the polymer, combined with the NMR results, implies that heating of the polymer, such as occurs during drying, thermally degrades the

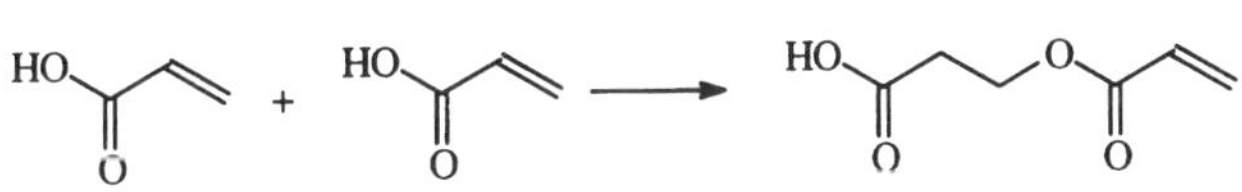

Scheme 2.1

Table 2.2. Effect of neutralization on acrylic acid dimer formation rate

	Dimer Formation Rate (ppm/day)	
Extent of Neutralization (%)	23°C	40°C
0	100	Not measured
62	450	2,700
80	125	770
90	31	180
100	0	0

incorporated dimer by a reverse Michael reaction to regenerate acrylic acid, as shown in Scheme 2.2. The temperature dependence of this result suggests that the fraction of co-polymerized dimer cleaved depends on the polymer drying temperature. A number of patents discuss controlling this thermal breakdown in superabsorbent polymers to minimize residual monomer formation.[16,17]

Minimizing or eliminating dimer in the monomer feed is clearly desirable in order to produce superabsorbent polymers with low levels of residual acrylic acid. At least two means of accomplishing this are possible. First, if the superabsorbent plant is connected to an acrylic acid plant, the time for dimer to increase can be eliminated by immediate usage. Second, a means of eliminating dimer can be incorporated into the process for manufacture of superabsorbents.

Acrylic acid manufacturers have converted dimer (and higher oligomers) to monomer using heat treatment.[18,19] In another patent, the dimer is "decomposed into acrylic acid in the evaporation and decomposing steps."[20] These represent a reversal of the Michael addition reaction (Scheme 2.1). The temperature in this step is 120°–220°C, preferably 150°–200°C. The residence time at these elevated temperatures may be rather short, 1–2 min at 200°C.[21] Additionally, purification steps are used to minimize the amount of dimer and other impurities in the acrylic acid for sale.

It has also been proposed that acrylic acid dimer can be decomposed with base hydrolysis at elevated temperatures or by using strong base at lower temperatures, as shown in Scheme 2.3.[14] This is not a simple reversal of the dimerization but a hydrolytic cleavage yielding both acrylic acid and β-hydroxypropionic acid (β-HPA, also known as 3-hydroxypropionic acid or as hydracrylic acid). Analysis of dimer before and β-HPA after the base hydrolysis of dimer in 65% neutralized monomer mix has shown very nearly 1 mol of β-HPA for each mole of hydrolyzed dimer.

The reaction is quite sensitive to the pH of the monomer solution, as shown in Figure 2.4.[21] This reaction may be utilized either during acrylic acid neutralization, as a continuous process,[22] or more simply by adding the acrylic acid to sodium hy-

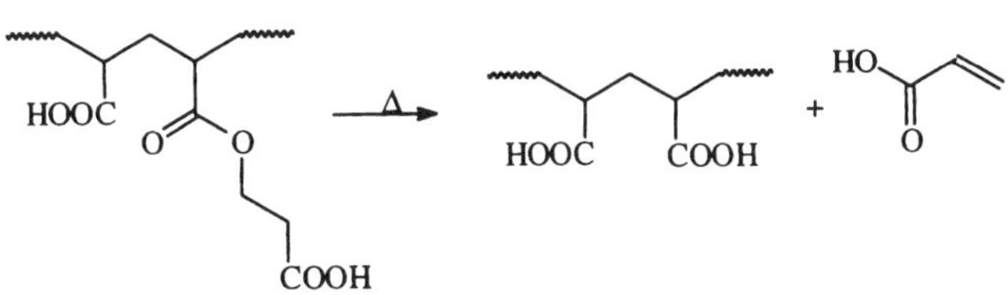

Scheme 2.2

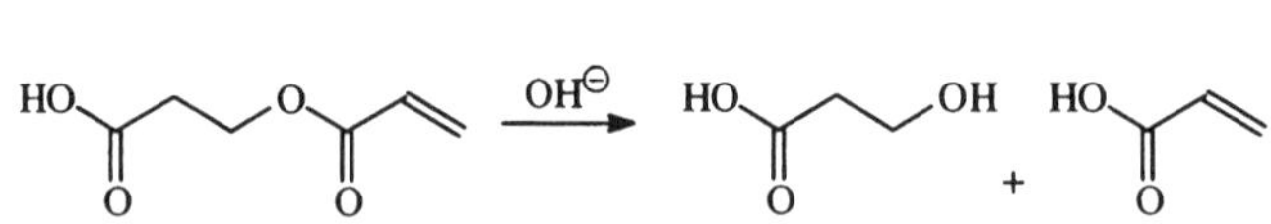

Scheme 2.3

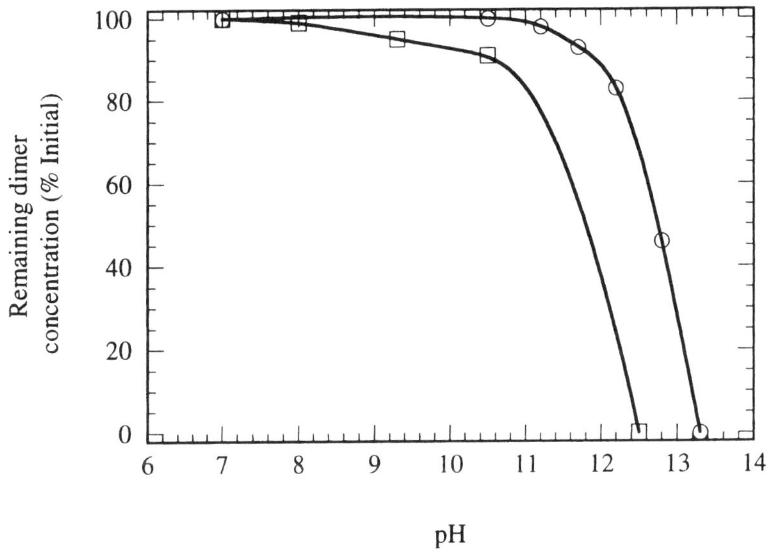

Figure 2.4. Acrylic acid dimer remaining after 15 min at the indicated pH at 23°C (○) or 55°C (□).

droxide slowly enough that the dimer is destroyed in the portion of acid that is 100% neutralized. If sodium carbonate is used as the base, however, the neutralized acid can only reach a pH of about 9.5–10, too low to effect the destruction of the dimer. This problem may be overcome by neutralizing with carbonate to the extent of around 90%–95% neutralization, purging the evolved carbon dioxide from the monomer solution, and then adding enough sodium hydroxide to raise the pH sufficiently to destroy the dimer. In either case, the resulting polymer product has reduced levels of residual acrylic acid, corresponding to the proportion of dimer destroyed.

Although the measured level of acrylic acid in the final product is reduced, β-HPA is produced and its fate must be considered. Most of the β-HPA will pass on through the process, including drying steps, and remain in the product. However, dehydration of β-HPA to form acrylic acid occurs at elevated temperatures. Thus, additional residual acrylic acid can result from having had dimer in the starting acid, even after a base hydrolysis step.[16] This was confirmed by making two polymer samples from acrylic acid originally containing 7,000 ppm dimer, one in which about 65% of the dimer had been hydrolyzed and the other in which it had not. After low temperature drying, the samples were first heated to 175°C and then additionally heated to 200°C for times ranging from 15 to 60 min. The results are shown in Table 2.3. The difference in the amount of unreacted acrylic acid found for the samples dried at 100°C and then heated to 175°C is attributed to the effect of temperature on cleavage of co-polymerized dimer. The increase in residual acrylic acid, after heating to 200°C, for the sample in which the dimer had been hydrolyzed was

Table 2.3. Effect of heat history on residual acrylic acid in superabsorbent product

Heat History	No Treatment (ppm, based on polymer)	Dimer Hydrolysis (ppm, based on polymer)
100°C drying	155	95
175°C, 30 min	1,330	920
200°C, 15 min	1,570	1,500
200°C, 30 min	1,670	1,640
200°C, 60 min	1,740	1,800

attributed to thermal cleavage of β-HPA. Note that this heat treatment brings both samples to essentially the same level of unreacted acrylic acid, thus eliminating any effect of dimer hydrolysis.

This mechanism is consistent with earlier work that shows the reversibility of the dehydration reaction of β-HPA. Zil'berman et al.[23] describe the preparation of β-HPA via the hydration of acrylic acid in the liquid phase at temperatures of 180°C and above. This also shows that the reaction to form β-HPA is reversible and hence confirms that β-HPA can be dehydrated at elevated temperatures to re-form acrylic acid, as noted by Sato et al.[20]

2.3. EXPERIMENTAL TECHNIQUES FOR STUDYING POLYMERIZATION

2.3.1. Standard Methods: Sampling, Dilatometry, and Thermal Measurements

Determining the rate of polymerization is a necessity for plant design. A standard approach for data gathering is to sample the reaction at various times, stop polymerization in the sample, and analyze the mixture for polymer, remaining monomer, or, preferably, both. Methods of analysis are numerous, and section 4.3 in Chapter 4 describes procedures appropriate for the polymer systems described in this book.

One general method for studying polymerization kinetics makes use of the significant change in specific volume typically associated with polymerization. Dilatometry accurately measures the volume changes that occur upon polymerization. For most monomers, including acrylic acid, the volume of the reaction mixture contracts as monomer is converted to polymer. Several of the reports of polymerization kinetics of acrylic acid cited below make use of this technique. One limitation, however, is that the technique cannot be used directly for crosslinked systems, since viscous bulk flow is necessary for the precise measurement of the volume change. Nevertheless, much information of importance for superabsorbents can be learned from dilatometry studies of the uncrosslinked polymerizations of acrylic acid, because actual crosslinker concentrations are typically quite low and have been shown to have little impact on overall polymerization rate.[24]

Another experimental method is based on the fact that polymerizations of vinyl monomers inevitably result in significant heat evolution. Indeed, this fact makes constant temperature studies at the high monomer concentrations used commercially difficult to perform. However, this situation can be turned to advantage by studying adiabatic polymerizations. In an adiabatic reaction, heat is neither absorbed from nor given off to the surroundings; hence, any heat generated (or consumed) in the reaction causes a temperature change. In the absence of a phase change, the heat capacity should be reasonably constant during polymerization, and, hence, the temperature change is approximately proportional to the extent of reaction. Accurate, automatic temperature measurement is routinely available today, and this greatly facilitates data gathering.

2.3.2. Nuclear Magnetic Resonance Technique

Typical polymerization studies, including much of the historical work on acrylic acid discussed below, employ monomer concentrations below 10%. This practice minimizes thermal and viscous effects. Kinetic studies in the concentration regime of industrial importance, however, are largely absent from the scientific literature. Relatively high polymer concentrations are particularly important for superabsorbent production, since drying large amounts of water from the mixture is energy intensive and thus expensive. At monomer concentrations of about 20% or greater (see Chapter 3, Fig. 3.2), the heat generated during polymerization is considerable, requiring quite small samples, i.e., high surface area, to allow maintenance of isothermal conditions. Furthermore, as noted above, systems of commercial interest are crosslinked, and this confounds some methods of continuous monitoring such as dilatometry. A significant step in the study of polymerization kinetics of superabsorbent systems has been the development of NMR methods.[24]

Experimental details of this approach can be found in Chapter 4, section 4.3.1. The method uses integrated line intensities from the ^{13}C-NMR spectrum to determine the relative fraction of monomer and polymer in the reaction mixture. With a sensitive spectrometer, this measurement can be made on a sample small enough to be held at constant temperature, even during rapid polymerization. Isothermal polymerization data on concentrated solutions of acrylic acid, critical for the comparison of theory to experiment for commercially useful reaction mixtures, has been largely unavailable until the development of this method.

An additional advantage of this technique is that the possibility exists to separately follow the co-polymerization of crosslinking monomers. To be successful at the relatively low levels of crosslinking found in commercial superabsorbents (<1%), the crosslinkers must be specially synthesized with enriched levels of ^{13}C. Particular examples are discussed in Chapter 4, section 4.3.2.

2.3.3. Initiator Decomposition

While transformation of monomer to polymer is the major kinetic event during polymerization, the (largely) independent reaction that drives this process is the decomposition of initiators to form free radicals. To have sound reinforcement of theoretical explanations of polymerization kinetics, it is most helpful to have an independent measure of initiator concentration during the course of polymerization. This is particularly true for the highly exothermic polymerizations of commercial practice, for which the common assumption of an unchanging initiator concentration (see below) is not realistic. When thermally activated initiators are used, as is routine for commercial processes, premature depletion or "initiator burn-out" is a typical concern with a rapidly rising reactor temperature.

Methods for the analysis of initiator remaining in a reaction mixture are discussed in Chapter 4, section 4.4.5. Because initiators are typically present at even lower levels than are crosslinkers, NMR techniques have not proved useful for monitoring their concentrations. Instead, independent analysis of remaining initiator must be made for a series of reaction mixtures subjected to related temperature histories. It is important to recognize that utilizing this approach presupposes that the decomposition of initiator is dependent on temperature alone. This is not always a satisfactory assumption, because induced decomposition of initiators is well known, particularly with compounds that can participate in redox systems. The commonly used persulfate di-anion is an example of such a species, and separation of this effect from the kinetic analyses discussed later has been addressed.[25]

2.4. MECHANISMS AND KINETIC EXPRESSIONS FOR TYPICAL RADICAL POLYMERIZATION

2.4.1. Initiation, Propagation, and Termination

Before the unique aspects of the polymerization of acrylic and methacrylic acids are discussed, the essentials of free-radical addition polymerization of vinyl monomers are summarized. Several excellent texts on this subject exist, and the reader is directed to one of these for more comprehensive information.[26]

Steps of a typical free-radical polymerization comprise initiation, propagation, and termination. Typical free-radical initiators are peroxides or diazenes, which thermally decompose to form two radical species:

$$I \xrightarrow{k_d} 2R\cdot \tag{2.1}$$

These radicals may add to monomer molecules (Eq. 2.2), forming a new radical:

$$R\cdot + M \xrightarrow{k'} RM\cdot \tag{2.2}$$

Additional monomer units add rapidly to this growing end, depicted in Eq. 2.3. This step is called *propagation*:

$$RM\cdot + M \xrightarrow{k_p} RM\cdot \tag{2.3}$$

Finally, two of these highly reactive radical chains find one another and combine, resulting in termination of the radical chain:

$$RM\cdot + RM\cdot \xrightarrow{k_t} polymer \tag{2.4}$$

Two possible termination reactions exist, recombination and disproportionation. Both are identical kinetically, but recombination results in the formation of a single polymer chain, while disproportionation results in two chains, each with half the molecular weight of the product from recombination.

We can define the indicated rate constants for these various processes, and a kinetic expression for the polymerization rate can be obtained with only a few assumptions. Propagation is the reaction that consumes monomer, and the rate of propagation, R_p, is given in Eq. 2.5, obtained directly from Eq. 2.3:

$$R_p = -\frac{d[M]}{dt} = k_p[M][RM\cdot] \tag{2.5}$$

Brackets around the symbols have been used to indicate concentrations in mol/L. For high-molecular-weight polymers, the amount of monomer consumed by Eq. 2.2 is small compared with the total, and Eq. 2.5 satisfactorily describes the disappearance of monomer.

Free radicals are very reactive species and tend to be consumed as quickly as they are formed. This situation generally results in the concentration of radicals quickly attaining a steady state in which $[RM\cdot]$ is relatively constant compared with the concentration of monomer. The rate of formation of $RM\cdot$, from Eq. 2.2, is balanced by the rate of termination, R_t, from Eq. 2.4, leading to the kinetic expression

$$\frac{d[RM\cdot]}{dt} = k'[R\cdot][M] - 2k_t[RM\cdot]^2 = 0 \tag{2.6}$$

Now, the species $R\cdot$ is in its own steady state, determined by the balance between Eq. 2.2 and the rate of initiation, R_i, from Eq. 2.1, as shown in Eq. 2.7, which can be rearranged to give Eq. 2.8:

$$\frac{d[R\cdot]}{dt} = 2k_d[I] - k'[R\cdot][M] = 0 \tag{2.7}$$

$$[R\cdot] = 2\frac{k_d[I]}{k'[M]} \tag{2.8}$$

Substituting into Eq. 2.6 and solving for $[RM\cdot]$ yields Eq. 2.9:

$$[RM\cdot] = \sqrt{\frac{k_d[I]}{k_t}} \tag{2.9}$$

When Eq. 2.9 is substituted into Eq. 2.5, Eq. 2.10 results:

$$-\frac{d[M]}{dt} = k_p[M]\left(\frac{k_d[I]}{k_t}\right) \tag{2.10}$$

Thus, for a typical radical polymerization, the rate of monomer consumption, $-d[M]/dt$, is first order in monomer and half order in initiator. The common situation, especially for kinetic studies, is that initiators are used at temperatures such that the polymerization is completed before significant depletion of the original amount of initiator occurs. In other words, the initiator concentration $[I]$ is typically constant during the course of polymerization (and equal to $[I]_0$). It is also important to note that, while the initiator decomposition rate $k_d[I]$ can typically be measured independently of the polymerization, there is no routine way to individually determine k_p and k_t. Instead, kinetic analysis of simple polymerization rate data gives only a value for $k_p/k_t^{1/2}$. The rotating sector technique, discussed by Odian,[26] is a method that involves rather complicated experiments with photoinitiators and has been used to obtain separate values for k_p and k_t in some systems.

The foregoing description is based on a thermally decomposing initiator. Redox couples are another class of radical initiators that are particularly useful for superabsorbents. In such cases an oxidizing agent and a reducing agent react to form at least one active radical, which then initiates polymerization:

$$Red + Ox \xrightarrow{k_{red}} R\cdot + byproducts \tag{2.11}$$

2.4.2. Kinetic Dependence of Molecular Weight

The molecular weight of a given polymer chain is determined by how much propagation occurs between initiation and termination of that chain. Closely related to the molecular weight of a polymer is the kinetic chain length, ν, defined as the average number of monomer units incorporated into a growing chain. When disproportionation is the primary termination mechanism, the kinetic chain length is equal to the degree of polymerization, which equals the polymer molecular weight divided by the monomer molecular weight. When recombination dominates, the degree of polymerization is twice the kinetic chain length. Using our standard system with a thermal initiator, Eq. 2.12 relates the kinetic chain length (and, therefore, polymer molecular weight) to kinetic parameters (see Odian[26]):

$$\nu = \frac{k_p[M]}{2(k_d k_t[I])^{1/2}} \tag{2.12}$$

Note that molecular weight does not generally increase as polymerization proceeds in a free-radical induced addition polymerization. In fact, frequently the molecular weight of polymer made late in the polymerization is lower than that made earlier, as is suggested by Eq. 2.12, because of the decreasing ratio of the concentrations of monomer to initiator.

2.4.3. Inhibition and Retardation

Certain substances, called *retarders*, reduce the rate of polymerization compared with the rate that is obtained with highly purified monomers. Another category of substances, *inhibitors*, essentially prevents polymerization from occurring at all until they are consumed. After an induction time, the reaction proceeds at its normal rate. Both retarders and inhibitors react with initiator fragments or growing chains to terminate radicals that would otherwise lead to polymer formation. Retarders are simply less efficient and do not terminate all the radicals.

The basic kinetic analysis of inhibited and retarded polymerizations begins with the addition of an inhibition reaction (Eq. 2.13) to the initiation, propagation, and termination reactions just discussed (Eqs. 2.1–2.4). In this description, Z represents the inhibitor or retarder species.

$$RM\cdot + Z \xrightarrow{k_t} RM\ (polymer) + Z\cdot \tag{2.13}$$

Assuming that the inhibitor radicals $Z\cdot$, neither reinitiate polymerization nor reform the inhibitor Z, the rate of change of concentration of growing radicals, $d[RM\cdot]/dt$, is represented at steady state by Eq. 2.14, which is similar to Eq. 2.6 except that now there are two paths for disappearance of the species $RM\cdot$:

$$\frac{d[RM\cdot]}{dt} = k'[R\cdot][M] - 2k_t[RM\cdot]^2 - k_z[Z][RM\cdot] = 0 \tag{2.14}$$

The substitution for $[R\cdot]$, defined in Eq. 2.8, can still be made, and the rate of propagation is still given by Eq. 2.5, which leads to Eq. 2.15 wherein the rate of propagation $-d[M]/dt$ is denoted as R_p and the rate of initiation as R_i:

$$\frac{2R_p^2 k_t}{k_p^2[M]^2} + \frac{R_p[Z]k_z}{k_p[M]} - R_i = 0 \tag{2.15}$$

Elaboration on both the derivation and utilization of this expression is given, for example, by Odian.[26] For our purposes here, note that the rate of polymerization is inversely proportional to z, defined in Eq. 2.16 as the ratio of the rate constant for inhibition, k_z, to the rate constant for propagation, k_p. Powerful inhibitors have large values of z.

$$z = k_z/k_p \tag{2.16}$$

2.4.4. Chain Transfer

Being a highly reactive species, the growing free radical can enter into reactions other than normal propagation and termination. An important class of such reactions is chain transfer, in which an atom, generally hydrogen, is abstracted from any of several sites in the polymerizing system by the growing radical. The growth of the original chain stops, but a new free radical is left at the site of the leaving atom, and this new site may then begin a chain. (If this new site is *not* sufficiently reactive to propagate, then the event would be categorized as inhibition or retardation rather than chain transfer.) The interesting aspect of chain transfer is that, in general, there is little effect on the overall rate of polymerization, but, since the chain is interrupted, there may be a profound impact on chain structure, i.e., branching, and molecular weight.

Chain transfer reactions are typically identified by the target moiety such as transfer to monomer, initiator, or added chain transfer agents (CTA). Incorporation of these reactions into a description of molecular weight leads to the rather complicated Mayo equation.[26] The case for added chain transfer agents allows a simplification of the Mayo equation to Eq. 2.17:

$$\frac{1}{\nu} = \frac{1}{\nu_0} + C\frac{[CTA]}{[M]} \tag{2.17}$$

C is the ratio of the rate constant for addition to the CTA to the rate constant of propagation, k_p; it is termed the *chain transfer constant.*

Another target of chain transfer may be the polymer backbone itself. Because polymer growth is maintained, the molecular weight is not directly affected. However, chain transfer to backbone is of particular interest in superabsorbents because it results in branching, and branching reactions can lead to crosslinking if two branches terminate by recombination. Evidence that this takes place in poly(acrylate) polymerizations are patent citations concerning the manufacture of superabsorbents without the use of any primary crosslinker.[27] Chain transfer to polymer reactions only become important when there is considerable polymer present, i.e., late in the course of polymerization. Quantifying the extent of this reaction is difficult, and no general values are available for chain transfer constants of poly(acrylic acid) under the conditions of interest.

2.5. UNIQUE ASPECTS OF ACRYLIC ACID AND METHACRYLIC ACID POLYMERIZATION

2.5.1. Effect of pH and Ionic Strength

Most polymers, and thus most polymerization studies, are made in organic solvents. Experiments in water reflect many of the unusual features of this ubiquitous yet extraordinary solvent. In the particular case of acrylic and methacrylic acids, the monomeric acids differ greatly from their salts in reactivity, and thus the course of polymerization is a strong function of pH. For example, in 1952 Pinner[28] reported that the potassium persulfate initiated polymerization of methacrylic acid proceeded much more slowly as pH was increased. Similar results had been seen by Melville[29] with photosensitized initiation. Blauer[30] reported data on the same system with the sparingly water-soluble, nonionic initiator, 2,2′-azobis(isobutyronitrile). He further noted the potential effect of termination rate changes with pH.

Other early studies quantifying the significant effect of pH, determined by the extent of neutralization, on the polymerization rate of acrylic acid were those of Ito et al.[31] and Kabanov et al.[32,33] These workers observed a minimum in the rate of polymerization around pH 7, approximately the pH at which most superabsorbent products are sold. Figure 2.5 is a plot of selected data from these reports illustrating the dependence of rate on pH. Polymerizing an electrolyte like sodium acrylate requires bringing the charged monomer into close proximity to the like-charged growing radical, and this electrostatic repulsion, not present in the case of the weak, largely undissociated acidic monomers, was assumed to slow the propagation reaction in partially neutralized systems compared with unneutralized systems.

It was also observed during the early development of superabsorbents at The Dow Chemical Company that unneutralized acrylic acid was much easier to polymerize into useful gels than were preneutralized systems. Increasing the monomer concentration lessened this difference, presumably because the higher monomer concentration suppressed dissociation, resulting in a smaller fraction of charged species. These higher monomer concentrations resulted in greater heat release during polymerization, however, and hence presented greater difficulty with temperature control. This problem was ultimately solved using evaporative cooling.[34,35]

An increased polymerization rate at increased total concentration suggests an effect of ionic strength. Reports on the addition of salts to aqueous polymerizations are rare but not unknown. Kabanov et al.[32] added NaCl to some of their kinetic runs, in conjunction with their pH study cited above. Salt concentrations were much higher than the relatively low acrylate concentrations. They saw a rate increase that appeared to plateau at $[Na^+]$ above 1 mol/L. They also speculated that at pH 5, where the macromolecular radicals, but not the monomer radicals, are undissociated, there should be no effect of added salt.

For polymerization in dilute solution, sodium acrylate would be expected to be

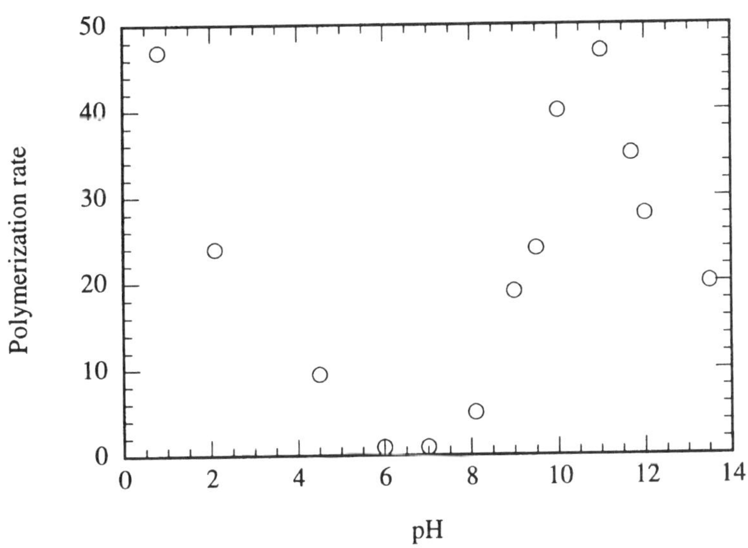

Figure 2.5. Polymerization rate as function of pH. Selected data from Blauer[30] and Ito et al.[31]

dissociated, and this charged ion would surely be repelled by the similarly charged growing chain. Although the monomer is a stronger acid than the polymer, both *salts* would be highly dissociated. Kabanov et al.[33] provide these pK_a values and explain their rate data as the sum of two reaction rates, one for free ions (dissociated) and a much faster one involving ion pairs.

2.5.2. Monomer Concentration Dependence of Polymerization Rate

As noted earlier, the simplest radical polymerization model leads to a rate dependence that is first order in monomer and half order in initiator. Unfortunately, as noted by Ito et al.[31] and Kabanov et al.[32,33] this model did not adequately fit the polymerization rate data. Data by Cutié et al.[36] from isothermal polymerizations of partially neutralized acrylic acid followed by NMR are presented in Figure 2.6 to illustrate this point. A first-order reaction would show the same time dependence for *fractional* conversion even at different initial monomer concentrations. In fact, the polymerization is close to 3/2 order in monomer, as shown in Figure 2.7, but still half order in initiator, as shown in Figure 2.8.

A plausible explanation for this monomer dependence was presented by Cutié et al.[36] It makes use of Eqs. 2.18 and 2.19, a mechanistic scheme essentially similar to that proposed by Ishige and Hamielec[37] for acrylamide:

$$I \underset{k_{-d}}{\overset{k_d}{\rightleftharpoons}} (2R\cdot)_c \qquad (2.18)$$

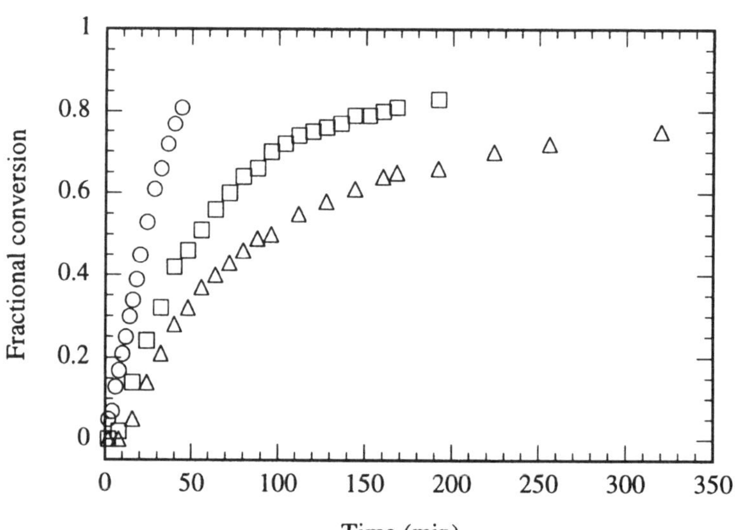

Figure 2.6. Fractional monomer conversion as a function of polymerization time for initial monomer concentrations of 4.56 mol/L (○), 2.29 mol/L (□), and 1.17 mol/L (△).

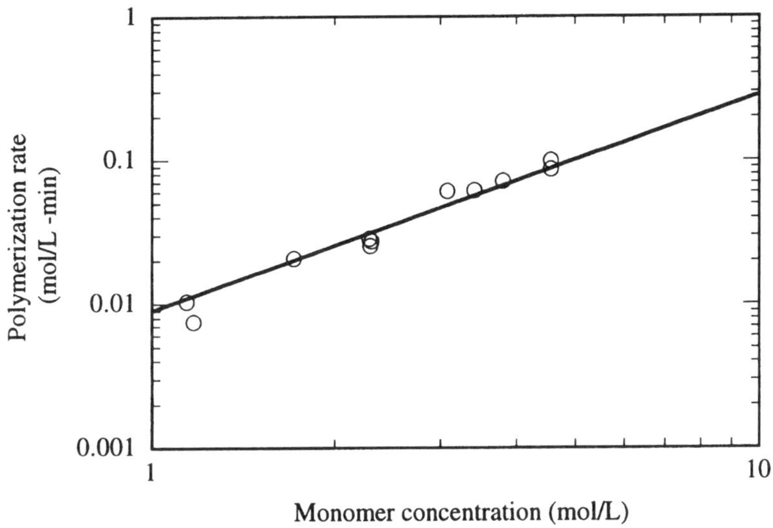

Figure 2.7. Dependence of polymerization rate of 65% neutralized sodium acrylate on initial monomer concentration. Slope is 1.5.

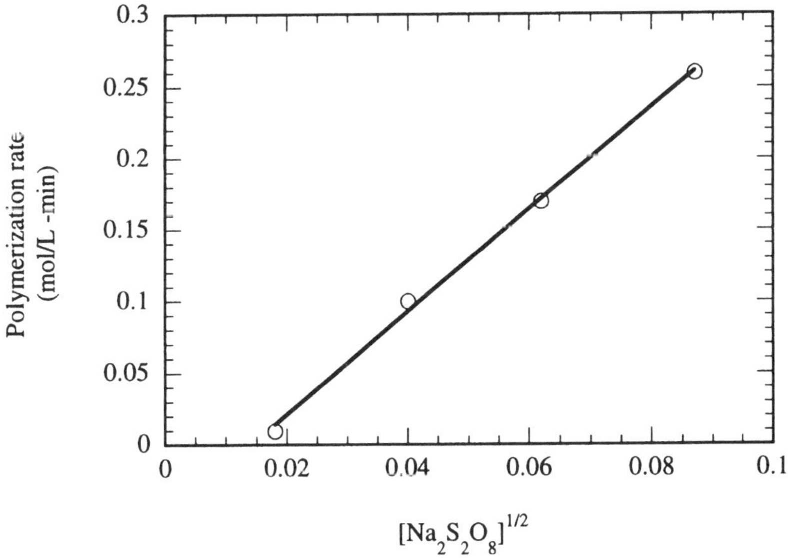

Figure 2.8. Linear dependence of polymerization rate on the square root of initiator concentration.

$$(2R\cdot)_c + M \xrightarrow{k_x} RM\cdot + R\cdot \tag{2.19}$$

The remainder of the polymerization follows Eqs. 2.3 and 2.4 outlined earlier. In Eq. 2.18, $(2R\cdot)_c$ represents two radical species trapped in a solvent cage. Only when these caged radicals are intercepted by monomer, Eq. 2.19, does polymerization ensue. Manickam et al. proposed direct reaction between monomer and persulfate (i.e., $I + M \rightarrow$ *growing radical*) for both methacrylic acid[38] and acrylic acid.[39] While this would give similar kinetics, the rate dependence of the independently measured initiator decomposition rate would be expected to depend directly on monomer concentration, which was not seen.[25]

Since $RM\cdot$ is still the primary propagating species in this description and Eq. 2.5 still defines the rate of propagation, determining the rate of polymerization reduces to finding an expression for the steady-state concentration of $RM\cdot$ from Eq. 2.20, which is analogous to Eq. 2.6:

$$\frac{d[RM\cdot]}{dt} = k_x[M][(2R\cdot)_c] - 2k_t[RM\cdot]^2 = 0 \tag{2.20}$$

From Eq. 2.18, a dynamic equilibrium, the concentration of caged radicals $[(2R\cdot)_c]$ can be defined as in Eq. 2.21:

$$[(2R\cdot)_c] = \frac{k_d}{k_{-d}}[I] \tag{2.21}$$

Substituting Eq. 2.21 into Eq. 2.20 yields Eq. 2.22:

$$[RM\cdot] = \sqrt{\frac{k_x k_d}{2k_t k_{-d}}[M][I]} \tag{2.22}$$

Utilizing Eq. 2.5 results in a propagation expression, Eq. 2.23, that describes the observed concentration dependence of the rate:

$$-\frac{d[M]}{dt} = k_p\left(\frac{k_x k_d}{2k_t k_{-d}}\right)^{1/2}[M]^{3/2}[I]^{1/2} \tag{2.23}$$

The effects of pH and ionic strength can be incorporated as empirical values for the overall rate of propagation. It is also important to consider the variation in k_d, the decomposition constant for sodium persulfate, as a function of the same parameters. The temperature dependence of the rate constant k_d has been determined under the same conditions of concentration as used in the studies above by Henton et al.,[25] and the temperature dependence for the overall decomposition for the data in Table 2.4 is based on that study.[36]

As discussed in section 2.4.2, the molecular weight of a polymer, in the absence of significant chain transfer, is essentially controlled by the ratio of propagation rate to termination rate (or initiation rate, since they are equal at steady state). The impli-

Table 2.4. Comparison of polymerization rate measurements

Reference	T(°C)	pH	$[M]_0$	Initiator type[a]	$[I]_0$	$R_p/[M]^{3/2}[I]^{1/2}$
Manickam et al.[38]	50	1	0.138	KPS	0.01	0.491–0.586
	50	4.2	0.172	KPS	0.01	0.236
	50	11	0.258	KPS	0.01	0.091–0.444
Kabanov et al.[32]	60	1	1.2	AIBN	0.005	0.14
	60	10	1.2	AIBN	0.005	0.19
	60	~7	1.2	AIBN	0.005	0.00278
Ito et al.[31]	50	2.4	0.5	APS	0.00285	0.90
	50	4.7	0.5	APS	0.00285	0.14
	50	7.2	0.5	APS	0.00285	0.045
Cutié et al.[24]	55	4.5	1.14–4.56	NPS	0.0023	0.126–0.239

[a]KPS, potassium persulfate; AIBN, 2,2′-azobis(isobutyronitrile); APS, ammonium persulfate; NPS, sodium persulfate.

cation for a batch polymerization is that the molecular weight will likely decrease as the polymerization proceeds, because the rate of propagation decreases while termination and initiation remain relatively constant. In Figure 2.9 the cumulative molecular weight data from identical isothermal polymerizations at 55°C are plotted versus the extent of conversion, illustrating this point.[36] Two notes should be made of these data: (1) The molecular weight decreases significantly with conversion, which, while anticipated, confirms the likelihood of forming a low-molecular-weight fraction late in the polymerization; and (2) the molecular weight values are large compared with those for many other polymers. The unusually large molecular weight of many commercial acrylic acid (and acrylamide) polymers is utilized in the application of the soluble (uncrosslinked) polymers as viscosifiers and flocculants.[40] This is probably a reflection of their high propagation rates, compared with those for other monomers, rather than low termination or initiation rates. For example, from a comparison of chemical structures one might expect the polymerization of acrylic acid to proceed at a rate similar to that for methylacrylate. In fact, at 60°C the propagation of aqueous acrylic acid, as determined from rotating sector experiments,[41] is faster than that of methylacrylate.[26] Results of Laborie[42] and of Chapiro and Dulieu[43] suggest that the high value for the propagation rate results from an interesting templating effect at high concentrations due to hydrogen bonding with water. In aprotic solvents, without such hydrogen bonding, the rates are much lower.

2.5.3. Inhibition

As noted in section 2.2.2.1, commercial monomer grade acrylic acid typically contains 200 ppm MEHQ, and it is recommended that it be stored under an air atmosphere. This is a high level of inhibitor when compared with that used in other vinyl

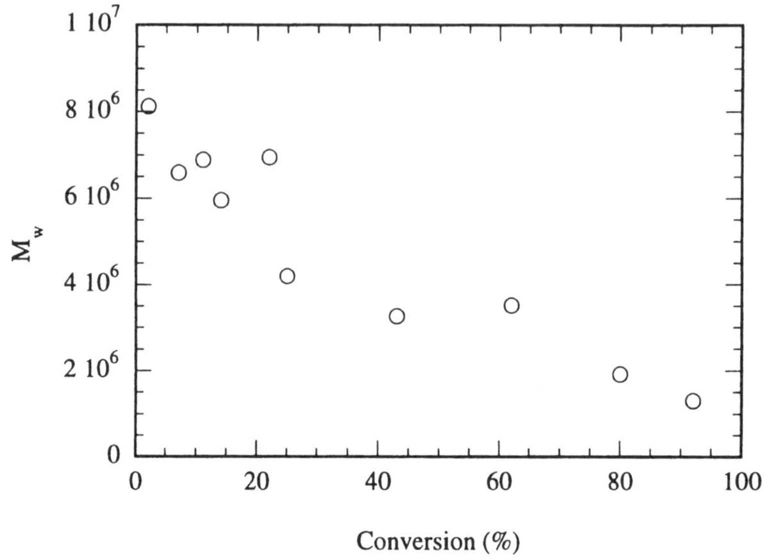

Figure 2.9. Effect of conversion on cumulative molecular weight (M_w). Polymerization was at 65% neutralization and 55°C.

monomers. Styrene, for example, is inhibited with 3–10 ppm of *tert*-butylcatechol. Reproducible polymerizations of acrylic acid generally begin with the deoxygenation of the system, and induction times vary roughly inversely with oxygen concentration.

The role of oxygen in the stability of vinyl monomers is complicated. For example, most purified vinyl monomers will not polymerize upon heating in the absence of oxygen (styrene being an exception), but virtually all vinyl monomers will readily form peroxides upon exposure to air. These peroxides have been found to break down and initiate polymerization at elevated temperatures.[44] At lower temperatures, where the peroxides are relatively stable, oxygen often reacts with other free radicals in the system, thereby functioning as an inhibitor. This delicate balance between inhibition and reinitiation is influenced by temperature, the oxygen content of the system, and the thermal stability of the peroxide formed from the particular monomer. In a closed system, oxygen is eventually depleted and polymerization proceeds. In contrast, when there is an unlimited supply of oxygen, such as in an open vessel, vinyl monomers will absorb a large quantity of oxygen, forming polyperoxides but not polymerizing.[45]

Even trace amounts of oxygen in otherwise very pure systems can suppress the polymerization of acrylic or methacrylic acids. Blauer[46] reported that even the small amounts of oxygen remaining in *purified* nitrogen would retard the polymerization of methacrylic acid if continuously bubbled through the solution during the reaction. Levy[47] has shown that, in the absence of oxygen, there is no induction time at 100°C for MEHQ-inhibited acrylic acid, and no detectable amount of MEHQ is consumed or reacts with the polymer radicals. By continuously monitoring the oxygen level in the reaction vessel, Kurland[48] determined the oxygen consumption during polymerization of neat acrylic acid at 50°C containing relatively high (106 ppm) and low (1.7 ppm) levels of MEHQ. At the lower level, a rapid increase in the rate of oxygen consumption was observed once the MEHQ was depleted. Initially about six molecules of oxygen were consumed per initiator radical generated, but, when the MEHQ was depleted, about 32 molecules per radical were consumed. Regardless of the MEHQ content, polymerization occurred when the oxygen content was reduced to about 0.5 ppm, clearly showing that the polymerization is inhibited by oxygen, not MEHQ. The rate of oxygen consumption, however, is much slower in the presence of higher levels of MEHQ.

Levels of MEHQ ranging from 16 to 151 ppm have been found in commercial superabsorbent polymers, as indicated in Table 2.5. The amount of MEHQ that survives the polymerization process may be the result of the efficiency of deoxygenation, the reaction temperature, and, most important, the nature of the polymerization catalyst. Persulfates are more powerful oxidants than azo compounds and more efficient at hydrogen abstraction; they may therefore convert more of the MEHQ to quinone.[24,49] The latter has been detected as a byproduct of the MEHQ-inhibited acrylic acid polymerization initiated with sodium persulfate.

The mechanism for inhibition/retardation presented in section 2.4.3 implies that, if the inhibition constant z (= k_z/k_p) is large, significant polymerization will not occur until the species Z is completely consumed (induction period), after which time the rate is normal (inhibition). If the effect is milder (z is smaller), polymerization occurs from the outset but at a reduced rate (retardation).[26] MEHQ in acrylic acid may not fit cleanly into either category. In a study of the isothermal persulfate-initiated polymerization of acrylic acid at various MEHQ levels,[24] the initial observations were the following:

1. In deoxygenated runs, polymerization began almost immediately.
2. The rate was lowered by the addition of MEHQ.
3. MEHQ was found in the polymer after essentially complete conversion.

Table 2.5. Residual MEHQ levels in selected superabsorbent polymer samples from eight manufacturers, over several years

Manufacturer Designation	MEHQ Content (ppm)
A	151, 67
B	109
C	32, 84
D	94, 46
E	46
F	54, 32
G	33
H	30, 34, 16

These three factors all indicate that MEHQ behaves like a retarder in this system. In other words, the reaction described in Eq. 2.13 would be significant but would not prevent considerable "normal" polymerization from occurring. Referring to Eq. 2.14, this would be a situation in which the third term $k_z[Z][RM\cdot]$ dominates the second term $2k_t[RM\cdot]^2$ sufficiently to reduce the rate by lowering the radical concentration. This in turn would imply that $[RM\cdot]$, and hence the polymerization rate, would depend on the *first power of initiator concentration.* In addition, one would expect chain termination by a small molecule (Eq. 2.13) to result in half the average molecular weight seen with normal recombination of growing polymer chains.

However, it was found that the rates of polymerization at various levels of MEHQ and persulfate depend approximately on the half power of the initiator, as would be expected for an uninhibited reaction. Furthermore, molecular weights show either little change with increased inhibitor level or actually appear to increase. These deviations from the expected retardation behavior can be accommodated by a novel mechanism described in Eqs. 2.24 and 2.25 (this approach was originally proposed to us by D.J. Arriola of The Dow Chemical Co.) If the radical originating from the initiator fragment $R\cdot$ has a noticeably different reactivity with the inhibitor or the monomer from that of polymer-based radicals, the following reaction scheme should be used:

$$R\cdot + Z \xrightarrow{k_z} inactive \tag{2.24}$$

$$R\cdot + M \xrightarrow{k'} RM\cdot \tag{2.25}$$

The reaction of $RM\cdot$ with Z (Eq. 2.13) does not occur to a significant extent, but the remainder of the polymerization follows the mechanism described in section 2.4.1.

The concentration of the initiator fragment $[R\cdot]$ is still at steady state, as in Eq. 2.26. Rearranging Eq. 2.26 results in an expression for the steady-state concentration of $[R\cdot]$, Eq. 2.27:

$$\frac{d[R\cdot]}{dt} = 2k_d[I] - k_z[R\cdot][Z] - k'[R\cdot][M] = 0 \tag{2.26}$$

$$[R\cdot] = \frac{2k_d[I]}{k_z[Z] + k_i[M]} \tag{2.27}$$

Substituting this expression for the steady-state concentration of initiator radicals into Eq. 2.14 yields Eq. 2.28:

$$[RM\cdot] = \left(\frac{k_d[I]}{k_t}\right)^{1/2}\left(\frac{k_i[M]}{k_z[Z] + k_i[M]}\right)^{1/2} \tag{2.28}$$

By inverting and squaring Eq. 2.28, Eq. 2.29 is obtained:

$$\frac{1}{[RM\cdot]^2} = \frac{k_t}{k_d[I]}\left(1 + \frac{k_z[Z]}{k_i[M]}\right) \tag{2.29}$$

Largely independent of polymerization mechanism, a pseudo-first-order rate constant for a polymerization (i.e., the slope of the ln$[M]$ vs. time curve) is equal to $k_p[RM\cdot]$, as apparent from Eq. 2.5. Different mechanisms primarily affect the value of $[RM\cdot]$. If one had pseudo-first-order rate constants at various inhibitor levels $[Z]$, but at a fixed series of initiator concentrations $[I]$, one could plot $1/(k_p[RM\cdot])^2$ vs. $[Z]$ for each $[I]$ and determine k_z/k_i. From experiments at fixed levels of inhibitor and varying, though irregular, levels of initiator, appropriate values of $k_p[RM\cdot]$ were interpolated for fixed initiator concentrations, and the plot described is shown in Figure 2.10. Values of k_z/k_i obtained from this plot are given in Table 2.6.

These values are remarkably constant. They are rather high compared with the values for the inhibition constant z ($= k_z/k_p$) given by Odian.[26] We should note, however, that k_i in this case is probably significantly smaller than k_p. If k_i were equal to or greater than k_p, there would be no apparent effect of MEHQ on the kinetics, because nothing would "intercept" the initiator fragment but monomer. Since k_i must be smaller than k_p, k_z/k_i would be larger than the value expected for k_z/k_p from Odian's table.[26] This is the case, and thus there is internal consistency. In summary, the data are adequately explained if the MEHQ inhibitor is inferred to primarily intercept the initiator radical rather than the growing polymer radical.

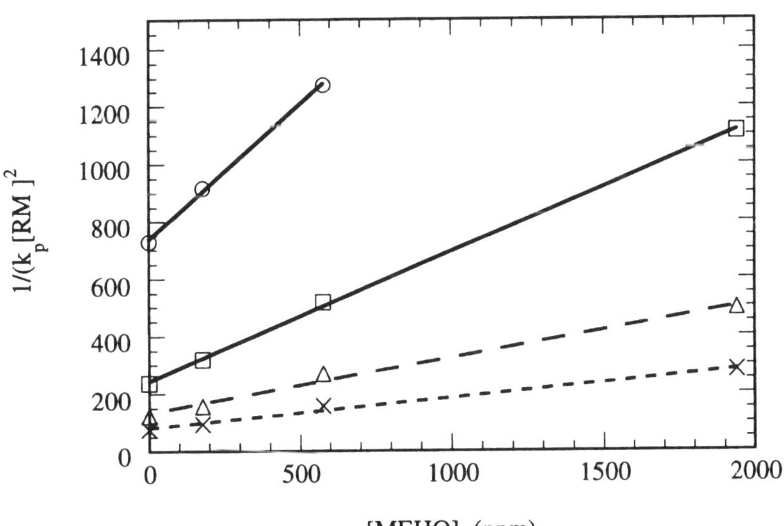

Figure 2.10. Dependence of inverse square of $k_p[RM\cdot]$ on the concentration of MEHQ for several values of $[I]$. $[I]$ = 0.0032 (○), 0.0064 (□), 0.010 (△), 0.0144 (x).

Table 2.6. Calculated values of k_z/k_i for various initiator concentrations

Interpolated [I] (mol/L)	k_z/k_i
0.0016	2,405
0.0036	3,506
0.0064	2,711
0.01	2,409
0.0144	2,523

2.5.4. Reactions Near Dryness

2.5.4.1. Continuation of Polymerization. Drying is an essential step in nearly every superabsorbent manufacturing process (see Chapter 3, section 3.2.5). Reactions normally studied in a polymerization reactor often continue during drying operations; events occurring in this regime may be particularly important in determining residual monomer levels in the final product. The impact on residual levels of the breakdown of co-polymerized acrylic acid dimer is discussed in section 2.2.2.2; but in the present section we consider the effect of the continuation of polymerization during drying. Both initiator and moisture levels are changing rapidly and independently of one another during drying, and this situation dramatically affects the polymerization rate of any remaining monomer. As an illustration of this interaction, a patent by Irie et al.[50] claims that the humidity of the drying air must be controlled so that polymer does not dry too quickly, thereby extending the polymerization reaction.

Studies of the rate of decrease in residual monomer concentration using nearly dry (down to 5% moisture) polymer samples held in closed vessels under nitrogen yield results that can be extrapolated from those obtained at reactor conditions. In other words, R_p is dependent on $[M]^{3/2}$ and $[I]^{1/2}$. In fact, at the increased solids content, the residual monomer can be reduced to undetectable levels under nitrogen. Similar studies of monomer reduction in closed systems in an air atmosphere show that oxygen inhibits this last bit of polymerization, as would be expected (S.S. Cutié et al., The Dow Chemical Co., unpublished results).

Very dry samples of sodium polyacrylate, for example, below 5% moisture, show reduced diffusion rates for moisture and oxygen. This can be interpreted as resulting from the transition of the polymer to a glassy state at reduced moisture. An inflection in the heat demand as measured by differential scanning calorimetry has been related to a glass transition temperature T_g. (see Chapter 4, section 4.5.2.1). A plot of this inflection temperature versus the weight loss obtained from thermal gravimetric analysis for a series of sodium polyacrylate gel samples is shown in Figure 2.11 (A.J. Pasztor, Jr. et al., The Dow Chemical Co., unpublished results). The effect of the state of the reacting polymer gel, whether glassy or liquid, on other reactions is under study.

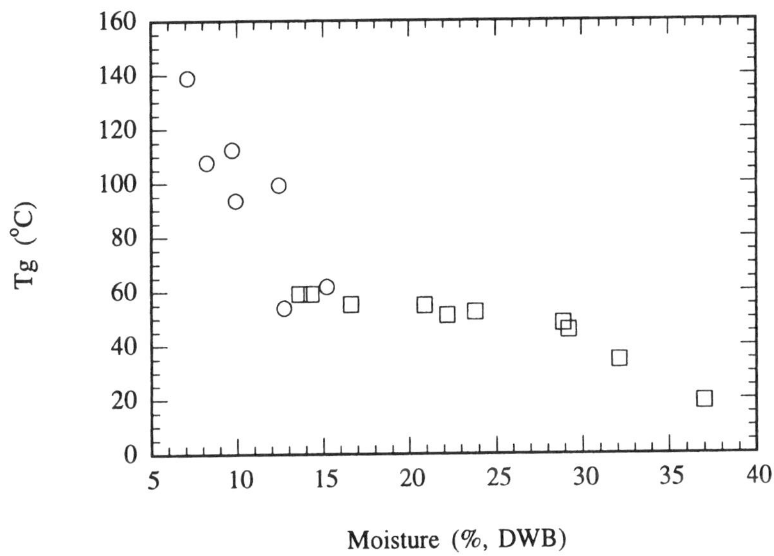

Figure 2.11. Glass transition temperature, T_g, as a function of moisture contained in the polymer, for a 65 % neutralized sodium polyacrylate, dried from the gel (○) or by adding moisture using a constant humidity atmosphere (□).

2.5.4.2. Anhydride Formation and Decarboxylation. The principal gaseous products evolved when superabsorbent polymer is heated to temperatures of 300°–400°C are water and carbon dioxide (see Chapter 4, section 4.5.3.2). These products were found under both aerobic and anaerobic conditions.[51–53]

Water may result from the formation of anhydride as neighboring carboxyl groups combine with the release of one mole of water (Scheme 2.4).[54] Because for small molecules hydrolysis of anhydride is more rapid at high pH, it may be expected that anhydride formation in polyacrylates would occur more readily, and the anhydride formed would be more stable, at low pH. The reaction can be followed using infrared spectroscopy and by monitoring water loss. In a polymeric system, intermolecular anhydride formation (but not intramolecular) may result in additional crosslinks. These added crosslinks will be susceptible to hydrolysis when the polymer absorbs water.

Anhydride formation has been observed in the infrared spectra of films of poly(acrylic acid) heated to 110°C[54] and in films of slightly neutralized polyacry-

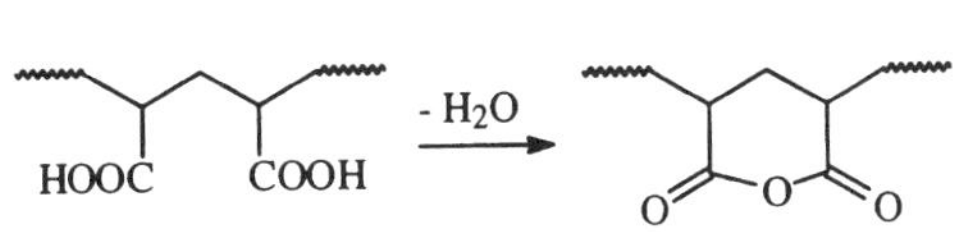

Scheme 2.4

lates heated to 250°C for 10 min.[55] Anhydride formation in other experiments at 250°C with polyacrylate polymers was found to be inversely proportional to the level of neutralization of the superabsorbent polymer (see Chapter 4, Fig. 4.19) (M.L. McKelvy, S.S. Cutié, The Dow Chemical Co., unpublished results). Temperatures above 225°C were necessary to initiate significant anhydride formation. Measurement of moisture loss versus heating time at 200°C suggested a two-step process consisting of an initial rapid rate of water loss followed by a slower process. However, at high extents of neutralization (50%–75%), evaporation of tightly bound water, not anhydride formation, was the suspected route for weight loss because sufficient acid functionality to support dehydration to anhydride is not present.

The decarboxylation process for small molecule organic acids and acid salts has been reviewed.[56,57] Decarboxylation of polyacrylates has been the subject of a number of recent studies, both in solution[58] and in the solid state.[55,59] The reaction has been studied by monitoring evolved gases using a variety of techniques, particularly thermogravimetry (TG) and pyrolysis combined with detection methods, such as infrared (IR) spectroscopy and mass spectroscopy (MS).

Published reports of CO_2 evolution in dry polyacrylates indicate that temperatures of 300°–400°C are required to observe significant decarboxylation.[51–53] Using TG/MS, however, low levels of carbon dioxide evolution have been observed when heating 60% neutralized superabsorbent in air at temperatures as low as 220°C (S.W. Froelicher, The Dow Chemical Co., unpublished results). Also, decarboxylation has been proposed as an explanation for results from heating polyacrylate films containing 10 mol% univalent metal ions at 250°C for 10 min.[55] IR absorptions at 1,795 cm^{-1}, attributed to anhydride, appeared initially. With further heating, the anhydride absorption bands disappeared. The chemical resistance of the films resulting from this process was attributed to the formation of crosslinks having ketone functionality, formed by intermolecular reaction with loss of CO_2. In contrast to anhydride crosslinks, ketone crosslinks will be stable to hydrolysis.

Buzanowski et al.[59] examined the products resulting from pyrolysis of polyacrylates neutralized to various extents. Cyclopentadiene was observed as a reaction product when samples neutralized in the range 20%–100% were rapidly heated to 700°C. The proposed reaction mechanism requires formation first of an anhydride, followed by decarboxylation to a cyclic ketone, and further reaction of the ketone to 1,3-cyclopentadiene. The proposed mechanism explains the acceleration of the decarboxylation reaction observed when some carboxylate ion is present. However, since little or no anhydride was detected in samples neutralized to 50% or greater and heated for 10 min at 250°C, as noted above, it seems likely that a different mechanism may be operating for the decarboxylation under these conditions. A mechanism has been proposed for the decarboxylation of barium acetate that does not involve anhydride intermediates but does lead to aldehyde functionality.[60] When applied to polymer salts at high neutralization, this mechanism would not lead to anhydride formation but could lead to ketone-type crosslinks and might be consistent with the formation during pyrolysis of cyclopentadiene, as observed by Buzanowski et al.[59]

2.5.5. Polymerization in Nonaqueous Solvents

Unneutralized acrylic acid is moderately soluble in numerous organic solvents, but its polymer is not. Soluble monomer and insoluble polymer allows the precipitating polymerization of unneutralized acrylic acid. This has been used for years in the production of branched poly(acrylic acid) thickeners such as the Carbopol resins of B. F. Goodrich, which have a similar structure to superabsorbent gels but are not crosslinked.[61] A particularly interesting medium for such work is supercritical carbon dioxide. This has been taught by Denzinger and Hartmann,[62] and recent work by Romack et al.,[63] has expanded the experimental range.

Several studies reported by Gromov, et al.[64] give polymerization rates for acrylic acid and related monomers in nonaqueous solvents. The general observation made by these authors is that the polymerization is first order in monomer, i.e., the typical behavior seen in free-radical polymerizations (Eq. 2.10), as opposed to the more unusual 3/2 power dependence seen in aqueous systems and discussed in section 2.5.2.

The alkali metal salts of both acrylic acid monomer and its polymer are virtually insoluble in organics. This limited solubility has allowed the commercial production of superabsorbents via suspension polymerization (see Chapter 3, section 3.3). This approach typically comprises dispersing an aqueous phase of neutralized monomers in a hydrocarbon, followed by initiation, either with an aqueous phase thermal initiator or often with a redox system using gaseous SO_2 as a reducing agent. It is important to note that a suspension polymerization is kinetically similar to a solution polymerization, each droplet acting as a small reactor.

2.6. CROSSLINKED GELS BY CO-POLYMERIZATION

2.6.1. Standard Co-Polymerization Kinetics

To be useful as superabsorbents, linear polymers must be crosslinked to form gels. Two general approaches exist for crosslinking. The first method is through co-polymerization of acrylic acid with a second monomer, which has multiple vinyl groups, to form a gel directly *during* the polymerization process. The second approach is via crosslinking of a polymer by reaction with a di-functional molecule, similar to a rubber curing reaction. The first approach is discussed in the following section and the latter in section 2.7.

Adequate description of a crosslinking co-polymerization requires at least a summary of co-polymerization in general. The model outlined here was developed by Alfrey and Goldfinger[65] and extended to systems containing many monomers by Walling and Briggs.[66] The model itself is known as the Mayo–Lewis model[67] and is the most widely used model for co-polymer systems. Standard texts (e.g., Odian[26]) expand on the summary given here.

The model is based on the concept that the nature of the very last monomer added to the growing chain controls its reactivity. In a simple co-polymerization of two monomers, M_1 and M_2, there are four possible propagation events. The growing chain with a monomer 1 moiety on the end $RM_1\cdot$ can react with another molecule M_1 with a rate constant k_{p11} or with a molecule M_2 with a rate constant k_{p12}. Likewise, when monomer 2 moiety $RM_2\cdot$ is at the end of the growing chain, it can add to M_1 with a rate constant k_{p21} or to M_2 with a rate constant k_{p22} as shown in Eqs. 2.30–2.33.

$$RM_1\cdot + M_1 \xrightarrow{k_{p11}} RM_1M_1\cdot \tag{2.30}$$

$$RM_1\cdot + M_2 \xrightarrow{k_{p12}} RM_1M_2\cdot \tag{2.31}$$

$$RM_2\cdot + M_1 \xrightarrow{k_{p21}} RM_2M_1\cdot \tag{2.32}$$

$$RM_2\cdot + M_2 \xrightarrow{k_{p22}} RM_2M_2\cdot \tag{2.33}$$

The reactivity ratios r_1 and r_2 are defined by Eqs. 2.34 and 2.35:

$$r_1 = k_{p11}/k_{p12} \tag{2.34}$$

$$r_2 = k_{p22}/k_{p21} \tag{2.35}$$

When $r_1 > 1$, radical $RM_1\cdot$ preferentially reacts with monomer M_1 (i.e., homopolymerizes), instead of monomer M_2. Very frequently $r_1 < 1$, which indicates $RM_1\cdot$ preferentially reacts with monomer M_2. This latter behavior can lead to alternating co-polymers. The reactivity ratio can also be zero, indicating that a monomer will not homopolymerize.

The composition of a co-polymer is controlled by both the reactivity ratios of the monomers and their concentrations. Analysis of the polymer composition formed *incrementally* is required for comparison to theory. Several approaches have been developed to determine the reactivity ratios of monomers.[68] The most frequently cited method is that of Kelen and Tudos.[69,70] Monomer feed mixtures of varying composition are polymerized to low conversion, thereby approximating incremental polymerization, and the composition of the resultant polymer is determined. The values of r_1 and r_2 are determined graphically from these data. This method requires multiple polymerizations as well as quantitative characterization of the co-polymer composition from each polymerization. This is difficult for systems with low levels of co-monomer (< 1%), as is the situation for crosslinkers in superabsorbents, and any analysis is more complicated in a gelled system.

Another approach utilizes NMR analysis to measure the fractions of, for example, M_1M_2 dyads and $M_1M_2M_1$ triads in the polymer chain rather than just the overall composition. The frequencies of these permutations can be used to calculate r_1 and r_2 values with data from fewer polymerizations than required by the Kelen-Tudos method.[71–74] However, with this method as well, there is little sensitivity in the sequence frequencies at low levels of co-monomer. For example, at low M_2 concentration most occurrences of M_2 in the backbone are $M_1M_2M_1$ rather than $M_1M_2M_2$, and thus there is little information to be gained by triad analysis.

Temperature typically has only a small influence on reactivity ratios. When a temperature effect is observed, increasing temperature causes the reactivity ratios to approach unity,[75] implying that the radicals are less selective at higher temperatures. Because activation energies are low for radical propagation reactions, temperature would be expected to have only a minor influence on their relative rates and thus the reactivity ratios.

Similarly, for polymerizations in organic solvents, reactivity ratios have been found to be only weakly influenced by the reaction medium.[76] However, for very polar monomers in polar solvents, larger changes in reactivity have been observed with changes in the solvent.[77–79] Indeed, ionizable monomers, such as acrylic acid, exist in both the protonated form and ionized form, and the two species exhibit significantly different reactivities. The reactivity ratio observed for acrylic acid reflects the relative amounts of the protonated and ionized forms present in solution. Both monomer concentration and the dielectric constant of the solvent influence the extent of ionization and thus the reactivity ratio of monomers such as acrylic acid.[80–81] Ponratnam and Kapur[82] reported a maximum for r_1 (0.92) at pH 2 and a minimum (0.33) at pH 7 in the co-polymerization of acrylic acid (M_1) and acrylamide. Ponratnam and Kapur[83] also observed a similar effect of pH in the co-polymerization of acrylic acid and *N*-vinylpyrrolidone. Reactivity ratios also change with conversion,[84] which can be viewed as a reaction medium effect. The reactivity ratio of acrylic acid with several monomers is shown in Table 2.7.

Although the reactivity ratios of many monomer pairs have been determined, many more have not, and new monomers are frequently prepared for which data are not available. Alfrey and Price[85] developed the *Q-e* scheme for estimating the reactivity ratios. This approach is based on establishing a pair of values (*Q* and *e*) for each monomer that take into account the resonance stability and polarization characteristics of the radical intermediates. Equations 2.36 and 2.37 give the reactivity ratios for two monomers M_1 and M_2 in terms of their respective *Q* and *e* values.

$$r_1 = \frac{Q_1}{Q_2}\exp[-e_1(e_1 - e_2)] \tag{2.36}$$

$$r_2 = \frac{Q_2}{Q_1}\exp[-e_2(e_2 - e_1)] \tag{2.37}$$

The *Q* and *e* values of a given monomer, as designated by the subscripts, can be derived from the experimental r_1 and r_2 values for various combinations with other monomers; many of these are tabulated.[90,91] Acrylic acid has a *Q* value of 0.83 ± 0.47 and an *e* value of 0.88 ± 0.23.[91] *Q* values range from 0.007 for 1-butene to 16 for 1-chloro-3-buten-2-one, and *e* values range from –2.2 for *tert*-butylvinylsulfide to 3.7 for maleic anhydride.

Table 2.7. Reactivity ratios of acrylic acid in radical co-polymerization

M_1	r_1	M_2	r_2	T (°C)	Solvent
Acrylic acid[86]	0.25	Styrene	0.15	60	None
Acrylic acid[87]	0.05	Styrene	1.6	60	DMF
Acrylic acid[87]	0.13	Styrene	0.30	60	Benzene
Acrylic acid[88]	3.25	Acrylonitrile	0.34	40	Water
Acrylic acid[89]	0.29	MMA	1.13	65	Benzene

2.6.2. Co-Polymerization in Gelling Systems

Copolymerization kinetics as outlined in the previous section was developed to explain composition drift in linear co-polymers such as poly(styrene-co-methyl-methacrylate), but additional considerations arise when co-polymerizing crosslinkers. Consider co-polymerization of a di-functional vinyl crosslinker; after incorporation of one double bond, a new species forms that was not present in the original monomer solution. This species has a pendent double bond, as well as some other unique characteristics that confound the kinetic analysis of a crosslinking co-polymerization. Most obvious is that the new species is now a macromolecular reactant, and so its diffusion to the growing radical chain is slowed more by the viscosity increase resulting from polymerization than is the diffusion of monomeric reactants. Another potential difference is a change in the reactivity of the remaining bond compared with that of the original di-functional molecule. Steric hindrance of the dangling bond is certainly a consideration, but other changes in reactivity, such as through loss of resonance, may occur as well.

The Flory-Stockmayer theory has been used to predict the build-up of a gel network during a crosslinking polymerization.[92–94] A key feature of this theory is the "gel point," which is the extent of conversion when continuous gel (i.e., at least one polymer molecule spans the entire volume) first forms. Side reactions such as intrachain cyclization decrease the efficiency of the crosslinker and result in both a gel point occurring later in the polymerization and a network that is less crosslinked than would be predicted from the amount of added crosslinker.

Deviation from this theory can be significant at high crosslinker concentrations. For example, Okay et. al.[95] determined that almost one-fourth of the double bonds of ethyleneglycol dimethacrylate (EGDM), in a polymerization of EGDM and methyl methacrylate, were consumed by internal cyclization reactions when the mole fraction of EGDM was 0.126 to 0.256. Cyclization reactions occur when a propagating radical adds to a pendent double bond that is part of its chain. This results in loops along the chain that may affect the swelling and elastic properties of the polymer. However, Landin and Macosko[92] showed that for levels of EGDM in methyl methacrylate below 2 wt%, with no other solvent, the rate of intramolecular cyclization was very low. Tobita and Hamielec,[96] however, found approximately 80% cyclization in aqueous co-polymerizations of acrylamide with bisacrylamide when the total monomer content of the solution was approximately 6%. Intramolecular cyclization increases with increasing crosslinker content, decreasing overall monomer content, and increasing conversion.

Efficiency of crosslinker incorporation controls the overall crosslink density in the final gel. The relationship between crosslink density and swelling capacity are discussed in considerable detail in Chapter 5. Another parameter of a gelled system is the ratio of "gel," the crosslinked continuous network, to "sol," that portion of the system that may be branched but is not attached to the network. At the completion of a vinyl gel polymerization, some portion is invariably left in this sol fraction, often referred to as *extractable polymer*. The impact of this fraction on performance properties is developed in Chapter 5, but it is generally detrimental. Higher levels of crosslinking, not surprisingly, minimize the amount of sol remaining at the end of polymerization. Some quantification of this fraction is discussed in the following section.

2.6.3. Polymerization of Acrylic Acid Gels

A wide variety of vinyl crosslinkers have been co-polymerized with acrylic acid to form networks. For polymerizations in organic solvents, almost any multi-vinyl–substituted monomer will cause crosslinking, including crosslinkers such as divinyl benzene and diallylphthalate. However, in the aqueous phase polymerizations typical of commercial processes, solubility of the crosslinker becomes a concern, and when the acid is partially neutralized, solubility of the crosslinker is even more limited. Figure 2.2 (above) shows structures for a number of crosslinkers cited in the literature. Consult Chapter 3 for a review of specific patents that discuss particular types of crosslinkers associated with various process technologies.

Discussion of the chemistry in this chapter focuses on multi-vinyl esters and allyl-substituted compounds. Examples of the esters include ethyleneglycol diacrylate and TMPTA.[9] Tetraallyloxyethane is an example of the latter group.[97] Allyl methacrylate, which combines both functionalities into one molecule, has also been used.[98] These crosslinkers are used at concentrations in the range of 0.01–2.0 wt%, and usually below 1.0%, to provide superabsorbents with both high swelling capacity and reasonably low soluble polymer content.

As a model crosslinker for superabsorbents, TMPTA has been studied extensively. The three double bonds of the TMPTA molecule are initially indistinguishable from one another and have an original reactivity similar to that of acrylic acid. But once incorporated into the polymer chain, it has been proposed that the remaining double bonds will be less reactive, primarily due to reduced mobility. In fact, it is believed that the third double bond is not reactive at all in preneutralized acrylic acid polymerizations, due to steric hindrance or shielding, and thus remains as a dangling double bond.[9] This conclusion was based on the observation that swelling and modulus values for gels crosslinked with TMPTA were similar to those for gels prepared from *divinyl* crosslinkers at comparable *molar* (but not *equivalent*) ratios. New data from a different approach, which leads to a substantially different interpretation, is discussed in detail below.[99]

Infrared spectroscopy,[96] NMR spectroscopy,[99] and titrimetry[99] have been used successfully to quantify high concentrations of co-polymerized crosslinker, but none are sufficiently sensitive to quantify the components at the low concentrations (<1%) normally used in crosslinked superabsorbent polymers. To study the reactivity of ester crosslinkers with acrylic acid at these low concentrations, a ^{13}C-labeled sample of TMPTA was prepared. Use of labeled monomer enhanced the NMR signal by a factor of 100 and allowed the polymerization process to be studied at low concentrations of crosslinkers. The crosslinker conversion was determined from the integrated area of the TMPTA carbonyl resonance at 164.5 ppm and the incorporated TMPTA carbonyl resonance at 174 ppm, while the acrylic acid conversion was determined from the area for the olefinic carbons of acrylic acid at 130 and 124 ppm and the incorporated acrylic acid backbone carbon resonances at 33 and 40 ppm[99] (D.E. Henton et al., The Dow Chemical Co., unpublished results).

The rate of incorporation of TMPTA into the polymer was much more rapid under all conditions than was the polymerization rate of acrylic acid, resulting in depletion from the monomer mixture of TMPTA relative to acrylic acid, as shown in Figure 2.12. This depletion occurred at lower conversion in the more highly neutralized systems. At 55°C and 65% neutralization, the crosslinker signal was below the NMR detection limits (ca. 0.01% based on mass of acrylic acid) at 70% conversion of the acrylic acid. A similar, but slower rate of incorporation was seen at 0% neutralization. It was explicitly observed that all three double bonds from the crosslinker reacted rather than just two, as inferred in an earlier study.[9] From the narrow line widths for the pendent double bonds and because the chemical shift had not changed from the original TMPTA position, it was concluded that significant mobility exists for the pendent bonds and that steric hindrance is not substantial. Some

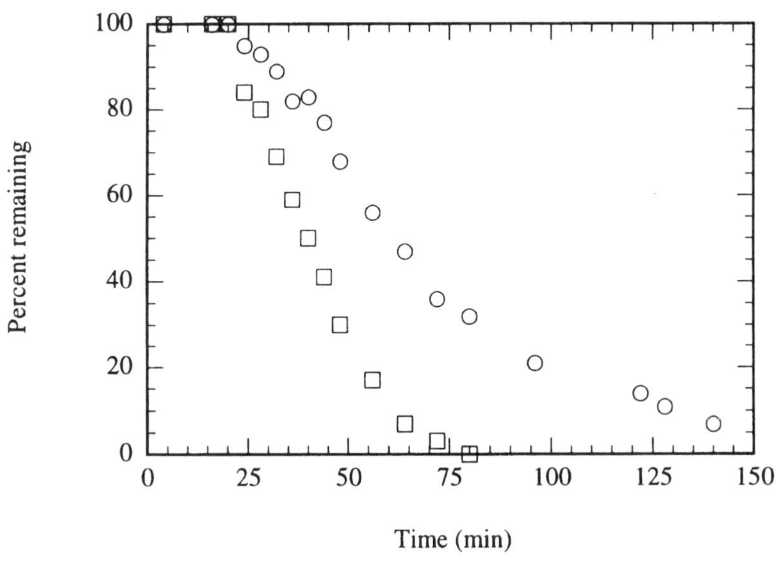

Figure 2.12. Percent of acrylic acid (○) and TMPTA (□) remaining during an isothermal polymerization at 55°C, 42% solids, and 65% neutralization.

reduction in mobility at the crosslink site appears to be present, however, based on the slight broadening of the saturated ester carbonyl peaks from the ^{13}C-crosslinker at 174 ppm (carbonyl groups adjacent to incorporated double bonds).

While this lack of chemical shift change for the pendent double bonds simplified the analysis, it prevented the determination of individual reactivity ratios for the second and third double bonds. The overall reactivity ratio r_1, representing all TMPTA double bonds whether free or pendent, was deduced by computer fitting of the data and values for different conditions, as given in Table 2.8. The value decreases from 0.77 at 0% neutralization to 0.30 at 65% neutralization. This is consistent with the trends discussed previously for various acrylic acid co-polymers upon neutralization. Because the other monomer species in the mixture are ionized, the growing polyelectrolyte chain, because of charge–charge repulsion, preferentially attacks the un-ionized crosslinker.

Temperature normally has little effect on reactivity ratios, and this was also the case in this study. From 55° to 80°C the value of r_1 increased only from 0.31 to 0.35. The reactivity ratio, r_1, increased from 0.44 to 0.77, however, when the monomer concentration, at 0% neutralization, was increased from 20% to 36%. This increase suggests a significant solvent effect on monomer dissociation.

The composition of the 65% neutralized acrylic acid feed mixture and the resulting polymer was compared with the predicted values using the integrated form of the co-polymerization equation and the measured value of $r_1 = 0.31$. An excellent fit of the data, as shown in Figure 2.13, is obtained throughout the entire conversion range with the use of the single reactivity ratio. This is consistent with the pendent double bonds having the same (similar) reactivity as the initial crosslinker.

From this study it is clear that acrylate ester crosslinkers are rapidly incorporated into the gel at a disproportionate rate at the beginning of the polymerization. In addition, the molecular weight of the poly(acrylic acid) naturally decreases with conversion (Fig. 2.14), because the ratio $[M]/[RM\cdot]$ decreases. These two factors reinforce one another, leading to the result that polymer chains made late in the polymerization are less likely to be incorporated into the network and are more likely to end up as extractable sol.

As an illustration, a simulation of the average distance between crosslinks for the network was generated using $r_1 = 0.31$, as determined for the polymerization of 65% neutralized acrylic acid with TMPTA. The plot in Figure 2.14 for 65% neutralized acrylic acid co-polymerized with 0.5 wt% TMPTA shows that the initial polymer chains formed have a M_n between crosslinks of about 5,500 g/mol or an average of one crosslink every 76 monomer units. If a number average molecular weight of these initial chains is taken at 5,000,000 g/mol, a plausible value, there are 65,790

Table 2.8. Effect of neutralization on reactivity ratio at 55°C

% Neutralization	% Solids	r_1
0	36	0.77
25	38	0.49
65	42	0.31

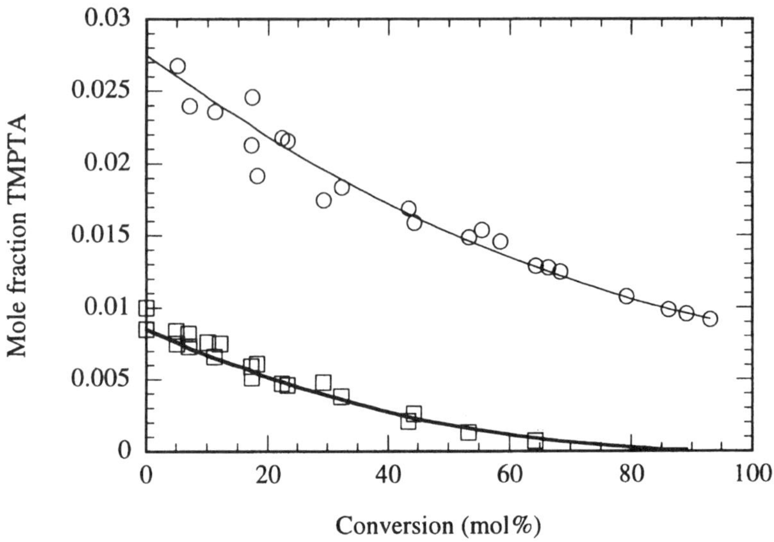

Figure 2.13. Experimental TMPTA content in the polymer (○) and in the 65% neutralized monomer feed (□) compared with the predicted content (lines), based on $r_1 = 0.31$.

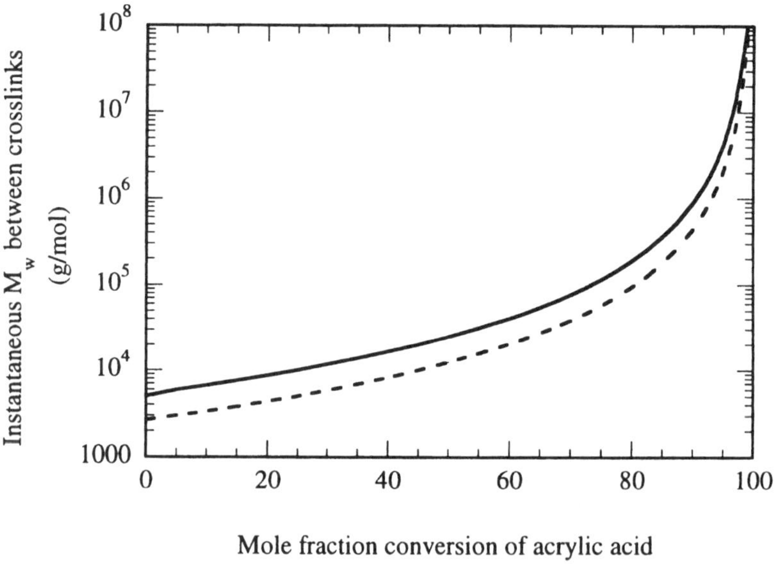

Figure 2.14. Predicted instantaneous molecular weight between crosslinks at 0.5 (—) and 1.0 (- - -) wt% TMPTA, based on $r_1 = 0.31$

crosslinks in each molecule. However, because of the rapid depletion of the crosslinker, by 90% conversion the distance between crosslinks would be 1,000,000 g/mol. In conjunction with the much lower M_n being formed at this time, the number of crosslinks per chain would be below one, on average. In other words, these chains would not be part of the network at all.

Superabsorbents with lower amounts of extractable polymer, and thus potentially better properties, could presumably be prepared by using a crosslinker that is more uniformly incorporated. The higher r_1 value for the polymerization of unneutralized acrylic acid and TMPTA implies a slower rate of incorporation for the crosslinker, thus leaving more available for reaction later in the polymerization. In a similar ^{13}C-NMR study of allyl end-capped polyethylene oxide crosslinkers by Smith et al.,[100] more uniform incorporation of the crosslinker than in the case of TMPTA was found.

At variance with the simulation above is the fact that rather low levels of extractable polymer can be achieved when preparing crosslinked poly(acrylic acid) superabsorbents, even though the crosslinker is depleted earlier in the polymerization. This may be due to grafting reactions onto the gel network by the polymer chains made later in the polymerization. Grafting onto a pre-formed polymer is well known in emulsion, suspension, and solution polymerizations.[101] Tobita[102] has addressed the issue of chain transfer to polymer in both emulsion and solution processes. Grafting is believed to occur at high conversion in solution processes, while in emulsion processes some grafting occurs even at low conversion. In a solution process, the high concentration of polymer and low concentration of monomer at the end of the polymerization favor grafting. The model of an inhomogeneous crosslinked network in the polymerization of crosslinked poly(acrylic acid) superabsorbents, resulting from the high reactivity of the crosslinker, is consistent with Tobita and Hamielec's modeling[103] of the crosslinked system of methyl methacrylate and EGDM. Organic azo compounds, such as azobis(isobutyronitrile) and 2,2′-azobis(2-amidinopropane) hydrochloride, are known to be rather poor hydrogen abstractors, which results in much lower efficiency of grafting.[104] Persulfates, on the other hand, are quite active in hydrogen abstraction.[105]

Extractable polymer can be formed by a variety of mechanisms besides the lack of incorporation of crosslinker. Chain transfer that prematurely terminates growing chains, chain degradation, and cleavage of crosslinks already formed are particular examples. Commercial superabsorbents typically contain 5%–20% extractable polymer, and significant attention is paid to maintaining a low level for performance reasons.[106] If the extraction is extended over a long time, the low-molecular-weight linear chains diffuse from the crosslinked particles first, and the high-molecular-weight chains follow, as illustrated by the data in Figure 2.15.

2.7. CROSSLINKING VIA CURING REACTIONS

2.7.1. Surface Versus Bulk Crosslinking

The other mode of network formation is curing or post-polymerization crosslinking, similar in principle to the processes used in the manufacture of rubbers. An agent, at

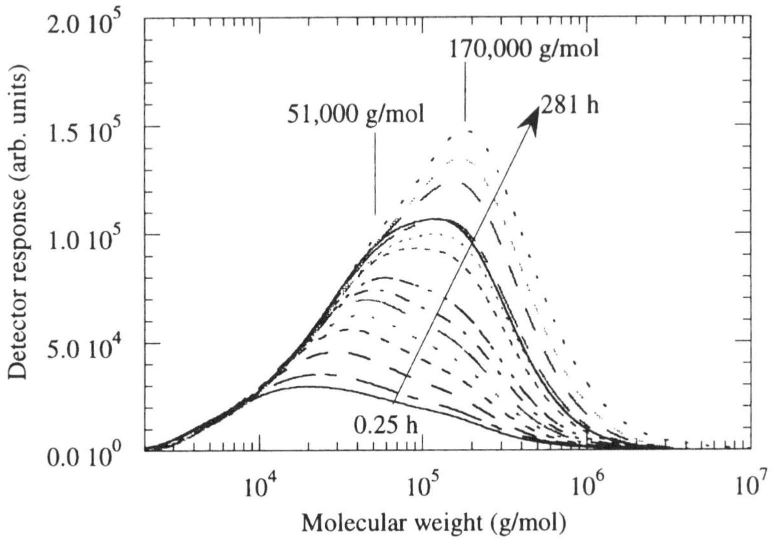

Figure 2.15. The molecular weight distribution of extracted soluble polymer as a function of extraction time. As extraction time increases from 0.25 to 281 h, the peak in the molecular weight distribution increases from about 10^4 g/mol to about 1.7×10^5 g/mol for a sample made by the process of Burgert et al.[118] (D. S. Allan, The Dow Chemical Co., unpublished results.)

least di-functional (or potentially so), is first mixed with pre-formed polymer chains, and a coupling reaction between the agent and functional groups on the polymer chains is then triggered. Decoupling the mixing from the curing reaction is frequently accomplished by low temperature mixing followed by heating.

Two different types of post-polymerization crosslinking agents have been used in the manufacture of superabsorbents, ionic and covalent, and these are discussed separately below. An important implication of post-polymerization crosslinking, in the case of absorbents, is the capability of increasing the crosslink density near the surfaces of gel particles. The advantages of doing this are discussed at length in Chapter 5. It is not uncommon for commercial products to be crosslinked in the bulk via co-polymerization and subsequently treated with either ionic or covalent agents to increase the crosslink density on the particle surface. Specific processes are described in Chapter 3, section 3.2.8.1.

2.7.2. Ionic Crosslinking

Multivalent cations can associate with more than one carboxyl group on polycarboxylated chains, forming at least a transient crosslink. For example, it is well known that a severe reduction in swelling capacity of polyacrylates occurs in the presence of calcium ion, as shown in Figure 2.16. Presumably the divalent cation in the external swelling liquid diffuses throughout the gel, and causes its catastrophic collapse by over-crosslinking. To diffuse, the cation dissociates from at least one

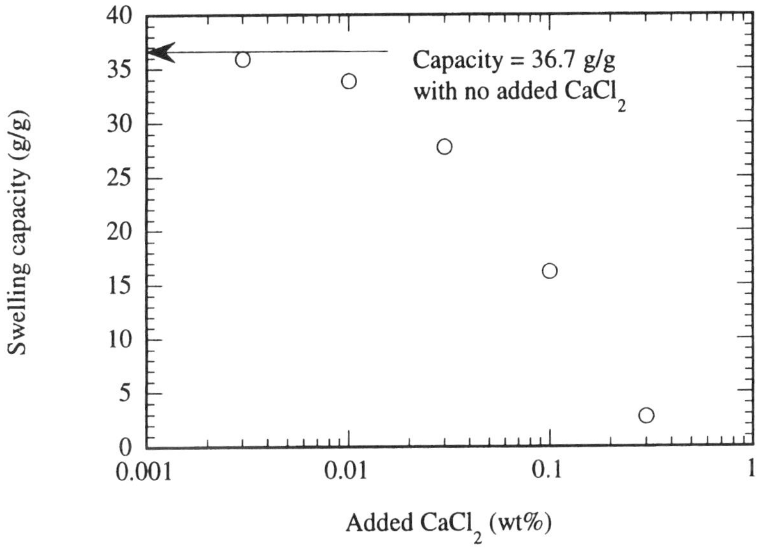

Figure 2.16. Impact of added calcium ions on swelling of crosslinked sodium polyacrylate in 0.9% NaCl.

site at a time and moves through the bulk of the gel, probably in a series of jumps. With trivalent ions, diffusion through a gel may be slower since at least two association sites must release to allow enough chain relaxation for the ion to move to a new location. Indeed, trivalent (or higher) ions such as aluminum can serve as useful crosslinkers. A review of much of this chemistry can be found in the thesis from Princeton University by Uhl.[107]

To use metallic ions as general purpose, post-polymerization crosslinkers requires contacting all portions of the gel mass uniformly. This is naturally difficult since the diffusion of Al^{3+} ions through the polyelectrolyte network is slow. High dilutions of polyelectrolyte and aluminum ion could be used, but this would require subsequent removal of large amounts of water since the final superabsorbent product must be substantially dry. Complexing agents such as citric acid have been used to delay association with the network relative to diffusion. Ionic crosslinking has been used for *pre*-polymerization crosslinking as well.[108] This procedure appears equivalent to preparing aluminum triacrylate *in situ* and incorporating it as a copolymerized crosslinker.

Surface treatment of polycarboxylated polymers with aluminum ions was first patented by Ganslaw and Katz in 1977.[109] The stated purpose was to improve dispersibility. An alcohol–water mixture was preferentially used to distribute the aluminum compound. Surface treatment of particulate gels with aqueous solutions of aluminum compounds has been shown to improve absorption under pressure.

Ionic post-treatments are essentially room temperature processes. As noted, separating the diffusion from the binding is difficult, and hence the crosslinking occurs primarily on the surface of the gel. An indication of the diffusion of aluminum can be inferred from the micrographs obtained by electron microprobe analysis (EPA) shown in Figure 2.17. This technique utilizes an electron microscope to map the concentration of individual elements across the sample. For these images, superabsorbent particles were swollen in aqueous aluminum sulfate solutions at a ratio of 10 g of solution to 1 g of polymer. The resultant gels were allowed to stand at room temperature for different time periods and then dried in a forced air oven at 165°C. Particles were mounted in epoxy, sliced, polished, and mapped for sodium and aluminum using EPA. The gel of Figure 2.17a was allowed to sit for 15 min before drying, while the gel of Figure 2.17b sat 16 h. The diffuse distribution of the aluminum ion in the latter sample is apparent.

Polyquaternary ammonium salts may also be used as crosslinkers. To be an effective crosslinker a species must contain at least two charged sites. The patent art suggests that multisited polyquaternary compounds perform best.[110,111] The preferred molecular weight used by Alexander et al.[110] is 300–4,000 g/mol; an example reagent is MAGNIFLOC 579C polyamine obtained from American Cyanamid Company. In this work the original polyacrylate substrate was already crosslinked, and the polyquat was added as a surface treatment to increase "both the rate of water absorption and the quantity of water absorbed and retained by the resin."

2.7.3. Organic Reactions With Carboxylic Acids

Esterification of the numerous pendent carboxyl groups on typical superabsorbents is another common strategy for post-polymerization crosslinking, illustrated generically in Scheme 2.5. An example is the use of polyhydroxy compounds as claimed in a patent to Tsubakimoto et al.[112] In the example, glycerol is added in an aqueous isopropanol mixture to the surface of dried and sized gel particles.

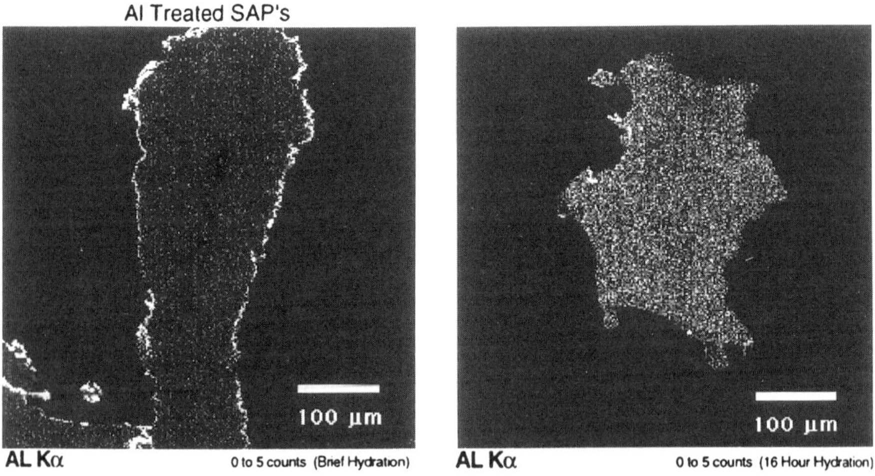

(*a*) (*b*)

Figure 2.17. Micrograph showing diffusion of aluminum into a superabsorbent polymer swollen in water for 1 **(a)** 15 min or **(b) 16** h. (C.K. Black, The Dow Chemical Co., unpublished results.)

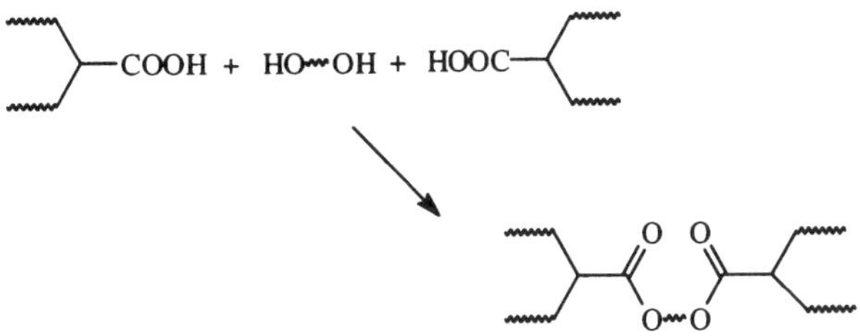

Scheme 2.5

This solvent mixture improves the uniformity of crosslinker distribution among the particles as well as modulating its penetration into each particle. The treated particles are then heated to around 200°C for several minutes. In addition to ester linkages, amide groups, formed from the reaction of polyamines with the carboxylic acid groups, may be used for crosslinking, but this reaction is more sluggish.[113] A more detailed discussion of the processes of these and the remaining methods of post-polymerization crosslinking are given in Chapter 3, section 3.2.8.1.

Dahmen and Mertens[114] have described the use of organic carbonates, especially ethylene carbonate in aqueous solution, for surface crosslinking; their approach avoids the use of flammable alcohols. A condensation reaction occurs producing carbon dioxide and a hydroxyethyl moiety that subsequently reacts with the available carboxyl groups to form ester bonds as crosslinks.[115] Both reactions are illustrated in Scheme 2.6.

Several patents disclose the use of diglycidyl compounds, particularly ethylene glycol diglycidyl ether (EGDGE), to effect surface crosslinking,[116,117] as shown in Scheme 2.7. A typical approach is to add a mixture of EGDGE and a solvent to the surface of dried and sized gel particles. These coated particles are then heated at approximately 150°C for generally at least 20 min. As with other systems, the depth of penetration, and thereby the crosslink density of the outer surface, can be controlled to some extent by the water concentration in the coating mixture.

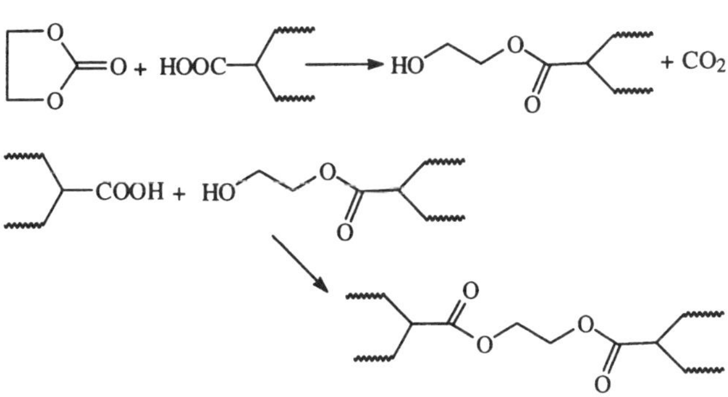

Scheme 2.6

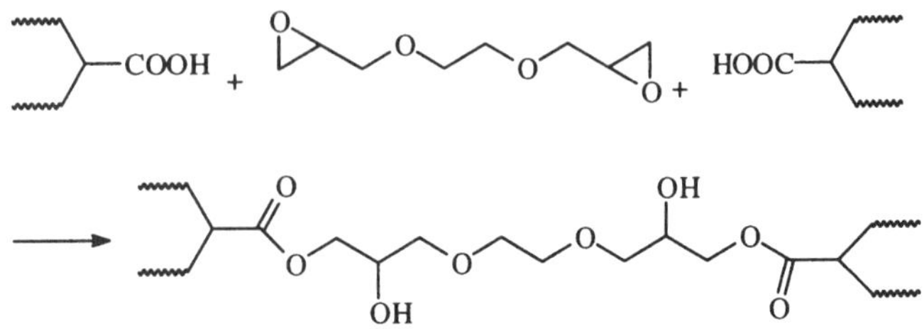

Scheme 2.7

Crosslinks internal to the gel particle, rather than on the surface, can also be obtained through post-polymerization esterification. This naturally requires the thorough incorporation of the reagent into the reaction mixture. An example of this approach is a process wherein polyethylene glycol is added to the polymerization recipe.[118] Separate from the crosslinking reaction, but also cited in the same application, is controlled degradation of the polymer network from oxidation by small amounts of sodium chlorate added to the polymerization mixture. This oxidizer is largely inert through the polymerization and drying operations but is activated during the heat treatment step that effects the esterification. This controlled breakdown of the polymer network in the center of the particle is thought to lead to a core/shell structure in the particle, which the authors speculate is responsible for the demonstrated improvement in modulus relative to swelling.

2.8. COMPUTER MODELING OF GEL POLYMERIZATION

The behavior of a complex system, such as the formation of crosslinked sodium polyacrylate gels discussed in this chapter, cannot be easily predicted with a single rate equation. Particular problems in this case are the extremely rapid propagation rate of acrylic acid, the early incorporation of crosslinker, and the inability of crosslinked gels to flow, all of which combine to make heat removal especially difficult. This situation would suggest that to maintain temperature control it would be advantageous to operate at low concentrations, but it is costly to remove large amounts of water. Interacting factors such as these are inducements to utilize elaborate computer models to predict behavior.

While development of a computer model for gel polymerization could begin with choosing among the many platforms available, there are elements that are common regardless of this choice. The discussion here will be sufficiently general to be applicable to many different hardware and software systems.

A model of gel polymerization could incorporate any of the reactions outlined in this chapter. However, central to any model is the numerical integration of the propagation equation from the initiation of polymerization to the time of interest, first shown in Eq. 2.5 and reproduced more generally in Eq. 2.38:

$$-\frac{d[M]}{dt} = k[M][growing\ radical] \tag{2.38}$$

Depending on the initiator system in use, the radical concentration can be approximated by the expressions developed earlier. For example, with thermal initiation by sodium persulfate, experiments showed that the chain radical concentration $[RM\cdot]$ depends on the half power of the monomer concentration (Eq. 2.22). The incremental decrease in monomer (Eq. 2.39), is calculated from this expression; the delta (Δ) is used to represent a small but finite increment of the variable quantity calculated by the program. Brackets ({}) are used to indicate the explicit instantaneous value of $-d[M]/dt$ calculated from substitution into Eq. 2.38. The other concentrations are incrementally changed at each time step as well.

$$-\Delta[M] = \{-d[M]/dt\}\Delta t \tag{2.39}$$

The size of the time step (Δt) is chosen such that the model output is well-behaved. For example, choosing too large a value for Δt has the effect of approximating a curve with a straight line and thereby missing the target value. These errors accumulate during the simulation and can lead to wildly fluctuating predictions and even nominally negative values for concentrations. Such behavior is obviously not a reflection of the real physical system, and a smaller value for the time step is called for. Very small time steps result in inordinately slow simulations, the actual computational time depending on the complexity of the simulation and the size of the computer.[119]

Accounting for the heat generated is required to predict the behavior of the real system, since only very high surface-to-volume ratio and hence generally quite small reactors are isothermal. The heat balance is accomplished in a general way by calculating the amount of heat generated from the incremental decrease in monomer ($\Delta[M]$), plus any heat flux from an external source, and using this heat in conjunction with the heat capacity and mass to calculate the resultant temperature change. Polymerization itself, i.e., monomer conversion to polymer, represents most of the chemical heat load; changes in the very low radical concentrations and the like are thermally insignificant. An Arrhenius expression can be used for each rate constant to calculate the parameter's value at the new temperature:

$$k = Ae^{-\Delta E/RT} \tag{2.40}$$

Here ΔE is the activation energy, and A is the pre-exponential factor. The process is repeated at a new increment of time, and the iteration continues until the time duration of interest, or until some other condition such as depletion of monomer is satisfied.

The number average molecular weight of the increment of polymer formed during the time step is the moles of monomer consumed per growing chain, $\Delta[M]/[growing\ radical]$. These incremental values may be accumulated during the course of polymerization simulation to obtain an estimate of the overall cumulative molecular weight distribution. Other average values (M_w, M_z, etc.) may be calculated and integrated over time.

Co-polymerization of crosslinker can be incorporated into the model through the use of r_1 and r_2 values, as were the results in Figure 2.13. It is important to note that in that study the composition as a function of conversion had been measured directly, and the model was run with trial r_1 values until the best fitting value was obtained.

A flow sheet for a simple model for nonisothermal polymerizations is given in Figure 2.18. This can serve as an outline; each step can be elaborately developed to

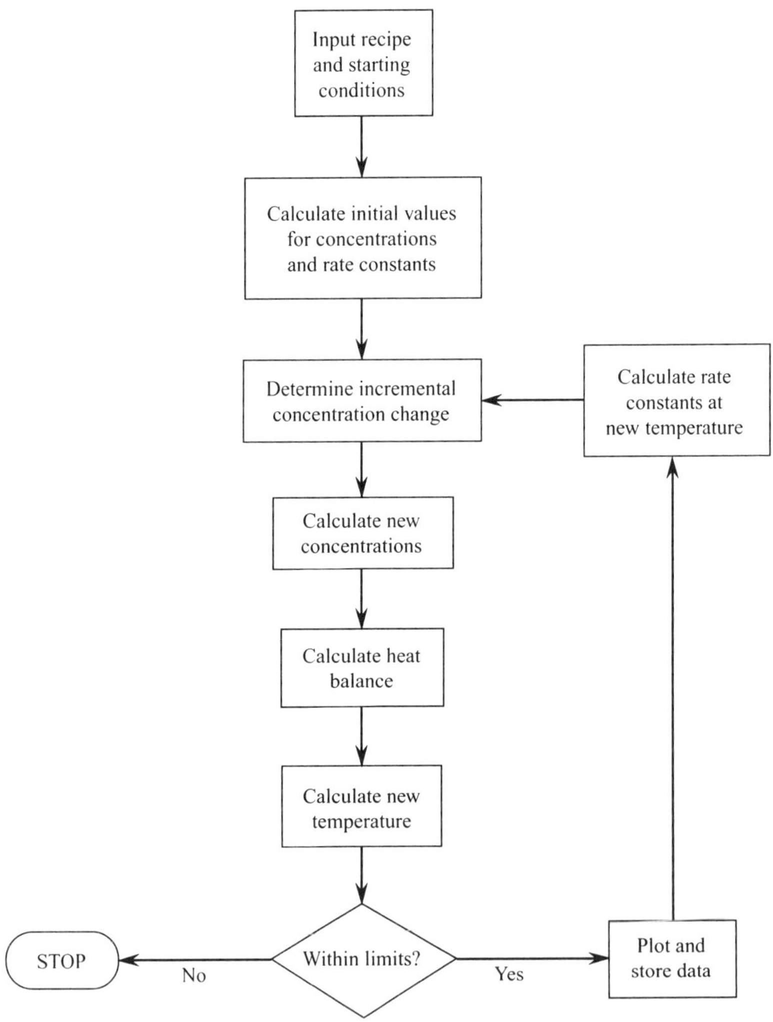

Figure 2.18. Schematic flow diagram for program modeling superabsorbent polymerization.

include the reactions discussed above, or others as are deemed important by later studies.

An example of the output from such a program for an adiabatic polymerization of acrylic acid, with the initial conditions indicated, is shown in Figure 2.19. Perhaps the first feature apparent is the rapid temperature rise. Note also the disappearance of monomer in relation to that of the initiator. Note further that the latter occurs after the polymerization is essentially complete ($[M] << 1\%$); with somewhat different starting conditions, the initiator could be depleted prematurely, leaving an incomplete polymerization, a condition sometimes termed *initiatcr burn-out.*

The value of models is to quantify the results expected on the basis of current knowledge. Insofar as observations differ from the model predictions, new features need to be included in the model. On the other hand, any realistic model can easily exceed the available computational resources. For efficiency, sufficient forethought must be given to asking only answerable questions of the model, such as time of polymerization, heat load, backbone molecular weight, molecular weight between crosslinks, and soluble polymer content. While total predictability is never completely realized, modeling leverages the knowledge gained from fundamental studies.

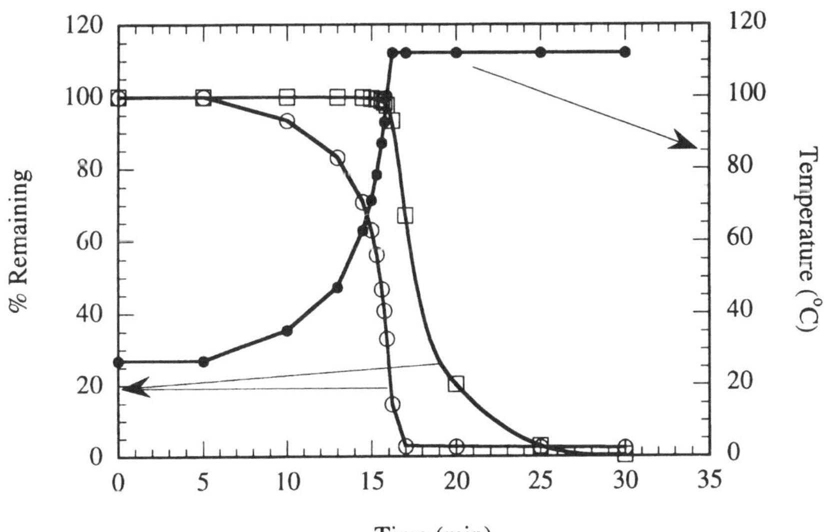

Figure 2.19. Results of simulation of an adiabatic polymerization of acrylic acid with sodium persulfate as the initiator. Monomer concentration (○) and initiator concentration (□) decrease as the temperature (●) increases.

REFERENCES

1. E. Doelker, *Adv. Polym. Sci.* **107,** 199–265 (1993).
2. A.J.K. Segeren, J.V. Boskamp, M. van den Tempel, *Faraday Disc. Chem. Soc.,* **57,** 255 (1974).
3. P. Aslani, R.A. Kennedy, *J. Controlled Rel.*, **42,** 75–82 (1996).
4. A.H. Clark, S.B. Ross-Murphy, *Adv. Polym. Sci.* **83,** 57–192 (1987).
5. A.H. Clark, In *Industrial Water Soluble Polymers,* C.A. Finch, ed., Royal Society of Chemistry, Cambridge, UK (1996).
6. N.A. Peppas, In *Hydrogels in Medicine and Pharmacy, Vol. 2,* N.A. Peppas, ed., CRC Press, Boca Raton, FL (1986).
7. N.B. Graham, In *Hydrogels in Medicine and Pharmacy, Vol. 2,* N.A. Peppas, ed., CRC Press, Boca Raton, FL (1986).
8. A. Apicella et al., In *Polymeric Drugs and Drug Administration,* ACS Symp. Ser. Vol. 545, American Chemical Society, Washington, D.C. (1994).
9. Y.-L. Yin, R.K. Prud'homme, F.W. Stanley, Jr., In *Polyelectrolyte Gels*, ACS Symp. Ser. Vol. 480, American Chemical Society, Washington, D.C. (1992).
10. M.L. Miller, In *Encyclopedia of Polymer Science and Technology, Vol. 1*, Wiley Interscience, New York, pp. 197–226 (1964).
11. W. Bauer, Jr., In *Kirk-Othmer Encyclopedia of Chemical Technology, Vol. 1,* John Wiley & Sons, New York, pp. 287–314 (1991).
12. *Acrylic Acid—A Summary of Safety and Handling,* Brochure by BASF Corp., Celanese Chemical Company, Rohm and Haas Co., and Union Carbide Corp., February (1980).
13. L.B. Levy, *Plant/Operations Prog.,* **6,** 188–189 (1987).
14. F.M. Wampler, *Plant/Operations Prog.*, **7,** 183–189 (1988).
15. S.S. Cutié, G.J. Kallos, P.B. Smith, *J. Chromatog.,* **408,** 349–355 (1987).
16. Y.F. Irie et al., *Eur. Patent Appl. 574,260* (1993).
17. T. Irie et al., *U.S. Patent 4,920,202* (1990).
18. N. Kutepow, H. Doerful, *U.S. Patent 3,086,046* (1963).
19. T. Horlenko, G.J. Fisher, A. Agullo, *U.S. Patent 3,868,410* (1975).
20. T. Sato, M. Baba, M. Okane, *U.S. Patent 4,317,936* (1982).
21. Anonymous, RD 37,631 *Res. Disclosures*, **376,** 538 (1995).
22. T. Shimomura et al., *Eur. Patent Appl. 372,706* (1990).
23. E.N. Zil'berman et al., *Zhur. Org. Khim.,* **19,** 33–37, (1983).
24. S.S. Cutié, D.E. Henton, C. Powell, R.E. Reim, P.B. Smith, T.L. Staples, *J. Appl. Polym. Sci.*, **64,** 577–598 (1997).
25. D.E. Henton, C. Powell, R.E. Reim, *J. Appl. Polym. Sci*, **64,** 591–600 (1997).
26 G. Odian, *Principles of Polymerization, 2nd Ed.*, John Wiley & Sons, New York, p. 246 (1981).
27. S. Aoki, H. Yamasaki, *U.S. Patent 4,093,776* (1978).
28. S.H. Pinner, *J. Polym. Sci.* **9,** 282–285 (1952).
29. H.W. Melville, In *Proceedings of the International Colloquium on Macromolecules, Amsterdam, 1949,* Centen, Amsterdam, p. 376 (1950).

30. G. Blauer, *J. Polym. Sci.*, **11,** 189–192 (1953).
31. H. Ito, S. Shoji, S. Shigenari, *J. Chem. Soc. Jpn. Ind. Chem. Sect.*, **58,** 194–196 (1955).
32. V.A. Kabanov, D.A. Topchiev, T.M. Karaputadze, *J. Polym. Sci. Symp.*, **42,** 173–183 (1973).
33. V.A. Kabanov, D.A. Topchiev, T.M. Karaputadze, L.A. Mkrtchian, *Eur. Polym. J.*, **11,** 153–159 (1975).
34. T. Tsubakimoto, T. Shimomura, *U.S. Patent 4,625,001* (1986).
35. J. Siddall, T. Johnson, *U.S. Patent 4,833,222* (1989).
36. S.S. Cutié, P.B. Smith, D.E. Henton, T.L. Staples, C. Powell, *J. Polym. Sci., Part B, Polym. Phys.* (in press).
37. T. Ishige, A.E. Hamielec, *J. Appl. Polym. Sci.*, **17,** 1479–1506 (1973).
38. S.P. Manickam et al., *J. Polym. Sci. Polym. Chem. Ed.*, **16,** 2701–2702 (1978).
39. S.P. Manickam, K. Venkatarao, N.R. Subbaratnam, *Eur. Polym. J.*, **15,** 483–487 (1979).
40. F.L. Buchholz, In *Ullman's Encyclopedia of Industrial Chemistry, Vol. A21,* VCH Publishers, New York, pp. 143–156 (1992).
41. N.I. Gal'perina et al., *Polym. Sci. (USSR),* **17,** 1670–1677 (1975).
42. F. Laborie, *J. Polym. Sci. Polym. Chem. Ed.*, **15,** 1255–1274, 1275–1288 (1977).
43. A. Chapiro, J. Dulieu, *Eur. Polym. J.*, **13,** 563–577 (1977).
44. C.E. Barnes, *J. Am. Chem. Soc.*, **67,** 217 (1945).
45. M.H. George, A. Ghosh, *J. Polym. Sci. Polym. Chem. Ed.*, **16,** 981 (1978).
46. G.G. Blauer, *Trans. Faraday Soc.*, **56,** 606 (1960).
47. L.B. Levy, *J. Polym. Sci.*, **23,** 1505 (1985).
48. J.J. Kurland, *J. Polym. Sci. Polym. Chem. Ed.*, **18,** 1139 (1980).
49. K.K. Georgieff, *J. Appl. Polym. Sci.*, **9,** 2009 (1965).
50. T. Irie et al., *U.S. Patent 4,920,202* (1990).
51. J. Lleras, S. Combet, *J. Chem. Phys. Physiochem. Biol.*, **69,** 1620 (1972).
52. J.W. Nicholson, A.D. Wilson, *Br. Polym. J.*, **19,** 67 (1987).
53. A. Frank, K. Weunscher, Int. Congr. Automat. Instrum. Pap., Rubber Plast. Ind. Pap., Antwerp, October 1966, *Chem. Abstr.*, **69,** 19737 (1968).
54. A. Crugnola et al., *J. Polym. Sci., Part C*, **16,** 4547 (1969).
55. J.W. Nicholson, A.D. Wilson, *Br. Polym. J.*, **19,** 449–452 (1987).
56. B.R. Brown, *Q. Rev. (Lond.)*, **5,** 131–146 (1951).
57. L.W. Clarke, In *The Chemistry of Carboxylic Acids and Esters,* S. Patai, ed., Wiley Interscience, New York, pp. 589–622 (1969).
58. A. Gurkaynak, et al., *J. Polym. Sci., Part A, Polym. Chem.*, **34,** 349–355 (1996).
59. W.C. Buzanowski, S.S. Cutié, R. Howell, R. Papenfuss, C.G. Smith, *J. Chrom. A,* **677,** 355–364 (1994).
60. J. Bell, R.I. Reed, *Nature,* **165,** 402–403 (1950).
61. J.A. Davidson, E.A. Collins, *J. Colloid Interface Sci.*, **55,** 163–168 (1976).
62. W. Denzinger, H. Hartmann, *U.S. Patent 4,748,220* (1988).
63. T.J. Romack, E.E. Maury, J.M. DeSimone, *Macromolecules,* **28,** 912 (1995).
64. V.F. Gromov et al., *Eur Polym. J.,* **16,** 529–535 (1980).

65. T. Alfrey, G. Goldfinger, *J. Chem. Phys.*, **12,** 205 (1944).
66. C. Walling, E. Briggs, *J. Am. Chem. Soc.*, **67,** 1774 (1945).
67. F.R. Mayo, F.M. Lewis, *J. Am. Chem. Soc.*, **66,** 1954 (1944).
68. G. Odian, *Principles of Polymerization, 2nd Ed.*, John Wiley & Sons, New York, pp. 440–441 (1981).
69. T. Kelen, F. Tudos, *J. Macromol. Sci. Chem.*, **A9,** 1 (1975).
70. T. Kelen, F. Tudos, B. Turcsanyl, *Polym. Bull.*, **2,** 71 (1980).
71. S. Charan, *J. Appl. Polym. Sci.*, **51,** 669 (1994).
72. P. Cheethan, T.N. Huckerby, B.J. Tabner, *Eur. Polym. J.*, **30,** 581 (1994).
73. L.M. Gan et al., *Macromolecules*, **27,** 6335 (1994).
74. F.A. Bovey, *Chain Structure and Conformation of Macromolecules* Academic Press, New York, pp 137–154 (1982).
75. F.M. Lewis et al., *J. Am. Chem. Soc.*, **70,** 1519 (1948).
76. C. Walling, F.R. Mayo, *J. Polym. Sci.*, **3,** 895 (1948).
77. R. Van der Meer, M.W.A.M. Aarts, A.L. German, *J. Polym. Sci. Polym. Chem. Ed.*, **18,** 1347 (1980).
78. E.L. Madruga, J. San Roman, J. Guzman, *J. Macromol. Sci. Chem.*, **A13,** 1089 (1979).
79. E.L. Madruga, J. San Roman, M.A. del Puerto, *J. Macromol. Sci. Chem.*, **A13,** 1105 (1979).
80. W.R. Cabaness, T.Y.-C. Lin, C. Parkanyi, *J. Polym. Sci. A1*, **9,** 2155 (1971).
81. T. Alfrey, Jr., C.G. Overberger, S.H. Pinner, *J. Am. Chem. Soc.*, **75,** 4221 (1953).
82. S. Ponratnam, S.L. Kapur, *Makromol. Chem.*, **178,** 1029 (1977).
83. S. Ponratnam, S.L. Kapur, *J. Polym. Sci. Polym. Chem. Ed.*, **14,** 1987 (1976).
84. C.Y. Chen et al., *J. Chin. Inst. Eng.*, **12,** 197–204 (1989).
85. T. Alfrey, Jr., C.C. Price, *J. Polym. Sci.* **2,** 101 (1947).
86. L.J. Young, In *Polymer Handbook*, J. Brandrup, E.H. Immergut, W. McDowell, eds., Wiley Interscience, New York, pp. 105–386 (1975).
87. S. Toppet, M. Slinckyx, G.J. Smets, *J. Polym. Sci.*, **13,** 1879 (1975).
88. P. Baja, D.K. Paliwal, A.K. Gupta, *J. Appl. Polym. Sci.*, **49,** 823 (1993).
89. G. Smets, R.Van Gorp, *Eur. Polym. J.*, **5,** 15 (1965).
90. L.J. Young, In *Polymer Handbook*, J. Brandrup, E.H. Immergut, W. McDowell, eds., Wiley Interscience, New York, p. 387 (1975).
91. R.Z. Greenley, *J. Macromol. Sci. Chem. (A)* **14,** 427 (1980).
92. D.T. Landin, C.W. Macosko, In *Characterization of Highly Cross-linked Polymers*, ACS Symp. Ser. 243, S.S. Labana, R.A. Dickie, eds., American Chemical Society, Washington D.C., p. 33 (1984).
93. P.J. Flory, *Principles of Polymer Chemistry*, Cornell University Press, Ithaca, NY (1953).
94. W.H. Stockmayer, *J. Chem. Phys.*, **11,** 45 (1945).
95. O. Okay, H.J. Naghash, I. Capek, *Polymer*, **36,** 2413 (1995).
96. H. Tobita, A.E. Hamielec, *Polymer*, **31,** 1546 (1990).
97. D.R. Chambers et al., *U.S. Patent 5,145,906* (1992).
98. F.L. Buchholz et al., *WIPO Int. Appl.* 93–05080 (1993).

99. D.J. Arriola, S.S. Cutié, D.E. Henton, C. Powell, P.B. Smith, *J. Polym Sci.* **63,** 439–451 (1997).
100. P.B. Smith, S.S. Cutié, D.E. Henton, C. Powell, J. Kosman, B.A. Howell *J. Polym. Sci. A: Polym. Chem.,* **35,** 799–806 (1997).
101. D.E. Henton, R.A. Bubeck, In *Polymer Toughening,* C. Arends, ed., Marcel-Dekker, New York (1995).
102. H. Tobita, *Polym. Reaction Eng.,* **1,** 357 (1993).
103. H. Tobita, A.E. Hamielec, *Polymer,* **33,** 3647 (1992).
104. S.L. Rosen, *J. Appl. Polym. Sci.*, **17,** 1805 (1981).
105. Y. Ogata, K. Tomizawa, K. Furuta, In *The Chemistry of Peroxides,* S. Patai, ed., John Wiley & Sons, New York, pp. 711–775 (1983).
106. K. Brandt, S. Goldman, T. Inglin, *U.S. Patent Re 32,649* (1988).
107. J.T. Uhl, *Rheological Studies of Water-Soluble Polymer Solutions With Interacting Solutes,* Thesis, Princeton University (1983).
108. F. Boardman, J.M. Lesniak, *U.S. Patent 4,808,637* (1989).
109. S.H. Ganslaw, H.G. Katz, *U.S. Patent 4,043,952* (1977).
110. W. Alexander, M. Anderson, B.R. Regan, *Eur. Patent Appl. 248,963* (1987).
111. M. Anderson, *U.S. Patent 4,954,562* (1990).
112. T. Tsubakimoto et al., *U.S. Patent 4,734,478* (1988).
113. J. March, *Advanced Organic Chemistry, 3rd Ed.,* John Wiley & Sons, New York, pp. 348ff, pp. 370ff (1985).
114. K. Dahmen, R. Mertens, *Ger. Offen. 4,020,780* (1991).
115. T. Yoshino et al., *J. Chem. Soc. Perkin I,* **11,** 1266–1272 (1977).
116. S. Obayashi et al., *Jpn. Kokai Tokkyo Koho* 57–44,627 (1982).
117. K. Nagasuna et al., *Eur. Patent Appl. 605,150* (1994).
118. J.H. Burgert et al., *WIPO Int. Appl.,* 94–20,547 (1994).
119. N.A. Dodson et al., *Polymerization Process Modeling,* VCH Publishers, New York (1996).

3

COMMERCIAL PROCESSES FOR THE MANUFACTURE OF SUPERABSORBENT POLYMERS

Andrew T. Graham, Larry R. Wilson

3.1. INTRODUCTION

Commercial superabsorbent polymers are partially neutralized polymers of acrylic acid that have been lightly crosslinked. The largest share of superabsorbent products is used in disposable baby diapers, with smaller quantities being used for adult incontinent devices and feminine hygiene pads. They are sold on the basis of their ability to prevent leakage in disposable personal hygiene articles. The polymer properties discussed most often for these applications are the swelling capacity of the polymer for urine (about 20–40 mL of urine per gram of polymer) and the modulus or ability of the polymer to resist deformation under an applied load (swelling under load).[1] Swelling and modulus have been treated,[2] based on the approaches of Flory[3] and Mark and Erman,[4] in terms of characteristic polymer properties.

The global market for superabsorbent polymers has grown and changed dramatically in the last 10 years as superabsorbents have replaced fluff pulp in diapers and other personal hygiene articles. Worldwide superabsorbent polymer production capacity (Table 3.1) grew from only a few million metric tons in 1985 to greater than 700 million tons in 1995.[5] The North American market is maturing, evidence for which is the desire to exit the business of one of the early major manufacturers in the United States, The Nalco Chemical Co.,[6] although market penetration is still

Modern Superabsorbent Polymer Technology. Edited by Fredric L. Buchholz and Andrew T. Graham.
ISBN 0-471-19411-5.

Table 3.1. Worldwide production capacity for producers of superabsorbent polymer (thousand metric tons [t])

Producer	1986	1989	1992	1995	1996	1998[g]
Allied Colloids	2.0	2.0	11.4	15.5	15.5	30.9
AMCOL Int. (Chemdal)	—	20.0	20.0	110.0	120.0	120.0
Arakawa	4.0	4.0	6.0	6.0	6.0	6.0
Atochem	—	5.0	8.0	9.0	12.0	12.0
Dow Chemical	22.7	46.4	60.0	90.0	90.0	120.0
U.S.	22.7	36.4	50.0	50.0	50.0	80.0[a]
Europe	—	10.0	10.0	40.0	40.0	40.0
Formosa Plastics	—	0	6.0	12.0	12.0	12.0
Hoechst	—	29.0	55.0	97.0	97.0	97.0
U.S. (Hoechst Celanese)	—	24.0	40.0	69.0	69.0	69.0
Europe (Casella)	—	5.0	15.0	28.0	28.0	28.0
Kao Soap	—	5.0	10.0	10.0	10.0	10.0
Kolon	—	—	—	12.0	12.0	24.0
Mitsubishi	—	0	10.0	10.0	10.0	10.0
Nalco[b]	—	20.0	40.0	45.5	45.5	45.5
Nippon Shokubai	15.0	37.9	84.0	131.0	141.0	141.0
U.S. (NA Industries)	—	6.0	12.0	30.0	36.0	36.0
Europe (BASF)	—	0.4	12.0	20.0	24.0	24.0
Japan	15.0	31.5	60.0	81.0	81.0	81.0
Nippon Synthetic[c]	—	3.0	4.5	6.8	10.0	10.0
Sanyo	8.0	17.0	20.0	50.0	55.0	55.0
Sekisui	—	0	3.0	6.0	6.0	6.0
SNF	—	0	0	2.0	7.0	22.0[d]
Stockhausen	—	50.0	79.5	110.9	110.9	110.9
U.S.	—	20.0	29.5	55.5	55.5	55.5
Europe	—	30.0	50.0	55.5	55.5	55.5
Sumitomo	3.1	5.6	8.0	22.0	32.0[e]	59.0[f]
Toa Gosei	—	0	10.0	10.0	10.0	10.0
Total	54.8	245.9	435.5	755.6	801.8	901.3

Data for 1992–1998 are from TranTech Consultants Inc. (TranTech began compiling their ChemPlan database for SAP in 1991.) Data for 1986 and 1989 are from The Dow Chemical Co. internal market studies.

[a]Includes 30,000 t potential capacity at low additional cost.

[b]Acquired by Stockhausen in 1996.

[c]Transferred technology and marketing rights to Mitsubishi in April 1996.

[d]Has capability to produce 15,000 t in Riceboro poly(acylamide) line.

[e]10,000 t of this capacity is suspension polymer.

[f]10,000 t of this capacity is suspension polymer. Total capacity includes 27,000 t/yr for Singapore for 1998.

[g]Estimated.

small in the major populated countries of Asia. Current manufacturers of acrylate-based superabsorbents in the United States include The Dow Chemical Co., Hoechst Celanese Corp., under a license from Sanyo Chemical Industries,[7] NA Industries (a subsidiary of Nippon Shokubai Co., Ltd.), AMCOL International (Chemdal), and Stockhausen GMBH. AMCOL has recently added significant manufacturing capacity both in Europe and in the United States. European superabsorbent manufacturers include AMCOL, Stockhausen GMBH, Dow, Hoechst Casella, Allied Colloids, Ltd., and BASF Aktiengesellschaft in partnership with Nippon Shokubai.[8] Japanese producers include Nippon Shokubai, Sanyo, Mitsubishi Petrochemical Co., Ltd., Kao Soap, and Sumitomo Seika. There are also a number of smaller producers in the United States, Europe, and the Far East. Elf Atochem S.A., Sumitomo Seika, and Kao Soap have been the three most prominent users of a suspension process to make superabsorbent polymers.

The synthesis and use of crosslinked, partially neutralized, polyacrylate superabsorbents have been the subject of several reviews[1,2,9–11]; however, very little information about manufacturing processes has been given. Perhaps the most unique, and arguably the first, commercial process is the polymerization technology patented by Sanyo[12] and licensed[7] to the Hoechst companies. Acrylic acid, in aqueous solution, was simultaneously polymerized, crosslinked and grafted to starch, and then neutralized and dried. Over the last 10 years, the dominant commercial technology for manufacturing superabsorbents has been solution polymerization of partially neutralized acrylic acid in the presence of a crosslinker using free-radical initiators, typically without a graft substrate. The resultant hydrogel is dried and ground to the desired particle size. Reverse-phase suspension polymerization is also practiced, but volumes produced are small relative to those from solution polymerization.

In this chapter, aspects of polymerization technology are reviewed, emphasizing commercial practice. While the review of the patent documents is intended to be thorough, it is not exhaustive. Our hope is to provide some insight into the interaction of engineering and chemistry in the design and implementation of commercial processes for the manufacture of superabsorbent polymers.

3.2. SOLUTION POLYMERIZATION: UNIT OPERATIONS AND THEIR EFFECT ON PRODUCT QUALITY

3.2.1. Introduction to Solution Polymerization Processes

3.2.1.1. Scale-Up Considerations. The initial development of a new superabsorbent polymer typically takes place at the laboratory scale, employing from a few grams up to several hundred grams of reactants. As with other polymeric products, it is convenient to think about the chemistry of the superabsorbent polymerization process as chemistry at the parts per thousand or parts per million scale (ppm). Polymerization initiators and crosslinkers are used at levels ranging from a few tens of ppm to several thousand ppm. Impurities in the monomer can have a strong effect on both the polymerization process and product quality. For example, the inhibitor methyl ether of hydroquinone (MEHQ), acrylic acid dimer discussed in Chapter 2, or iron as an (unwanted, variable) catalyst for an initiator system each affects the process when present in very small quantities.

Early patents cite examples involving small-scale experiments, while later patents focus on problems associated with commercial scale processes. Typically, as the size or volume of the equipment or process increases, the impact on the final product properties of variables or processes (unit operations), taken for granted in small-scale laboratory work, becomes significantly more important. The larger quantities of raw materials and product handled at a commercial scale require greater and more detailed knowledge of the chemistry and engineering variables.

At the laboratory scale, reactants are typically weighed to several significant figures and then mixed in open beakers either on a bench top or in a fume hood. At the manufacturing scale, liquid reactants are handled in closed containers of up to several thousand gallons and moved from one container to another in piping made of appropriate materials of construction. Quantitation of reactants may be accomplished by load cells attached to a vessel, by a level measurement in a vessel, or by integrated flow measurements. In each of these cases, process considerations affect the ability to hit the desired target quantity. For instance, the time needed to open or close a valve in a large-scale manufacturing facility affects the amount of material that is transferred through that valve and must be accounted for in the plant operations. Small-quantity additives may be weighed with great accuracy using an analytical balance in the laboratory. However, at a commercial scale they are added either continuously and quantitated using in-line flow meters, weigh-cells mounted on one of the vessels, or batch-wise using a small closed vessel ("shot pot") that may weigh 10–10,000 times more than the quantity of reagent being added from or to that container. Other variables affecting scale-up of most processes include mixing and heat transfer. Also of concern in a commercial process are releases of small amounts of vapors or liquids to the environment, which, on a laboratory scale, would be handled, for example, in a fume hood. The need to process large quantities of raw material, in-process intermediates, and product, and the manner in which the process actually functions, are affected by these considerations. These factors should be considered in the discussion that follows.

Typically chemists and engineers are most familiar with chemical reactions that take place in the liquid state because liquid-state reactions are the most easily managed at any scale. Applying the principle that research effort should be focused on the most obvious issues first, typically, at least initially, the primary reaction step is considered most fully. As the development of superabsorbents has proceeded, the problems associated with the polymerization reaction have been solved, and the importance of process steps (unit operations) beyond the polymerization reaction has become more evident. These other process steps include preparation of the wet hydrogel for drying, drying of the wet hydrogel, sizing of the dried particles to a size useful for the diapers, and the addition of additives for property enhancement.

3.2.1.2. Overview of Unit Operations. Figure 3.1 is a block flow diagram of a generic solution polymerization process. This diagram represents laboratory practice as well as commercial processes of the major superabsorbent polymer manufac-

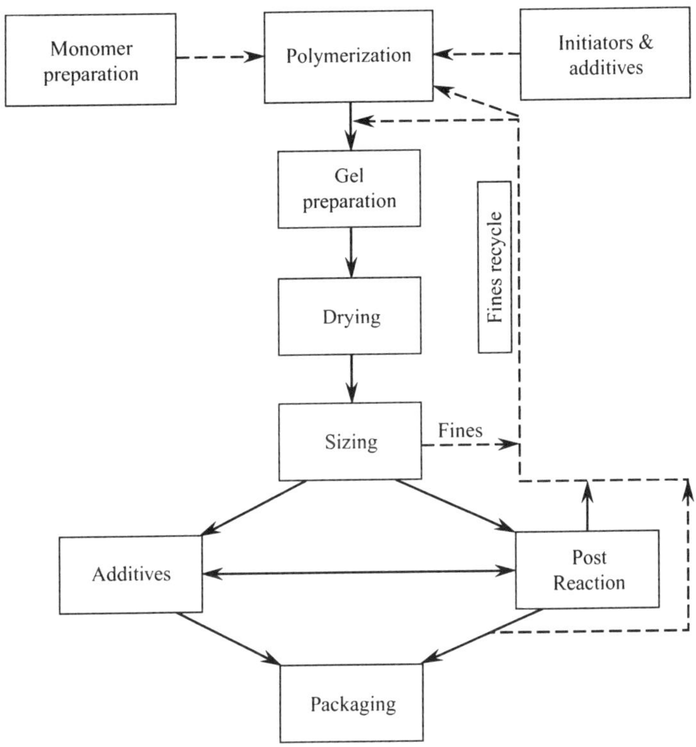

Figure 3.1. A block flow diagram of a generic solution polymerization process.

turers. A monomer mix is first prepared (see section 3.2.2) from acrylic acid, water, a crosslinking monomer, and, in many cases, a base, and then transferred to a reaction vessel. Contacting the monomer mix with an inert gas, either in-line during the transfer or in a batch process, for example, by bubbling with nitrogen, removes oxygen from the liquid. Addition of free-radical initiators, for example, sodium persulfate, or redox couples, for example, ascorbic acid and hydrogen peroxide, causes the monomer mix to polymerize (see section 3.3.3) with the evolution of heat (~77.4 kJ/mol acrylic acid).[13] The reaction mix begins to gel at just a few percent conversion as a result of crosslinking and the high backbone molecular weight. During the polymerization process, the gel may be broken up to produce a rubbery particle that is either dried directly or further reduced in size to facilitate drying. High monomer conversions (90%–95%) are attained in reaction times as short as 30 min. However, the polymer intermediate is held for an extended period of time at elevated temperature, in the reactor or in a separate hold tank, to achieve the ultrahigh conversion required in the dried commercial products.

Various types of dryers, including through-circulation conveyor dryers, fluidized-bed dryers, and rotating drum dryers (see section 3.2.5), are used to dry the polymer. A combination of grinding and screening the dried gel gives the final product particle size distribution. The presence of extremely fine particles ("fines"), typically less than 140 μm in diameter, has been unacceptable in the commercial product for diapers. Several means of converting the fine material to other saleable products, or recycling it in the current process, are described in section 3.2.7.

Technology for preparing products with a higher degree of crosslinking on the surface has long been known and is useful as a means of increasing absorption rate. More recently, superabsorbent products with an improved combination of swelling capacity and absorption under a compressive load (see Chapter 5) have been realized using such structured particle technology. Often, a crosslinker is added to the desired particle size distribution and is reacted to yield the structured product. Patents describing the chemistry indicate that a variety of blenders and heating devices are suitable.

Most unit operations in a commercial superabsorbent polymer plant have the potential to release small quantities of acrylic acid, other organics, or particulates to the environment. For example, as the polymerization reactor is filled with liquid, the vapor in the vessel is displaced and must be treated to remove any acrylic acid vapor before being released to the environment. The hot, moist gel fed to a dryer may release small quantities of unreacted acrylic acid into the surrounding air, creating an industrial hygiene problem for plant personnel. Also, in addition to containing unreacted acrylic acid, the air used in a through-circulation dryer to transfer heat to the polymer may contain fine particles of dry polymer product. These would need to be removed before the air from the dryer is returned to the environment. Enclosing the operations allows the air to be treated to remove contaminants. All major U.S. and European manufacturers treat their air streams to reduce emissions.

3.2.2. Raw Material Preparation

In this step, the ingredients to be added to the reactor, acrylic acid, water, an optional graft substrate, for example, starch[12] or poly(vinyl alcohol) solution, and a base, are prepared,[14–18] usually in a batch-wise fashion. The decisions taken in this step affect how the rest of the process operates, particularly the polymerization reaction, and are the primary determinants of the network structure of the dried particle. Monomer concentration (polymerization solids), the degree of primary crosslinking, and percent neutralization of the final product are typically set in this step. The degree of crosslinking affects the final product performance properties, swelling, and modulus (see Chapter 5). The degree of neutralization also affects the swelling behavior of the final product. However, when the pH of the product is to be maintained in a range safe for contact with human skin,[18] the most desirable percent neutralization is from 60 to 90 mol%. The monomer concentration in the reaction step determines, throughout the reaction, the demands on the heat removal system and the rheology of the partially polymerized hydrogel. The modulus, swelling capacity, and amount of extractable polymer of the polymer gel, determined in the polymerization step, affect its subsequent handling.

The primary monomer is high purity acrylic acid alone or in combination with its partial sodium or potassium salt. Typical impurities in acrylic acid include the dimer of acrylic acid, the polymerization inhibitor MEHQ, and small amounts (ppm) of several other materials. In at least one case, purification by distillation, just prior to use, of the acrylic acid monomer removes the inhibitor MEHQ and removes acrylic acid dimer to prevent the formation of β-hydroxypropionic acid.[19]

The formation of diacrylic acid (acrylic acid dimer) from acrylic acid during storage has long been known. The rate of formation is temperature dependent and dependent on the amount of moisture present as an impurity in the acrylic acid; this has been thoroughly documented.[20] The dimer of acrylic acid can be decomposed by heating[21] at temperatures from 120° to 220°C, yielding 2 mol of acrylic acid.[22,23] Acrylic acid dimer, because it contains a double bond, may also be incorporated into the polymer backbone.[20] A portion of the incorporated dimer may be cleaved during the drying of the polymer and apparently leads to 1 mer of polymer backbone and 1 mer of acrylic acid. The percentage of dimer cleaved depends on the heating temperature.[24] The unreacted acrylic acid is detectable as residual monomer. To minimize the amount of acrylic acid detectable in the product, the quantity of dimer in the acrylic acid feed is generally minimized.

Acrylic acid containing dimer may be treated with strong aqueous base prior to polymerization.[20,24,25] The reaction product of dimer hydrolysis is 1 mol of acrylate ion and 1 mol of β-hydroxypropionic acid. The β-hydroxypropionic acid may be trapped in the polymer matrix (although not polymerized) and will not be detected as residual acrylic acid by standard techniques (see Chapter 4, section 4.4.1). The β-hydroxypropionic acid may also be dehydrated to acrylic acid during a heating step.[26] However, it is our experience that higher temperatures are required to dehydrate β-hydroxypropionic acid than are required to thermally cleave acrylic acid dimer incorporated in the polymer backbone.

MEHQ is added to acrylic acid during its manufacture and acts in combination with oxygen as a polymerization inhibitor during storage and handling. It is normally present in acrylic acid at 200 ± 20 ppm. MEHQ, in combination with the thermal initiator persulfate ion can contribute to color formation in the final product.

A primary crosslinker is used at concentrations of about 500 ppm to 1 wt% (0.01–0.2 mol%), based on acrylic acid, so that the reaction product will be a crosslinked gel of the desired swelling capacity. Suitable divinyl crosslinkers include trimethylolpropane triacrylate, methylenebisacrylamide, diethyleneglycol diacrylate, triallylamine, allyl methacrylate, tetraallyloxyethane, and the like.[17,18,27–29] The choice of crosslinker,[30] through the crosslinker reactivity ratio, in combination with other polymerization parameters, will affect the amount of soluble polymer produced. Additional minor components in the reaction mix include chelants and dispersants,[28] or surfactants,[18] to aid in the solubilization of the hydrophobic crosslinkers.

Lower monomer concentrations have been chosen to improve end-use properties[18,31] or to facilitate handling. However, operating at the highest monomer concentration possible improves reactor throughput, in mass of product per unit time, and economics. However, as monomer concentration increases, a higher adiabatic temperature rise is possible (Fig. 3.2). Thus, higher monomer concentrations, in combination with temperature-sensitive initiator systems, lead to higher tempera-

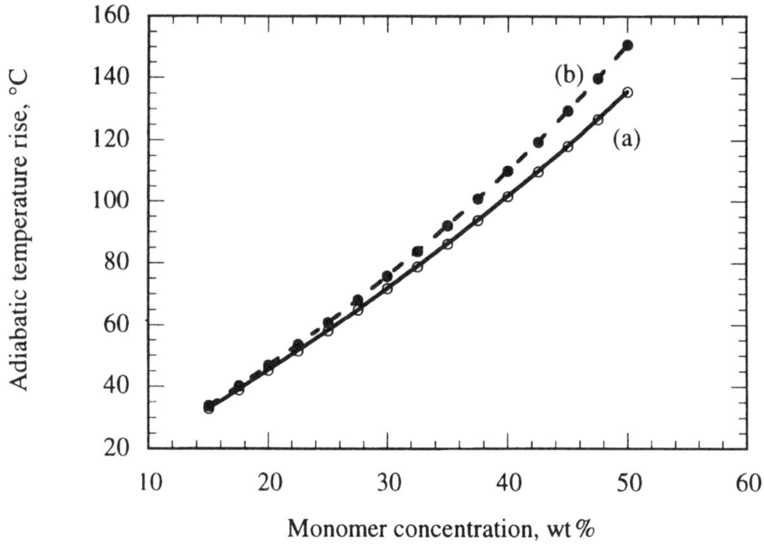

Figure 3.2. Adiabatic temperature rise for polymerizations of crosslinked partially neutralized acrylic acid as a function of monomer concentration. The heat of polymerization of acrylic acid was taken as 77.4 kJ/mol, and the heat capacity of the polymer was taken as either 0.50 **(a)** or 0.35 cal/g°C **(b)**.

tures and higher initiation rates near the end of the polymerization when monomer concentration decreases. As a net result, lower molecular weight chains are produced near the end of the polymerization. If the crosslinker concentration is constant or decreasing, which depends on the relative reactivity of the crosslinker and the monomer, the shorter chains are less likely to be crosslinked into the network structure. Therefore, larger quantities of undesirable[31] extractable polymer (lower molecular weight polymer, not crosslinked into the network structure) are produced toward the end of the polymerization. Commercially useful monomer concentrations range from about 15% to 30% for the polymerization of acrylic acid[12] and from 30%–43% for the polymerization of partially neutralized monomer.[16,17,25,28,29,32,33]

The degree of neutralization (mole basis) used for polymerization ranges from 0% (unneutralized)[31,34,35] to 80%. When acrylic acid is polymerized, the polymer is subsequently neutralized to an extent of 50%–80%, most typically from 60% to 75%, by adding a solution of sodium hydroxide or sodium carbonate to the acidic gel polymer. At the commercial scale, neutralization of the monomer may be accomplished either by simply adding the aqueous base to the acid monomer or by more involved processes. In one specific process[25] the monomer is neutralized to greater than 100.1 mol% and held at that composition for up to 2 hours to hydrolyze acrylic acid dimer to β-hydroxypropionic acid.[19,25] The pH of the monomer mix must be greater than about a pH of 12 to effectively hydrolyze the acrylic acid dimer.[24] This process results in a reduction in the residual monomer content of the final product.

Several orders of addition of reactants are possible, including the order specifically described earlier in discussing the neutralization process. The order of addition of ingredients may also be adjusted to accommodate factors such as crosslinker solubility or crosslinker sensitivity to hydrolysis. The crosslinker may be added to the acrylic acid prior to neutralization or to the fully neutralized monomer mix. If desired, the oily crosslinker may be dissolved in a portion of the acrylic acid prior to addition of the mixture to the balance of the neutralized monomer mix. Other orders of addition are described in the examples of the references of this section. After preparation, the monomer mix is moved to the polymerization step.

3.2.3. Polymerization Vessels and Systems

In the polymerization process, the monomer mix prepared in the previous process step is converted to a hydrated gelatinous rubbery polymer (hydrogel) with the evolution of heat. The polymerization vessel used for this process has substantially different requirements placed on it compared with a vessel for a liquid system because of the rubbery, semisolid nature of the crosslinked polymer hydrogel. In liquid systems, mixing in a reaction vessel is often of primary concern as a means of both uniformly mixing reactive ingredients and improving heat transfer required to maintain a uniform temperature in a vessel. However, essentially no mixing can take place after gelation, which occurs at only a few percent conversion, because of the high monomer concentrations, the high molecular weights achieved, and the presence of crosslinker.

The temperature increases in a reaction system if the rate of heat generation (rate of reaction) is faster than the rate of heat removal. Adiabatic or nearly adiabatic polymerizations of acrylic acid or partially neutralized acrylic acid can be very rapid, taking only 10–30 min to go from a few percent conversion to over 90% conversion (Fig. 3.3). This situation is favorable economically because it permits a high space–time yield for the capital dollars invested in reaction equipment. The kinetics of adiabatic polymerizations of acrylamide[36] and methyl methacrylate[37] have been reported and are similar in many features to those observed for acrylic acid or partially neutralized acrylic acid (see Chapter 2). Smaller molecular weight at high conversion (and higher temperature) results from lower monomer concentration, increased probability of chain transfer reactions, and increased initiation rate from the thermally activated initiators used for the polymerization. Therefore, to control the temperature of polymerizations with relatively large adiabatic temperature rises, and to avoid polymer degradation at high temperatures,[18,31] an efficient means of heat removal is required. However, the ability to remove heat is also affected by gelation at low percent conversion. When the surface-to-volume ratio is low, efficient mixing is required to transfer heat to a cooling surface such as a reactor wall.

Several polymerization systems have solved these problems. In one case, polymerization of acrylic acid is conducted at only about 20 wt% acrylic acid, thereby reducing the need for heat removal by limiting the total heat released. Also, the

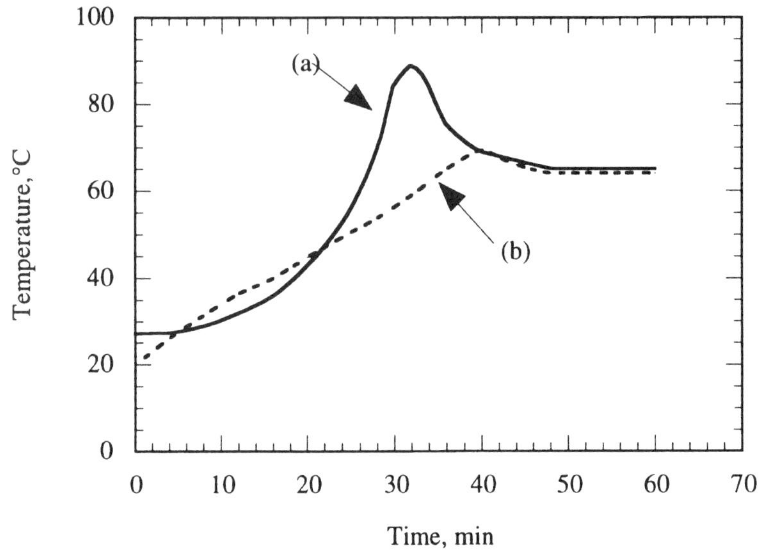

Figure 3.3. Temperature versus time for polymerizations in a 2-L laboratory reactor with no cooling except through heat losses to the atmosphere. Monomer solution 1,200 g was loaded to the reactor: **(a)** 32 wt%, 65% neutralized acrylic acid[28]; **(b)** 20 wt% acrylic acid.[35]

acidic hydrogel is more easily broken up than is the product that is neutralized prior to polymerization. In a second case, batch or continuous reaction systems have been designed that will grind the hydrogel into smaller particles, which may be easily removed from the reactor and which will permit heat removal from the rubbery polymer hydrogel. This system can be used for the polymerization of either acrylic acid or its partially neutralized salt. However, commercially, it is used for the polymerization of partially neutralized acrylic acid. In a third case, a continuous polymerization process has been developed in which the monomer is polymerized on a continuous belt. The thickness of the polymer layer is about 6–12 cm and may be several meters in width. The rate of heat loss from the relatively thin strand to the surroundings (or to the cooled belt) is apparently great enough to avoid problems with elevated temperatures in the reacting gel. The resulting strand may be successfully fed to a gel grinding device. This process is used commercially to polymerize partially neutralized acrylic acid.

In the first case above,[35,38] in which the concentration of acrylic acid and the need for heat removal are limited, solutions of acrylic acid at concentrations of 17–25 wt% have been polymerized to yield a high viscosity gel that can subsequently be extruded into strands. At these lower monomer concentrations, the adiabatic temperature rise is about 50°C (Fig. 3.2) so that a polymerization initiated at 20°C reaches high conversion by about 70°C. The patents specifically describe a batch polymerization process in which an "adiabatic reaction" is initiated at between 5° and 20°C, with a peak temperature of not greater than 90°C. "The peak temperature is controlled, if necessary, so as not to exceed 90°C," and is preferably about 70°C;

however, no information is given as to how the temperature control is accomplished.[35]

A reactor for the solution polymerization of acrylamide has been described in which the high-molecular-weight and high-viscosity, soluble, linear polymer is pushed or extruded from the reactor by a piston.[39,40] In the patent examples for crosslinked poly(acrylic acid) superabsorbents the polymerization vessel is only described as a stirred vessel, but a high-viscosity, gel-like product is pumped from the reactor and extruded (through a die) into strands. The patent descriptions seem to suggest a process that is an extension of the technology cited above for aqueous solution polymerization of uncrosslinked acrylamide or acrylic acid. The strands are chopped into small pieces so that subsequent gel handling and drying steps may take place satisfactorily.

The use of the crosslinker tetraallyloxyethane, the initiator system consisting of the redox pair ascorbic acid and hydrogen peroxide, and the thermal initiator 2,2′-bisamidinopropane dihydrochloride are specifically claimed for polymerizations of acrylic acid.[35] In another example, acrylic acid was polymerized at a concentration of about 25% and initiated with either a sodium sulfite or, preferably, sodium thiosulfate redox pair with *tert*-butyl hydroperoxide. The thermal initiators sodium persulfate and 2,2′-bisamidinopropane dihydrochloride were also used.[41]

In the second case, the need for improved heat removal led to several different reactor designs in which the rubbery, semisolid hydrogel could be broken into small pieces. The design of these vessels addresses, first, the need to control the peak polymerization temperature[18,31,35] and therefore removal of the heat of polymerization and, second, the need to size the polymer gel prior to drying.[18] These problems have been solved for reaction vessels mounted either vertically or horizontally (see below), and at least some designs may be used for either a continuous or a batch process A common feature of these vessels is an agitation system designed to knead or tear the polymerizing rubbery mass into small pieces. As conversion (polymer concentration) increases, the modulus and strength of the partially converted rubbery gel increases. Under the stress of the agitation system, in a manner similar to that of taffy as it is being pulled, the rubbery hydrogel fractures to finally produce a collection of rubbery gel particles, each of which has a consistency somewhere between that of gelatin and an automobile tire.

The size of the resulting pieces depends on the agitator configuration, especially on clearances between stationary and rotating members, and is characterized as a few millimeters to several centimeters. These smaller pieces have several advantages. First, smaller pieces are easier to handle than chunks the size of a commercial reactor (3–40 m^3). A reactor agitation system can move smaller pieces around within a reactor so that, over time, more hot polymer surface area will be able to contact a cold, jacketed, reactor wall and improve cooling. Also, the larger ratio of surface area to volume of small polymer pieces improves any cooling as a result of vaporization of water under an applied vacuum using a vacuum/condenser system.[28,32] Drying of the wet hydrogel is also more rapid using the smaller sized gel particles.[18]

A number of polymerization vessel types have been described in some detail: vertical agitated vessels[17,28,42] and several types of horizontal vessels.[32,43–45] The horizontal vessels may have either a single shaft or several mutually parallel rotary stirring shafts[32,33] (Fig. 3.4). The agitators on both vertical and horizontal vessels have high torque agitation systems. The shear forces between the rotors and stators of the agitation system provide the energy to tear the polymerizing hydrogel into small pieces. Careful design of clearances between the agitator members and the reactor walls is required to produce the desired small hydrogel pieces and to avoid mechanical problems (for example, contact of rotor members with the vessel wall) that could lead to eventual vessel damage. Polymerizations at lower monomer concentrations require less shear energy to tear the hydrogel and, consequently, a less-robust agitation system, because, as monomer concentration is reduced, fewer carbon–carbon bonds must be broken to produce a unit of surface area in the polymerizing hydrogel. In all cases, the agitator members are designed to resist the mechanical stress of moving and tearing the semisolid rubbery hydrogel, and both vertical and horizontal vessels deliver a finely divided gel product.

In addition to careful design of agitator clearances for particle sizing, suitable agitation systems are all designed so that the combination of rotors and stators "scrape" all the surfaces of the vessel and the rotating members. The scraping action minimizes deposition of polymer on the vessel walls and on the agitator itself. A surface coating for the reactor walls such as tetrafluoroethylene resin helps prevent or minimize build-up of polymer.[32] A combination of very smooth reactor surfaces and vessel jacket temperature less than 70°C is also used to minimize fouling of the reactor wall and agitator.[46]

Either vessel configuration is suitable for the polymerization of either unneutralized[32,33] or neutralized[33] acrylic acid and may be used in a continuous[32] or a batch[33] polymerization process. Single shaft, two-arm, and three-arm "kneaders" are described. Where multiple agitator shafts are used, the shafts are rotated in opposite directions and may be rotated at unequal rates. Rotary stirring shafts may use blades of various types, including S-type, sigma, fish tail, and Banbury type blades. Efficient use of the capital invested in the polymerization vessels requires that the majority of the vessel contents are emptied from the reactor after each batch polymer-

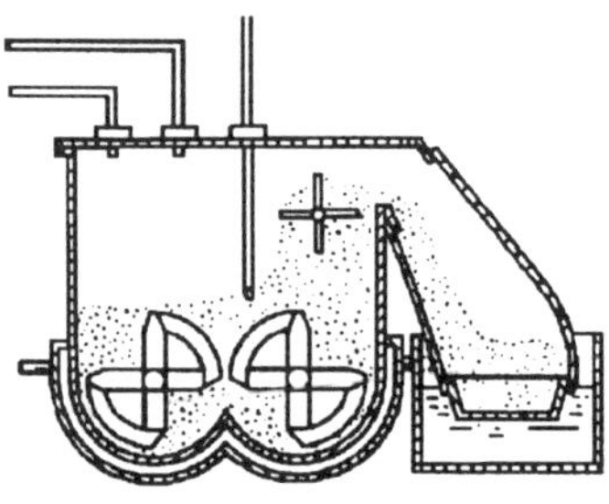

(a) Reactor

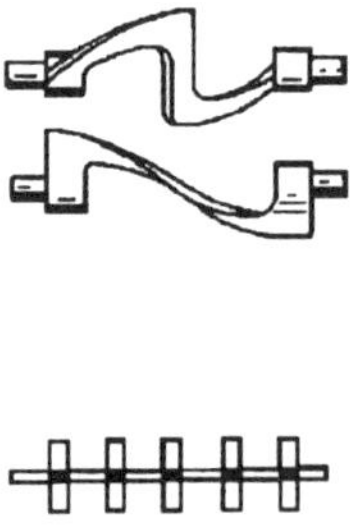

(b) Two alternate stirring shafts

Figure 3.4. Dual-shaft jacketed kneader reactor and two other possible rotary stirring shafts. (Drawn from Tsubakimoto and Shimomura[32]).

ization. In one case, a "self-purging effect of more than 80%" is required.[43] Appropriate dimensions and blade arrangements have been briefly described.[43] In practice, either horizontally or vertically mounted reaction vessels can be used to polymerize superabsorbent polymers.

Polymerization vessels may be provided with a jacket through which hot or cold water may be circulated either to heat or to cool the reactor and contents. The polymerization vessel may also be connected to a condenser that provides heat removal during the polymerization.[28,32] Typically, condenser systems are used in which the condensate is drained back to the reaction vessel. In the early stages of polymerization,[43,47] reduced pressures may be used to fix the temperature and to remove heat by vaporization of water. In a second stage the temperature and pressure may be increased to 100°–170°C and 1–8 bar, respectively, to complete the polymerization.

Advantages of the kneader reactors in continuous service are those that would be expected. These include an improvement in operational efficiency in that once the reaction has reached steady state virtually no manual work is required to maintain the reaction and a more constant generation of heat of polymerization. Another advantage claimed for a kneader reactor, either in a continuous process or in a batch process, is that the reaction product is broken up into small pieces of a few millimeters in diameter with essentially no large chunks.[32]

Commercially, batch processes are used with polymerization vessels of the above types. A batch monomer mix, prepared as described above, is about 55%–80% neutralized and has a monomer concentration of about 30%–43%.[16,17,25,28,29,32,33] Recycling of dry ground fine particles to the polymerization by addition of the fines to the monomer mix is described in more detail in section 3.2.7.

To prevent excessively long reaction times, oxygen, required as part of the inhibitor system for storage of acrylic acid monomer, must be removed from the monomer solution prior to polymerization. Oxygen removal may be accomplished by the in-line addition of nitrogen as the monomer mix is transferred to the reactor[48] or by addition of inert gas (typically nitrogen or carbon dioxide) to the monomer either in the polymerization vessel or in a monomer storage vessel prior to the polymerization.[27] Possible mixing devices for in-line deoxygenation include static mixers of various types, aspirators, or perforated plates.[48] After addition to the reaction vessel and deoxygenation of the monomer mix, the free-radical polymerization of the monomer mix is initiated with redox pairs such as bisulfate/persulfate or ascorbic acid/persulfate.[16,25,49,50] Redox initiators are used so that the polymerization may begin at temperatures lower than those that are accessible using thermal initiators alone. Low initial temperatures, for example, 20°–25°C, are preferred for at least two reasons. First, low temperature initiation reduces the amount of heat that must be removed toward the end of the polymerization, and, second, polymerization at lower temperatures avoids some chain transfer reactions (see Chapter 2, section 2.4.4), which can reduce the average backbone molecular weight of the polymer. Thermal initiators such as sodium persulfate or 2,2′-bisamidinopropane dihydrochloride, are also required for optimum reduction of residual monomers toward the end of the polymerization.[51]

A belt reactor for the polymerization of water-soluble monomers, such as acrylamide and acrylic acid, was described in 1973.[52] An improved belt polymerization apparatus[53] and process[27] for using that apparatus were subsequently patented. The improved apparatus (Fig. 3.5) consists of a continuous belt whose shape can be changed by strategically placed rollers. The belt first provides a trough for the containment of liquid monomers that are added to the belt. As the polymerization proceeds, the monomer solution thickens and gels; the belt is caused to flatten and pull away from the polymerized gel mass. When the polymerization is complete, the gel strand is pulled over a guide roller, onto a spiked roller, and carried forward for further processing. Strand thicknesses of 6–12 cm and belt speeds of 5–12.5 cm/min are cited. A continuous, high monomer concentration process using two belts in series has also been described.[54] Although the claims of the patent are specifically limited to a high solids process,[54] this latter process is also expected to be adaptable to lower solids polymerizations and their consequent lower polymerization temperatures and higher product quality.

A comparison of patent examples shows that the chemistry used with the continuous single-belt process[27,53] is not significantly different from that described above for batch polymerizations. A batch monomer mix is first made up, as described above. The acrylic acid is about 55%–80% neutralized, and the monomer concentration ranges from 30%–43%.[27,29] Recycling of fine particles to the polymerization by addition of fines to the monomer mix, is also employed with this process. Catalysts can be redox pairs, photochemical initiators, thermal initiators, or combinations and are contained in the monomer mix as it is added to the belt. The catalysts are mixed "in a mixing tube" prior to adding the monomer mix to the belt. The

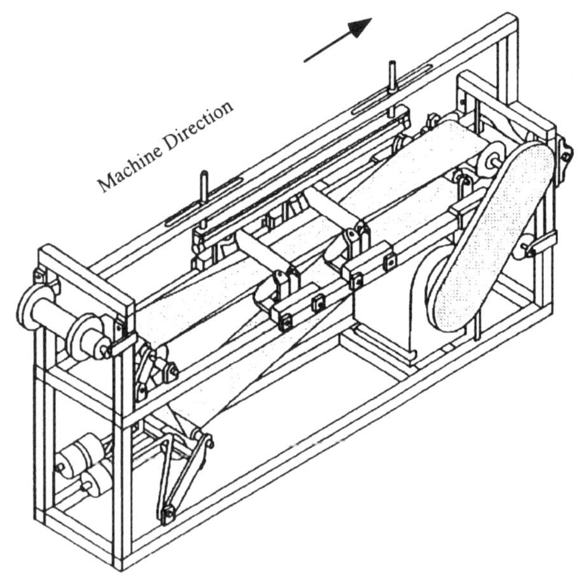

Figure 3.5. Continuous belt reactor with trough to contain liquid monomer. (Drawn from Chmelir and Paven[27]).

monomers have been deoxygenated prior to the addition of the catalyst. The monomer mix is then fed to the belt polymerizer.[27]

It is not clear whether the atmosphere above the belt is maintained oxygen free, as would be normally required to avoid inhibition of typical monomer systems. However, enclosing the belt in a hood might serve to minimize the volume that would need to be maintained oxygen free.

Polymerization temperatures above about 80°C adversely impact the product properties.[18,31] Higher adiabatic temperature rises are associated with higher monomer concentration combined with limitations on heat removal. Therefore, limitations on heat removal with increased strand thicknesses would constrain the combination of monomer concentration and strand thickness that could be used to obtain the desired product properties. Cooling and heating devices are referred to, but no details are given. In a subsequent patent, temperature control in a continuous process[55] is obtained by polymerizing the monomer mix between two continuous thermally conducting belts that are adjustable to provide a thickness of 1–30 mm. The belt surfaces are heated or cooled to maintain the desired temperature. Similar, modified, continuous processes may also be used to polymerize poly(acrylic acid) onto a substrate, for example, a fabric.[56] Deoxygenated monomer mix containing initiators is applied to a porous substrate. The monomer-impregnated substrate, sandwiched between two polymerization-inert surfaces, is polymerized by heating to 100°C.

In 1985, polymerization technology for a batch reactor without agitation or conventional heat removal was disclosed.[57] Monomer concentrations greater than 70% by weight are taught. Potassium hydroxide alone or in combination with ammonium hydroxide is used to neutralize the monomer. These bases are used instead of sodium hydroxide or sodium carbonate because of the higher solubility in water of potassium and ammonium acrylate than sodium acrylate. Initiation is provided by the redox system sodium bisulfite/sodium persulfate. As the heat of polymerization is liberated, water contained in the monomer mix boils, providing some temperature control to the polymerization. An advantage cited for this process is that a substantially dry (<15% by weight of water) product may be obtained without any special drying equipment.

A continuous process based on the previous batch process was also developed.[58] Monomer mix is pumped into the bottom of an inverted conical vessel, simultaneously with initiator solution. As the material polymerizes, it is forced out of the narrow top of the cone onto a moving belt. In one example, employing a 70% neutralized, 73% monomer concentration reaction mixture,[58] the maximum reaction temperature was 130°–135°C. Some cooling can be provided by circulating water in the jacket of the conical reactor. Additional cooling can be obtained by including 1%–10% of an organic solvent with a boiling point range of 55°–120°C. The polymerization is completed on the belt to yield a substantially dry (<15% by weight of water) product. As with the earlier batch process, cooling is provided by the evaporation of water and organic solvent as the heat of polymerization is liberated. The very high temperatures reached during the polymerization and their negative effects on the desired combination of product properties (low extractables, high modulus)[18,31] are disadvantages of this process. Advantages include the substantially dry product (<15% by weight of water) and the fast absorption rate of the rather foamy product commercially produced by this process.

3.2.4. Gel Handling

3.2.4.1. Post-Reactor Chemistry. After the polymerization in the primary reactor has been substantially completed, as determined by the presence of a measurable exotherm, the conversion is expected to be in the range of 93%–97%. Additional time must be allowed for the reaction to proceed to higher conversion, typically about 98.5%–99.9%, prior to drying. This higher conversion is reached by holding the polymer at an elevated temperature for an extended period of time. In one case, after reaching the peak reaction temperature of about 70°C, the polymer is kept in an insulated vessel for 2–12 h so that the residual monomer is reduced below about 1,000 ppm.[35] In another process, "the reactor is maintained at a desired hold temperature for a period of time necessary for high monomer conversion to be achieved."[17] In another example, a two-stage reaction process is described. The first stage is used to set temperature and to remove heat by vaporization of water. The polymerization is completed in a second stage at a higher temperature and pressure.[43,47] These references imply that all of the reaction chemistry is accomplished in the polymerization vessel. However, for optimum capital utilization a second, separate vessel might be used while the reaction proceeds to very high conversions.

The time and temperature needed to reach high conversion was not specified by the examples cited above. The temperature of the product and the amount of time ("hold time") the product must spend in this step to obtain the desired properties are defined by the reaction kinetics (see Chapter 2) and the desired process throughput. The rate of monomer reaction at conversions greater than 90% is independent of the concentration of persulfate. However, we would expect any chain scission process to be dependent on the concentration of any persulfate or azo-initiator. Thus a lower temperature (Table 3.2) or a reduced initiator concentration would tend to reduce the rate of formation of extractable polymer. This, in turn, through the reaction kinetics, affects the operator's selection of vessel temperature and the amount of thermal initiator, for instance, sodium persulfate, that is actually added in the primary reaction step. At this stage, it is still desirable to maintain the process under nitrogen because of the negative impact of oxygen on free-radical processes. These considerations, with the desired process throughput rate, allow the process engineer to determine the required vessel volume for this process step.

The kinetics described in Chapter 2 describe the rate at which unreacted monomer will decrease in this step. However, the effects of chain scission on swelling capacity and extractables are not described and can only be inferred from the material presented. To better illustrate these effects, a series of samples was made in the laboratory using 2,000 ppm of trimethylolpropane triacrylate as the crosslinker. The recipe, reactor, and process were as described in Comparative Example A of Graham and Gartner[17] except that the reactor contents were held after the peak reaction temperature for 3 h at the indicated hold temperature. The gel was dried at 100°C in a circulating air oven and the properties measured using the methods described in Chapter 4. As hold temperature increased, extractables and swelling capacity increased as shown in Table 3.2. (Tables II and III of Graham and Gartner[17] show similar data, but also include data on modulus and unreacted acrylic acid.) Increasing hold time and increasing amounts of sodium persulfate, as can be inferred from the kinetic results presented in Chapter 2, also increase the swelling capacity and extractables (data not shown).

3.2.4.2. Post-Reactor Gel Preparation. The next logical unit operation in the flow sheet might be viewed as drying. However, preparation of the hydrogel prior to drying is essential for economical operation. The desire to optimize drying (reduce dryer size and cost) is often mentioned in patents that disclose polymerization process technology. One means of increasing drying rate is by reducing the particle size of the gel. This takes advantage of Fick's second law of diffusion[59] that relates diffusion-controlled drying time to the square of the particle radius.

Two methods of sizing absorbent hydrogels prior to drying are discussed in the literature. They are gel sizing in the reactor and gel sizing in a separate unit operation. When sizing in the reactor, the agitator must also be designed as an effective shearing device (section 3.2.3). The descriptions of some of the preferred polymerization vessels,[32,33] teach the need to produce a finely ground material suitable for drying. The action of the vessel agitator produces a ground gel material with 98% of the product finer than 3 cm and the majority of particles between 0.5–1 cm. Improved grinding of gel may be obtained if a hydraulic pressure head is maintained on the reactor during the gel grinding process.[60] Attempting to use an agitator in this fashion no doubt increases the power and torque requirements of the agitation system and may make it excessively costly and prone to maintenance problems. Therefore, a separate unit operation is usually chosen to make the particle size of the dryer feed as uniform as possible.

Preparing dryer feed for through-circulation belt dryers, a type of dryer commonly used for superabsorbents, is well-known technology.[61] Some types of materials can be dried without any special shaping or sizing. However, pasty materials are often extruded to form spaghetti-like pieces, about 6.3 mm in diameter. Gel grind-

Table 3.2. The effect of hold temperature on swelling capacity and extractable polymer

Hold Temperature (°C)	Swelling Capacity[a] (g/g)	% Extractables[b] (4-h swelling time)
60	34.3	7.3
70	36.0	11.8
78	35.9	12.2
90	39.3	18.9

[a]Method of Chapter 4, Section 4.6.1.2.
[b]Titration method of Chapter 4, Section 4.3.4.

ing and extrusion have also been used to prepare dryer feed when the product is dried on drum dryers.[35]

An extruder[18,35,62] can reduce the particle size of the rubbery gel in a cost-effective manner because the shear developed between the stationary extruder wall and the moving screw grinds the rubbery particles. The efficiency of the extruder allows particle sizing at lower cost than with a reactor agitator.[63] Standard references discuss the principles of operation.[64,65] In summary,[63] the extruder's screw speed, screw design, and die design and the polymer's temperature, crosslink density, and water content all affect the degree of particle size reduction. The majority of the size reduction occurs in the screw as a result of back pressure from the die rather than of the rubbery hydrogel being reduced in size as it is squeezed through the holes of the die. Following this logic, either a single screw extruder or a double screw extruder could be made to grind the rubbery superabsorbent gel. Also, the exact geometry of the die holes is not as important to effective grinding as is the back pressure generated by the screw design and the combination of hole geometry and percent open area chosen for a given die design.

The reaction kinetics previously described also influence the selected operating parameters for the gel grinding equipment. The hydrogel, prior to grinding, is at the temperature of the holding vessel, and reactions are still proceeding in the gel as they would have in this vessel. Then, mechanical energy, as power supplied to the extruder screw, is added to the sticky, rubbery hydrogel in order to effect the grinding process. The result will be an increase in polymer temperature. The magnitude of the temperature increase depends on the amount of energy input per mass of hydrogel and affects the reaction processes taking place in the gel. Thus, some designs of this system can increase swelling capacity and extractables of the resulting polymer while decreasing absorbency under load (Table 3.3).[63]

The discussion of the desirability of hydrogel grinding, the parameters affecting the grinding process, and the chemistry occurring in the gel during the grinding step may serve as a starting point for understanding the development of grinding technology for superabsorbent hydrogels. Initial descriptions of hydrogel sizing de-

Table 3.3. Properties of the dried product after extrusion of the gel through a die with circular die holes of the indicated diameter. Data reproduced from reference 63.

Die Hole Size (mm)	Absorption Capacity[a] (g/g)	Absorbency Under 2.0 kPa Load[b] (g/g)	Extractable Polymer[c] (%)
2	34	23	11.6
4	29	25	9.3
6	29	25	8.8
8	29	25	9.2
10	29	25	8.7

[a]Method of Chapter 4, Section 4.6.1.2.
[b]Method of Chapter 4, Section 4.6.4.2.
[c]Method of Chapter 4, Section 4.3.4.

scribed cutting or extruding the wet hydrogel such that the surface area per unit volume was at least 9 cm^2/cm^3 (a sphere of 3.3 mm diameter would have this ratio of surface to volume).[18] In the patent examples, a small single-screw extruder having a die with 1.5 mm holes was used. If the information is taken at face value, the particle diameter was twice the hole size so that substantial energy input into the gel must have been required to force it through the holes. In later patents assigned to the same company, the descriptions of preferred polymerization vessels[32,33] teach the need to produce a finely ground material. It is preferred that 98% of the product is finer than 3 cm in diameter and the majority of particles be 0.5–1 cm in diameter. In a subsequent optimization, a polymer gel in the temperature range of 45°–90°C is extruded through a perforated plate containing holes in the range of 6.5–18 mm. The most preferred final gel particle size is 0.5–3 mm, the same final particle size range as first described.[18] Now, the particle size is smaller than the hole size, suggesting lowered energy input into the gel and reduced property degradation. Thus, extrusion through a die is used[14] so that the gel is formed into "shapes having sufficient surface area for drying." At least for one manufacturer of superabsorbents,[62] the hot hydrogel (50°–70°C) is extruded through a die with holes whose size may range from 8 to 15 mm. The acceptable percent open area of the die covers a broad range, 30%–90%. This set of conditions yields a ground hydrogel of highly uniform particle size, with an average particle diameter of 0.5–3 mm with a high swelling capacity and low extractable polymer content.

An additional benefit of a separate unit operation for gel is that mixing of additives may be accomplished in the equipment during chopping/grinding.[35] When acrylic acid is polymerized, the polymerized and ground hydrogel may be neutralized in a gel grinding step using sodium hydroxide solution, typically at 50 wt% concentration.[35] If an ester crosslinker such as trimethylolpropane triacrylate is chosen for the polymerization, the neutralization would preferably be accomplished with a sodium carbonate solution to avoid hydrolysis of the crosslinker. Other additives may be added in this step with, if needed, additional gel mixing and grinding. Additives include recycled fines,[15] an additional crosslinking reagent such as ethylene glycol diglycidyl ether,[35] sodium bisulfite for reduction of residual monomer,[66] or an inhibitor to suppress the formation of unwanted soluble polymer that may be produced during later processing steps.[66]

3.2.5. Drying

The water present in the hydrogel intermediate (added during the preparation of the monomer) is removed by evaporation using convective or conductive heating devices. Because the monomer concentration ranges commercially from about 16% to about 43% on a weight basis, about 2–5 kg of water must be removed for 1 kg of polymer produced. The moisture concentration at the end of the drying step will be from about 1 to 5 wt% as measured by the 105°C oven test described in Chapter 4, section 4.5.3.1. Fundamentals of the drying process are discussed in a number of chemical engineering references. Drying rate is controlled by the dryer design, chemical properties of the hydrogel, such as the diffusion coefficient of water through the matrix, and physical characteristics of the bulk material on the dryer, such as particle size, porosity, and uniformity of the hydrogel. The polymer can be dried in any number of commercial dryers.[61] However, through-circulation belt[32,51] and rotating-drum dryers[35] are both specifically mentioned in the literature for superabsorbents.

In addition to water removal, additional monomer conversion takes place in the drying step by means of a free-radical process (Chapter 2, section 2.5.4.1). Optimization of this process depends on the product temperature; moisture concentration; polymer glass transition temperature and diffusion coefficients of all reacting species, monomer, initiator, polymer chains, etc.; hydrogel particle size (because of the moisture concentration and nonuniformity in the particle); and initiator concentration. The unreacted monomer will be reduced in this step from about 3,000–30,000 ppm to commercially acceptable quantities of 200–1,000 ppm.

From a description of a pilot-scale drying process to reduce monomer, it is not obvious that a chemical process is taking place.[51] However, both in the description of the pilot-scale dryer and in a second process,[67] the presence of a thermal initiator, either a persulfate salt or 2,2′-azobis(2-amidinopropane)-hydrochloride, is necessary to reduce the amount of unconverted monomer. The monomer reduction is most satisfactory when the product temperature range during drying is from greater than 60°C to about 100°C. (The product temperature, when significant moisture is being lost, will be nearly the same as the dew point of the drying air after it exits the polymer bed.) Of course, the selected thermal initiators are very active in this temperature range. Preferred residence times are about 4–5 min.[67] In both cases, monomer reduction is more efficient in the presence of some moisture. The lower moisture limit is about 20% and the upper limit about 60%. At these temperatures and moistures, the polymer is above its glass transition temperature (Chapter 2, section 2.5.4). Thus, mobility of the polymer chains may play a role in monomer reduction in the dryer. The presence of an upper moisture limit implies that monomer reduction does not occur at a commercially satisfactory rate if the moisture content is too great. No explanation for this behavior has been given previously. However, it may reflect the rate of oxygen diffusion, as a function of moisture concentration, into the particle. As in more dilute systems, the rate of monomer reduction in the dryer is much faster when oxygen is rigorously excluded (Chapter 2, section 2.5.4.1).

In addition to the reduction in unreacted monomer, presumably by a free-radical mechanism, swelling capacity and extractable polymer are also affected by the drying process. Samples of gel made by the process of Sidall and Johnson[28] were dried in a commercial pilot-scale through-circulation tray drier at several different air temperatures with the results shown in Table 3.4. The air velocity was 65 m/min for sample A and 100 m/min for sample B. The gel bed depth was about 6.5 cm. The dried samples were ground. The swelling capacity and extractables were determined using the methods of Chapter 4. As the inlet air temperature increased, extractables increased considerably and swelling capacity increased slightly. The residual monomer reduction, the swelling capacity increase, and the extractables increase may all be interpreted, consistent with the discussion of reaction kinetics in Chapter

Table 3.4. The effect of drying air temperature on swelling capacity and extractable polymer

Sample	Drying Air Temperature (°C)	Swelling Capacity[a] (g/g)	% Extractables[b] (16 h)
A	120	30.1	5.7
A	150	31.0	7.9
B	150	30.8	7.3
B	180	33.1	9.8

[a]Method of Chapter 4, Section 4.6.1.2.
[b]Titration method of Chapter 4, Section 4.3.4.

2, section 2.5.4.1, as resulting from the production of free radicals from the decomposition of persulfate in the drying step.

A single-stage through-circulation belt dryer, similar to ones used by many superabsorbent polymer manufacturers, is shown in Figure 3.6a, and a sectional view is shown in Figure 3.6b. Two-stage, through-circulation belt dryers may also be used.[68] Both single-stage and two-stage designs are well known commercially, and a general description of through-circulation belt dryers is available.[61]

The wet hydrogel is spread evenly at one end of the dryer on a continuous, perforated, typically stainless steel, belt. The result is a porous bed of wet material across

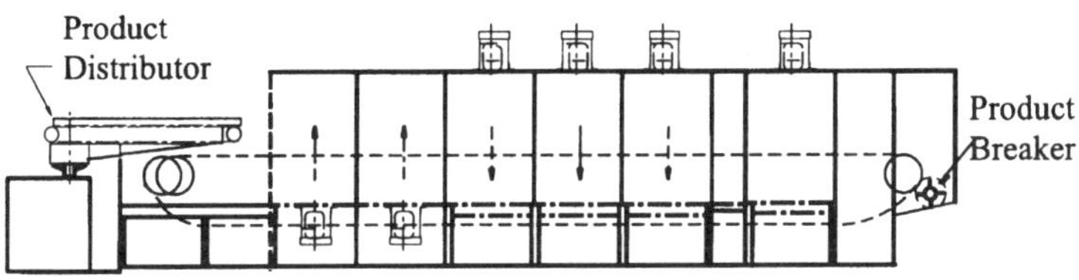

(*a*)

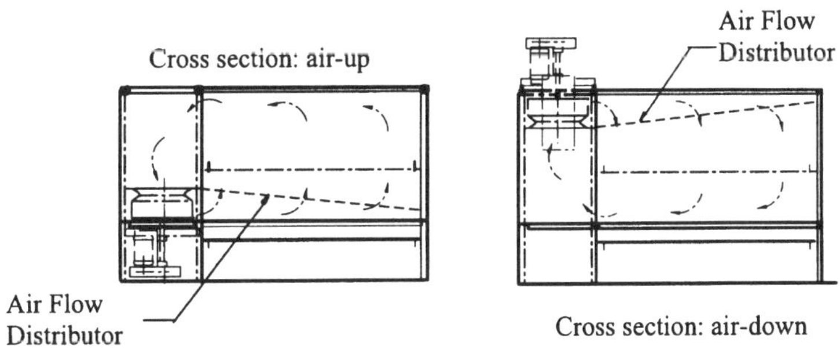

(*b*)

Figure 3.6. (a) Through-circulation, single-conveyor, single-stage dryer. The product distributor is of the oscillating-belt type, and a product breaker is shown at the dryer exit. **(b)** Cross-sectional views of air-up and air-down sections of a through-circulation single-conveyor dryer. (Courtesy Procter & Scwartz Inc.)

the width of the dryer bed at a desired depth. The particle size of individual pieces of polymer vary over a range. Bed characteristics, such as bed porosity, bed bulk density, and leakage around the edges of the bed, also vary. When dry, the bed may consist of individual dry particles or a mass of dry particles stuck together to form a porous sheet.

The drying air is circulated at 1–3 m/s, first through a heat exchanger and then through the product and the belt as the belt moves down the length of the dryer (Fig. 3.6a). A portion of the air may be continuously vented to the atmosphere (through the environmental control system) so that recirculation of the air in the dryer is in the range of 60%–90%.[61] The fraction of recirculated air affects the relative humidity and the drying rate in a particular dryer section. A dryer will have several sections, each with its own circulating fan and heat exchanger, arranged in series to form a tunnel through which the belt travels (Fig. 3.6a). Individual dryer sections can be controlled independently. Air may be circulated upward through the bed in the wet end and downward in the dry end. Naturally, at the wet end of the dryer, the air humidity is relatively high, especially since the percentage of recirculated air is large to minimize energy costs. Air dew points in the range of 50°–70°C are common in through-circulation dryers.[61]

Through-circulation dryers work best with particles or granules.[61] Special pretreatment to make materials suitable for through-circulation drying, such as the gel grinding used for superabsorbent polymers, is common. The belt can be loaded directly by oscillating belt or vibrating type feeders, as is shown in Figure 3.6a. The belt speed and the rate at which product is loaded to the dryer determine the depth of the product on the belt. If the permeability characteristics of the product bed at any chosen spot are constant over time, the thickness of the bed may be estimated, from pressure drop measurements,[69] using equations related to Darcy's law (Chapter 1, Eq. 1.1).[70] In one case, a large diameter pipe, which serves as a distribution vessel, has been mounted at one end of the dryer, perpendicular to the direction of travel of the belt.[71] Holes in the bottom of the pipe are located so that the soft, deformable hydrogel is evenly distributed across the width of the dryer belt. Spreaders are used to level and uniformly distribute the wet, sticky product across the width of the dryer belt. In one case, a multiprong rake has been attached to the end of an oscillating belt to more evenly distribute gel.[72] As a result of the increased product uniformity, an improvement in drying efficiency is obtained.

A broad range of drying temperatures is inherently possible and would be expected in the operation of a through-circulation dryer. Greater amounts of moisture are removed at the beginning of the drying cycle; thus, the temperature of the recirculating air tends to be lower at the beginning of the dryer. At the end of the dryer, little heat is required to remove moisture from the nearly dry polymer and the temperatures are higher. Naturally, reduced average temperatures require longer drying times. In one case, 160°C air (inlet air temperature for one-pass air) was used.[32] Elsewhere,[68] the most preferred range of drying air temperature has been given as 150°–170°C. A detailed description of a pilot-scale through-circulation drying process is given.[51] The hot air has been humidified (60°–100°C dew point), and drying temperatures are from 80°–250°C. However, the range of 100°–180°C is preferred for producing a product with a low content of residual monomer. The temperature ranges and dew point ranges are consistent with the desire both to dry the polymer in a short time and to reduce the monomer content to an acceptably low level.

Three drying stages are observable in the drying curve for a typical polyacrylate superabsorbent (Fig. 3.7). The first is constant-rate drying; the second is a transition stage between constant-rate drying and diffusion-controlled drying; and the third is falling-rate or diffusion-controlled drying.

In the first sections of a through-circulation dryer, drying rate is typically limited by the rate at which moisture can be removed from the surface of the particles and drying rate is constant. Moisture diffuses to the polymer surface faster than it can be removed by the hot, moist, recirculating air. The rate of moisture removal depends on the heating gas temperature, humidity, and flow rate. Characteristics of the bed of hydrogel on the dryer belt, including particle size and shape and bed porosity, also affect drying rate. The actual product drying rate, in this part of the dryer, is a complicated function of these variables. A mathematical model of a through-circulation dryer is available describing the constant-rate portion of a drying curve.[73] The design of through-circulation dryers has been discussed,[74] and design criteria are available, also primarily based on the constant-rate portion of the drying curve.

As a result of the distribution of particle sizes and the variable porosity of the gel bed on the dryer belt, a transition region in the drying behavior between the constant-rate period and the diffusion-limited, falling-rate period (Fig. 3.7) is typically observed. One view of the transition stage of the drying process is that a portion of

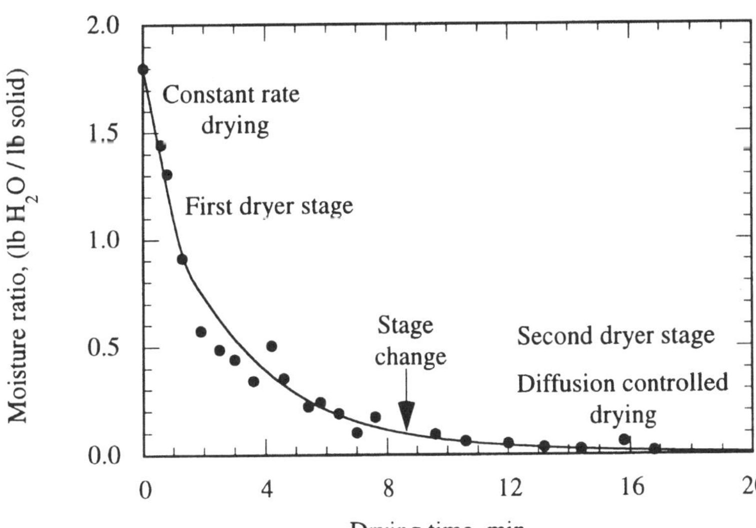

Figure 3.7. A typical drying curve of a 65% neutralized superabsorbent polyacrylate obtained on a commercial two-stage through-circulation dryer. Drying was essentially completed in 15 min.

the polymer is being dried under the constant-rate regime while another portion of the polymer has entered the diffusion-controlled regime. The overall drying rate changes from constant-rate drying to diffusion-limited drying as the available "surface" moisture is depleted for material in the whole range of particle sizes and bed porosities.

Drying in the diffusion-limited, falling-rate period is expected to follow Fick's second law of diffusion.[75] In spherical coordinates, where diffusion is independent of the angular coordinates, the governing equation is

$$\frac{\partial C}{\partial t} = \frac{1}{r}\frac{\partial}{\partial r}\left\{ rD\frac{\partial C}{\partial r} \right\} \tag{3.1}$$

where C is the concentration of the diffusing species (water), D is the diffusion coefficient, t is time, and r is radial distance in a spherical particle. Solutions to this differential equation may be found where the diffusion coefficient is either constant or linearly dependent on the concentration of the diffusing species.[75] Equation 3.1 shows that when the drying rate is diffusion controlled, the drying rate depends on the concentration of the diffusing species. Therefore, as the polymer dries, the drying rate decreases, hence the name "falling-rate" for this drying stage.

Moisture diffusion coefficients for 65% neutralized crosslinked sodium polyacrylate have been estimated from drying curves over the range 130°–180°C (M.A. Fialkowski, The Dow Chemical Co., unpublished results). At 150°C and moistures of 0.17 kg water/kg polymer or less, the diffusion coefficient was found to depend linearly on moisture concentration (Eq. 3.2). The diffusion coefficient has units of m^2/s, and the concentration of water is given in weight of water per weight of solid polymer on a dry basis:

$$D = 1.15 \times 10^{-11} + 68.3 \times 10^{-11}[H_2O] \tag{3.2}$$

The temperature dependence of the diffusion coefficient was included using an Arrhenius-type approach. Combining Eq. 3.2 with exponential temperature behavior leads to an expression for the diffusion coefficient of water in 65% neutralized polyacrylate as a function of both temperature T (K) and moisture:

$$D(T, [H_2O]) = (1.15 \times 10^{-11} + 68.3 \times 10^{-11}\,[H_2O])\exp\left(\frac{-9{,}470}{T} + \frac{9{,}470}{150 + 273}\right) \tag{3.3}$$

The exponential factor was not determined at exactly constant moisture conditions and should be used with some caution. Also, the equation used to calculate the diffusion coefficient from the experimental data assumed spherical particles, only a crude approximation to the geometry of the particles in the wet polymer bed.

A brief, general description of drum dryers, which consist of two rolls rotating in opposite directions to form a nip, is available.[76] Drum dryers rely on conductive heat transfer to provide the energy necessary to evaporate moisture. To have short, economical drying times, and to avoid removing a sticky, moist product from the drum rolls, the hydrogel is limited to a thickness on the drum that can be dried in about three-fourths of a revolution of the drum roll. Air flow is across the surface of the drying gel rather than through it. Where steam provides heat to the dryer rolls, the drying temperature may be regulated by the steam pressure supplied to the rolls. The hydrogel throughput rate depends primarily on the surface area of the rolls, the thickness of the hydrogel layer (controlled by the gap, or nip, between the rolls), the speed at which the rolls are turned, and the desired final product moisture. Some means of evenly spreading the granular gel across the width of the rolls would be required for efficient drying.[35] To have a thin enough gel layer to dry efficiently, the hydrogel must be of low enough modulus to be deformable in the dryer nip. Patent examples use concentrations of polymer in the gel feed to the drum dryers of 22–30 wt%, although we have successfully drum-dried polymer concentrations as high as 36 wt%. Drying temperatures from 100° to 200°C are taught. However, the drying temperature was 105°C in one example.[35]

The product is removed from the rolls with flaking knives, yielding a flaked product rather than the granular product resulting from the through-circulation dryer. Our experience suggests that drum drying leads to a product of higher surface area and faster liquid absorption rate than that achieved with oven drying or through-circulation belt drying.

3.2.6. Handling of the Dried Material: Particle Sizing

After drying, the superabsorbent product would not usually be in the correct particle size for a particular application. The product from the dryer may range in character from individual particles to a porous sheet of dried polymer, the dimensions of which may be the width of the dryer and the depth of the wet cake added to the dryer (reduced by the drying process). A two-stage dryer may yield chunks of dried sheet at the end of the drying process. If a two-stage dryer is not used, some means of breaking up the sheet of material exiting the dryer must be provided before attempting to feed the product to the particle sizing step. Dryer manufacturers often supply this equipment as part of the dryer. In all cases, the particle size distribution of the final product may be obtained by a combination of comminuting (milling) and screening to give a particle size distribution ranging from about 200 to 800 μm. The reader is referred to standard references[61,77] for descriptions of suitable process equipment.

To achieve the rather narrow particle size distribution advocated for use in diapers,[78] two-stage milling is used in combination with product screening and recycling of the oversize stream back to the grinding step. This process may consist of two dissimilar or two identical milling stages, typically selected for a gentle grinding action. Knife mills, attrition mills, roll crushers, and the like are suitable for use in the first stage. The first stage of sizing prepares the polymer for the second stage and narrows the overall distribution of particle sizes fed to the second sizing step.

The requirements on particle sizing in the second stage are more stringent because of the desirability of producing a narrow particle size distribution with minimum fines generation. Roll crushers (Fig. 3.8) are commonly used for this portion

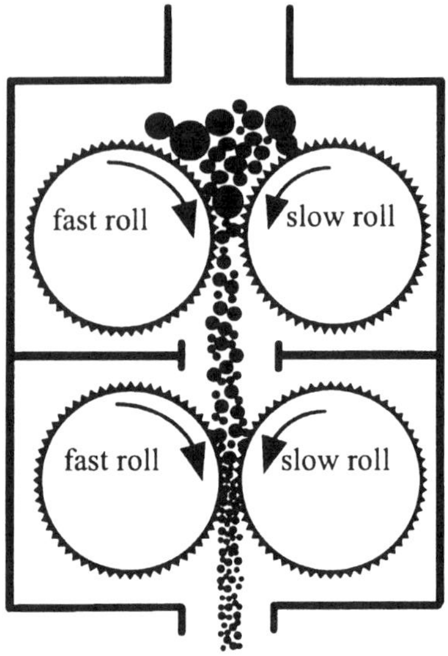

Figure 3.8. Schematic diagram of a two-pair high roll crusher.

of the particle sizing process. A particle classification step, usually screening, is used in combination with the grinding step. The polymer granules pass through a series of screens of decreasing opening size. The fraction of particles that is larger than the desired particle size is recirculated to the grinding step, while the on-size fraction is forwarded to the next unit operation. A screening operation may be used in combination with one or both of the grinding steps. Undersize or fine material may be separated from the desired particle size fraction at this step.

Figure 3.8 is a schematic of a two-pair-high roll crusher. The rolls are typically corrugated, with the corrugation pattern strongly affecting the particle size distribution of the ground product. The corrugation design includes the groove size and spacing between grooves. The angles associated with the shape of the grooves may also be varied and have been optimized by the equipment manufacturers. In combination with the selection of equipment for the first stage of the grinding process, roll groove design, roll operational parameters (rpm of the rolls and differential rpm between the rolls), and gap between pairs of rolls are chosen to give the desired particle size distribution. The primary variable in setting the particle size distribution for a given set of rolls, however, is the gap between the rolls. Increasing the gap reduces the amount of grinding that is performed, to produce larger particles. The gap between the rolls also acts as an internal classifier in the mill, allowing only particle sizes larger than the gap setting to be crushed. Particles that are smaller than the desired size pass through the mill without further size reduction, thus minimizing the quantity of undesirable fine particles. However, feeding the mill at too fast a rate may result in a broader particle size distribution as a result of particle–particle grinding. Guidance is available on choice of equipment for comminuting and screening[61] and for minimizing the width of a particle size distribution.[79]

The desired product particle size cut is then transported to the next operation in the process. This may be merely dropping the product vertically through a pipe, or it might require specialized solids transporting equipment. Choices include belt conveyors, bucket conveyors, screw conveyors, vibratory conveyors, and dilute or dense-phase pneumatic conveyors.[80,81]

The choice and design of a solid transport system depends on plant layout and the results expected of the transport process. Commercial processes are designed to minimize particle attrition for at least two reasons. First, it is desirable to minimize the generation of quantities of particles smaller than about 140 mesh (100 μm), as noted below. Second, degradation of the product properties may result from attrition of the particle surface (see section 3.2.8.1).[82] For short transport distances, low-attrition mechanical conveyers are desirable. For longer transport distances, pneumatic conveyor systems are preferred. Since particle attrition depends on the velocity at which the product is transported, lower velocity systems are advantageous.

Only limited information is available concerning particle size distributions for the final product. One reference specifies a final "mass median" particle diameter between 400 and 700 μm, with the amounts of very small and very large particles kept below a minimum concentration.[78] The presence of fine particles (fines), typically below about 100 to 150 μm in diameter, is unacceptable in the product sold to the customer.[14,78,83] Several means of converting the fine material to a saleable product or reusing it in the current process are described in section 3.2.7.

The choice of sizing and handling equipment, including gel grinding, also affects the bulk density and flowability of the product (for measurement of these characteristics, see Chapter 4, section 4.5.6). Microscopic examination of particles led to the conclusion that generation of fines can occur as rough edges of the product particles are rounded off in milling or transporting of the product. A higher product bulk density results from the elimination of void volume arising from irregular particle shape. The presence of amounts of fine particles also alters product flow characteristics. Product flow characteristics, in turn, affect particle segregation in storage hoppers (and, therefore, the design of the storage hoppers) and product flow in feeding and metering equipment associated with diaper manufacturing. For further information in this area, the reader is referred to standard materials on the handling of solids.[61]

3.2.7. Fines Recycle

The production of particles over a range of sizes is an inevitable result of the grinding, screening, and solids handling operations used to handle dried superabsorbent polymers. Fine fractions in superabsorbent polymers or, indeed, any solid water-soluble or water-swellable product are undesirable to the extent that "gel-blocking" occurs when the product contacts an aqueous fluid. Gel-blocking occurs when the outer surface of a tightly packed mass of fine particles swells too quickly, preventing penetration of the liquid into the interior particles, whether by diffusion or by capillary action. It is primarily this effect that leads to reduced product performance in superabsorbent products containing too great a fraction of fine particles.

Even when the comminution and transport operations are selected to minimize the production of unwanted fine-particle size fractions, their handling and disposition must be considered. The patent record shows that superabsorbent manufacturers realized, after the fact, that the unwanted fines fraction needs to be managed. One solution is to simply remove the fines from the product by screening. That, however, only postpones the question of what to do with them. Two paths are discussed.

The first path attempts to transform the fines into a second product or to include them as an inert or active filler in a second product.[84–89] This path is not discussed in this volume. The second path is to recycle the superabsorbent fines into the primary superabsorbent product. In one case, fines, water, and fresh gel (direct from the reaction step) are physically blended to yield a uniform blended product.[14,15] About 15% fines, based on the weight of the final dried product, are blended. Fines of partially neutralized product can be rehydrated, without substantial agglomeration or gel-blocking, by using a high speed mixer, followed by mixing of the hydrated fines with fresh gel.[14] (Similar equipment is described for the addition of reactive additives for advanced generation products.[90–92]) Also, recycle of fines to a gel stream consisting of the fines, fresh gel (un-neutralized), and a few parts of water processes more easily than the fresh gel.[15] The mixture of hydrated fines and fresh gel is extruded through a die so that the gel is formed into "shapes having sufficient surface area for drying."[14]

Unless a reaction takes place between the hydrated fines and the gel, for example by using a reagent such as ethylene glycol diglycidyl ether, the fines are not chemically bound to the gel polymer. As a consequence,[17] the product may not retain its "unitary nature under the stresses imposed by hydration." Viewing the swelling of agglomerated particles under a microscope shows that the agglomerated larger particles break into its fine-particle components as the mass swells.

Recycle of superabsorbent fines to the polymerization step is also described. In this art, the fines are swollen with monomer or monomer solution and the mixture polymerized.[16,17,29,93] In one case,[16] the monomer is first added to the reactor, followed by addition of the fines. Although the viscosity of the monomer mix increases substantially with the addition of fines, no particular problems were noted. Equipment to mix fines and monomer is not described in these patents. However, equipment of the type used to swell fines with water might be useful at a commercial scale.[94] Equipment necessary for polymerizations containing fines[16] is similar to the reaction vessels previously described.[32,33] When polymerizing on a continuous belt, fines were added to the belt polymerizer as a part of the monomer mix.[29] In this case, the fines are first mixed with a nonswelling solvent, a nonswelling salt solution, or a nonswelling monomer such as acrylic acid and then added to an aqueous solution of monomer. The amount of fines that can be added may be as high as 50%,[16] but is probably more on the order of 20% or less.[16,17,29] When fines are recycled to the polymerization, product properties are nearly equivalent to properties of products made without fines recycle. However, in one case for polymers of the same swelling capacity, a slight reduction in absorbency under load was noted.[17]

3.2.8. Addition of Post-Treatments

3.2.8.1. Advanced Products. The processes described above produce a particle with crosslinking that may be assumed to be essentially uniform throughout the particle. As such, the swelling and modulus behavior of the product particles can be described by theories of network swelling and elasticity (see Chapter 5). One problem associated with these superabsorbent polymers was also recognized in a variety of water soluble polymers, e.g., poly(acrylamide), cellulose ethers,[95] or soluble poly(acrylic acid).[96] This problem is the tendency of the granular polymers to clump, "gel block," or form fish-eyes when aqueous liquids are added to them or they are added to aqueous liquids. The surfaces of the massed particles swell rapidly to form a soft, deformable layer. The resulting particle deformation and interparticle adhesion reduces interparticle porosity and limits the swelling rate of the polymer mass to the diffusion rate of liquid through the partially swollen mass.

The first attempts to solve this problem in highly absorbent anionic polyelectrolytes used multivalent cations to form a crosslinked surface layer that was more rigid than the original core polymer.[97] Subsequently, to give products with improved dispersibility and absorption rate, the surface of crosslinked polyacrylate particles was reacted with a variety of multifunctional organic compounds.[49,50,98,99] This structural improvement is shown schematically in Figure 3.9. Swelling capacity data as a function of particle size readily demonstrate that a shell of real, as opposed to infinitesimal, thickness is formed by surface crosslinking processes. The shell of higher crosslink density provides a more rigid surface layer during swelling and prevents the gel-blocking that would otherwise occur early in the swelling process (Chapter 5). As a result, liquid can flow through the bed of particles to each particle, increasing the effective surface area available for swelling and the apparent swelling rate. The swelling and modulus behavior of structured particles are discussed in Chapter 5.

Because a surface layer is formed on each individual particle, the crosslinking reaction, but not necessarily the reagent addition, must occur after achieving the desired particle size distribution. Any compound that can react with two or more functional groups on the polymer backbone can function as a surface crosslinking reagent. For superabsorbents, this requires reagents that can react with the car-

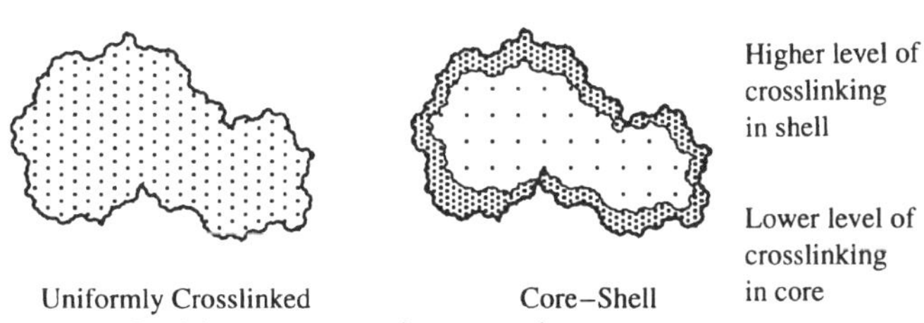

Figure 3.9. Schematic representation of a uniformly crosslinked superabsorbent particle and a structured particle with a surface layer of higher crosslink density.

boxylic acid or carboxylate moieties on separate, adjacent polymer chains. Quaternary amines have been used,[100] but the most common reagents add surface crosslink density as a result of formation of ester crosslinks. These surface crosslinking agents include polyhydric alcohols,[49,50] glycidyl compounds,[101–103] and alkylene carbonates.[104]

Blenders suitable for the addition of surface crosslinking agents include screw mixers, ribbon mixers, fluidized beds,[49] high-speed rotating paddle mixers,[50,90] vortex mixers, rolling mixers, or double screw mixers.[104] Spraying the surface crosslinking reagent into the blender, using small droplet sizes, preferably less than 300 μm, helps reduce the amount of liquid required to disperse the surface crosslinking reagent.[50] A high-speed paddle mixer used for this process operates at greater than about 600 m/min of a lead-end peripheral speed.[90] In one example, a Turbulizer, produced by Hoskawa Micron K. K., was operated at 3,000 rpm. The inner surface of the mixer is preferably coated with a material that minimizes adhesion of the wetted polymer to the walls of the mixer, for example, polypropylene or a poly(tetrafluoroethylene) resin. When a high-speed mixer is used, the amount of organic solvent used to help disperse the surface crosslinking reagent may be minimized or even completely eliminated.

A heating step is normally necessary for the reaction of surface crosslinking agents with the partially neutralized poly(acrylic acid) backbone. The time and temperature of heat treatment are chosen, considering the reactivity of the crosslinking reagent, to give the desired end-use properties. Naturally, longer reaction times and higher temperatures increase, for a reagent of any reactivity, the extent of reaction. Tsubakimoto et al.,[50] claim temperatures of 90°–300°C while Dahmen and Mertens[104] teach 150°–300°C. However, at temperatures greater than about 250°C, degradation of the polyacrylate by oxidation or decarboxylation may become significant (see Chapter 2, Section 2.5.4.2). More specifically, the reactive epoxy groups in ethylene glycol diglycidyl ether react in 30 min at 140°C.[105] The less reactive ethylene carbonate requires 1 h at 180°C or 15 min at 215°C.[104]

Suitable heating devices include fluidized-bed dryers, infrared dryers, agitated-trough dryers,[49] paddle dryers,[50] vortex dryers, or rotating-disc dryers.[104] Comparative data that would allow the selection of the most suitable device are not given. Both direct-contact dryers (rotating-disc dryers, paddle dryers) and fluidized-bed dryers are used commercially to provide heat for the reaction step. Direct-contact dryers deliver heat to the polymer particles by contact of the particles with a hot surface, while fluidized-bed dryers deliver heat to the particles with a stream of hot air of high enough velocity to suspend the particles in the airstream. Direct-contact dryers, for this application, must be designed keeping in mind the possibility of attrition of the particle surface (from the mechanical action of the paddles or other movable surfaces on the particles) and consequent degradation in product properties. The air-handling equipment associated with fluidized-bed dryers may make them expensive to operate.

Initially, structured particle technology was viewed as solving the problem of gel-blocking, as measured by a fish-eye test[49] or a similar test conducted without a compressive load. These early tests were only able to qualitatively distinguish between structured polymers made under different sets of process conditions. For example, Tsubakimoto et al.[50] showed that a water/alcohol carrier for the surface crosslinking reagent yields a structured polymer with less fish-eye formation than when the reagent is added undiluted. Later, more sophisticated tests (see Chapter 4, section 4.6.4) that measure swelling under a compressive load (absorbency under load) were able to quantitatively distinguish between structured polymers made under different sets of process conditions. Results of these tests suggest that penetration of the reagent into the particles and control of the dispersion of the crosslinking agent over the surface of the particles are critical process considerations for achieving success in the absorbency under load test.

Optimization of the surface crosslinking process to reduce gel-blocking in an absorbency under load test has been addressed in a number of ways. Surface crosslinking reagents have been added in aqueous solutions and other diluents such as alcohol or aqueous salt solutions.[102,103] However, there are several difficulties with these approaches.[106] Water swells the particle surface and lowers the glass transition temperature of the polymer at the surface, increasing penetration of the surface crosslinking agent into the polymer particle. However, excessive amounts of water cause the polymer to swell and can make handling difficult. Inorganic particulates, such as silicates, in addition to controlling the hydrophilic/hydrophobic balance of the treatment reagent, are used, in combination with high-speed blending of the reagents with the polymer, to control agglomeration during the reagent addition process.[50,90] The alcohol or salt solution helps disperse the surface crosslinking agents over the powder by diluting the crosslinker (larger volume) and helps control or prevent agglomeration of the powder by reducing the activity of water in the solution. The use of an organic solvent has the obvious disadvantage of requiring recovery and recycle. Also, for example with the lower alcohols, the low flash points require extra equipment and more safeguards to maintain safe operation.

Adding 0.2 wt% glycerol, in combinations of water and *n*-propanol, to a 65% neutralized polyacrylate made by the process of Siddall and Johnson[28] shows the effect of reagent penetration. One gram of liquid was added to 100 g of polymer powder. Twenty-five grams of the reagent treated polymers was heated in an aluminum dish for 15 min at 200°C in a preheated circulating air oven equipped with a steel plate as a heat sink. Table 3.5 shows the swelling capacity (method described in Chapter 4, section 4.6.1.2) and absorbency under load (described in Chapter 4, section 4.6.4.2) data. Swelling capacity decreases as the amount of water in the reagent solution increases. Similar results have been obtained by using aluminum acetate, ethyleneglycol diglycidyl ether, or ethylene carbonate as the surface crosslinking reagent. This has been interpreted as resulting from increased penetration of the reagent into the particle and increased crosslinking throughout the particle rather than just on the surface.

Other process choices also affect the resulting properties. Table 3.5 suggests that for glycerol the optimum weight ratio of water to *n*-propanol is about 30:70 (based on the values of swelling capacity and absorbency under load). The amount of water may be reduced by choosing a more hydrophilic organic solvent such as methanol. Increasing the salt concentration of the crosslinker solution also decreases the hy-

Table 3.5. The effect of solvent hydrophilicity on the properties resulting from surface crosslinking of a polyacrylate treated with 0.2 wt% glycerol and then heated at 200°C for 15 min

Solvent Composition Water/*n*-propanol (w/w)	Swelling Capacity (g/g)	Absorbency Under 2.0 kPa Load (g/g)
Starting material	39.0	9.0
0/100	36.4	23.2
10/90	35.0	21.6
20/80	34.7	29.0
30/70	28.8	29.3
40/60	30.5	29.3
50/50	29.3	28.6

drophilicity of the solution (reduced particle surface swelling) and has the expected effect on product quality. The amount of crosslinker solution and the concentration of crosslinker in the solution also impact the results. When too little crosslinker solution is used, for example to avoid agglomeration, the particle surface coverage can be incomplete, leading to increased gel-blocking in the swelling under load tests. When the concentration of crosslinker is too low, or an insufficient quantity of reagent is applied, the amount of additional crosslinking at the surface does not make the surface sufficiently rigid to give the required resistance to gel-blocking.

More recently, the desire for products with an improved combination of swelling capacity and absorption under load[107,108] has led to more complex surface crosslinking sytems.[100,106,109] In one case,[100] a shell structure is achieved by using two surface crosslinking reagents with different solubility parameters, one greater than 12.5 $(\mathrm{cal/g^3})^{1/2}$ and the other less than 12.5 $(\mathrm{cal/g^3})^{1/2}$. The combination of solubility parameter and low molecular weight of the crosslinking reagents (less than 350 g/mol) is said to optimize the penetration of the surface crosslinking reagent into the particle (compare row 4 to rows 2 and 3 of Table 3.6). In a second case, surface crosslinking with a specific class of high-molecular-weight (greater than 1,000 g/mol) azetidinium compounds leads to polymer particles with more ideal shell thicknesses (less than 10% of the particle radius).[109] The high molecular weight of the reagent limits the penetration of the crosslinking agent into the particle so that the swelling capacity is not reduced too greatly. A range of surface tension, about 40–50 dynes/cm, of the applied crosslinker solution promotes spreading of the crosslinker solution over the surface of the particles.[110] A two-step addition of surface crosslinking agent also leads to improved properties.[111] Another two-part process, in which a first crosslinker (hydrophilic polymer or metallic compound) is reacted with a second additive on the surface of a wet polymer (10–100 parts), has also been described. The extent of hydration of the polymer and crosslinker are chosen to give the desired penetration of the surface crosslinking reagents into the particle and the desired combination of properties. The process seems most suitable for suspension-made polymers, but its inventors claim its utility is for polymers made by solution polymerization. In one example, a coating of poly(ethyleneimine)

Table 3.6. A comparison of properties resulting from "optimum" process conditions for making structured particles

Reagent	Swelling Capacity[b] Before Treatment (g/g)	Swelling Capacity (g/g)	Absorbency Under Load[c] (g/g) 2.0 kPa	3.9 kPa	6.2 kPa
Aluminum acetate	39	35	25	18	—
Glycerol	39	35	33	21	12
EGDGE[a]	40	35	35	31	24
Glycerol/EGDGE (process of Date et al.[109])	45	35	35	31	26
Ethylene carbonate	50	36	30	25	21
Example 1 of Brehm and Hartan[111]	40	36	33	25	21
Process of Burgert et al.[112]	39	36	32	26	22

[a]Ethyleneglycol diglycidyl ether.
[b]Method of Chapter 4 Section 4.6.1.2.
[c]Method of Chapter 4 Section 4.6.4.2.

crosslinked with a polyglycerol polyglycidyl ether resulted in a superabsorbent resin with "excellent" stability (by which the inventors seem to mean gel modulus). This process is said to require less stringent temperature control to achieve adequate reaction.

Despite the differences in the various surface crosslinking reagents and the resulting differences in reaction conditions, products with very similar properties may be produced with a variety of surface crosslinking processes. Table 3.6 shows swelling capacity and absorbency under load results when a variety of surface crosslinking reagents were applied in a methanol/water solvent system to a 65% neutralized poly(acrylic acid). The reaction conditions (temperature and time) were chosen to give the highest values of absorbency under load at the target swelling capacity (35 g/g). The choices were based on our experience with these processes and on data in the indicated references.

The surface crosslinking process of Brehm and Hartan[111] and the alternate structured particle technology of Burgert et al.[112] (discussed in more detail below) also produce similar results (last two rows of Table 3.6). In each case, the absorbency under load was substantially improved compared with the material before surface crosslinking. Even so, property differences remain that depend on the individual surface crosslinking reagents or process, despite our attempts to optimize the process. Some of these may result from insufficient optimization of the process conditions, such as spreading of the reagent, or chemistry, as discussed in Chapter 2, section 2.7. However, a consideration of the several reagents leads to the speculation that the crosslinker structure might also be a factor in the performance of the structured particle. The distance between functional groups in the crosslinker might be such a factor. This is a question that needs to be resolved by future study.

Improved properties have also been achieved by forming a structured particle in a different way. Polymers made with certain crosslinkers or combinations of crosslinkers and heated in the presence of special oxidizing agents have dramatically improved absorbency under load and swelling capacity.[112] In this technology, all necessary ingredients are added prior to polymerization so that they are mixed uniformly in the resulting particles. A variety of primary crosslinkers may be used alone or in combination with a secondary crosslinker, also uniformly distributed throughout the particle. Primary crosslinkers that have been used include 1,1,1-trimethylolpropane triacrylate (TMPTA), ethoxylated TMPTA, allyl methacrylate, and various diacrylates and methacrylates based on poly(ethyleneglycol). Nonvinyl or bimodal secondary crosslinkers have been used. Examples of nonvinyl crosslinkers include glycerol, various molecular weight poly(ethyleneglycols), and ethyleneglycol diglycidyl ether. Bimodal crosslinkers include hydroxyethyl acrylate and acrylate or methacrylate esters of polyethylene glycols. The preferred oxidizing agent is sodium or potassium chlorate. The amount of chlorate affects the reduction in the level of unreacted monomer, the amount of extractable polymer in the product, and the final swelling capacity. Figure 3.10 is a graph of data from examples 78–86 of Burgert et al.,[112] in which polymer samples were heated for 20 min after reaching 200°C. As the level of chlorate increases, swelling capacity and extractable polymer increase. The optimum chlorate level, considering swelling capacity, extractable polymer, absorbency under load, and unreacted monomer, is about 200 ppm. The moisture remaining in the particle, after drying and grinding and prior to the reaction step, influences the magnitude of the observed oxidizer effect.

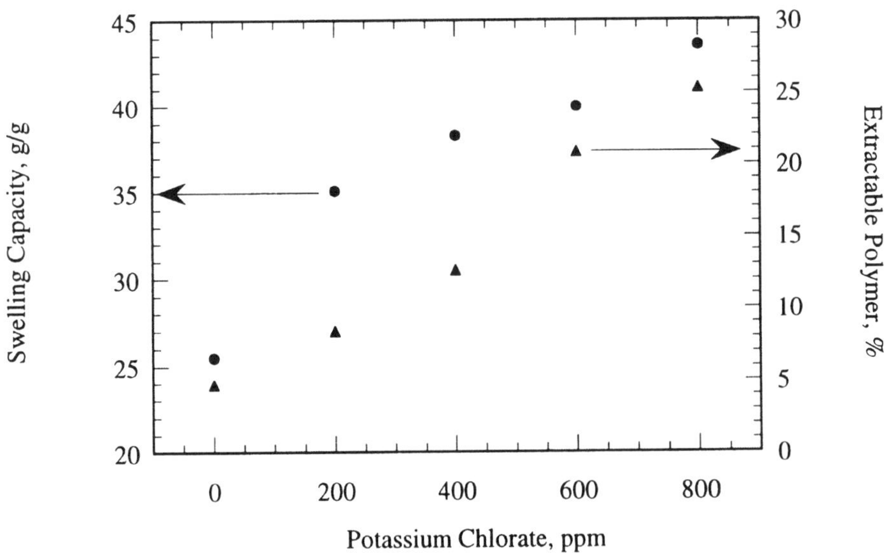

Figure 3.10. The effect of chlorate level on the swelling capacity (●) and extractable polymer (▲) of structured particle product made by the process of Burgert et al.[112]

The reaction temperature should be between 170° and 235°C, and the heating time should be less than 40 min. The best combination of product properties occurs when the heating temperature is between 210°–230°C. Heating times at 230°C of 10–30 min yielded product with nearly the same swelling capacity, extractable polymer, and absorbency under load. Combinations of swelling capacity and absorbency under load, even at high imposed loads, made by this technology (Tables 3.6) are nearly equivalent to properties achieved with surface crosslinking technology, presumably because gel-blocking is limited by a surface layer of higher crosslink density.

A consequence of the "core–shell" particle structure (Fig. 3.9) is a vulnerability to degradation of product quality by fracture of the particle surface.[82] High-quality methods for handling the solid product are required (see section 3.2.6) to avoid particle surface fracture and consequent degradation of product properties. Chemical means of preventing surface degradation have also been disclosed. For example, the addition of about 10 mol% of a monomer such as poly(ethylene glycol) acrylate improves the fracture resistance of the resulting superabsorbent product.[113] Coating the superabsorbent particles with a "non-reactive, water-insoluble, film-forming polymer" also reduces the dust level (small particle size fraction from abrasion of the particle surface in a ball mill test).[114]

3.2.8.2. Additives for Improved Handling. Techniques to reduce dust in the final product include agglomeration of the finer dusty particles via surface crosslinking, as described above, or the addition of low levels of hydrophilic liquid additives.[115] Both techniques reduce the concentration of air-borne respirable dust (less than 10 μm in diameter) by agglomerating the fine particles either to one another or to larger particles. When low levels of liquid additives are used, they are typically blended into the final product just prior to packaging. In one case, a nonreactive additive to reduce product dustiness may be added concurrently with any surface crosslinking reagents.[110]

Additives that reduce the rate of moisture absorption in humid environments and improve polymer flow characteristics are also of interest. One can appreciate the scope of this problem by simply exposing a quantity of superabsorbent polymer to humid air. In a matter of a few hours, the polymer begins to clump and will not flow as required in processing in a diaper plant. Dehumidified air may be used to alleviate or eliminate this problem, either in polymer or end-use manufacturing plants. However, particularly in the tropical Pacific markets, dehumidified air is not commonly available in diaper manufacturing plants, and additives must be used to allow adequate dry flow of the polymer. A number of such additives are known: oily materials for use with hygroscopic fertilizers,[116] polymeric soaps as drying aids for copolymers of acrylamide,[117] and particulate silica in combination with polyols or polyalkylene glycols as a flow aid for polymers and copolymers of poly(acrylamide).[118] Quaternary surfactants have also been found useful as anti-clump aids, alone or in combination with additives, to reduce product dustiness (a drawback of the silica anti-caking technology).[119] These chemicals may be added in a variety of types of commercial blenders.

3.3. SUSPENSION POLYMERIZATION

3.3.1. General Terminology

The phrase *suspension polymerization* as used here is compatible with the terminology described by Arshady[120] to mean a polymerization in which droplets of monomer or monomer-containing solution are dispersed in a continuous phase that is immiscible with the monomer-containing solution. The polymerization is carried out in the dispersed phase, and the monomer droplets polymerize independently of one another. There is no transfer of monomer to, or polymerization in, the continuous phase as is the case for a latex polymerization. The size of the resulting particles is typically greater than 1 μm. If the particles are smaller than this, the term *emulsion polymerization* is typically used. In principle, there is no upper limit on the product particle size. In practice, it becomes difficult to maintain the suspension if the monomer droplets are much larger than a few millimeters. Another feature of a suspension polymerization is that if the dispersed phase is a solution, the monomer and polymer are both soluble in the solvent. If the polymer precipitates as it forms, the term *dispersion polymerization* is used. The products described here are sodium polyacrylate gels, and the term *solvent* is used in the sense that the solvent is a good solvent for the uncrosslinked polymer. Uncrosslinked sodium polyacrylates have been made for many years by solution polymerization. Because the products are ultimately used in dilute aqueous solution as thickeners, and rapid dilution of the as-purchased polymer is desired, the products have been produced and sold as water-in-oil emulsions.

The essential features of a suspension polymerization are a continuous phase, a dispersed monomer-containing phase, and a suspension aid for stabilizing the dispersed-phase particles against recombination (agglomeration) during collisions. Sufficient agitation is supplied to initially break up the dispersed phase into droplets and then keep the droplets from accumulating at the bottom or top of the reactor as a result of any density difference between the two phases. Equation 3.4 has found utility in predicting the particle size[120]:

$$d = \frac{kD_v\eta_m\gamma R}{D_a\eta_c C_s N} \tag{3.4}$$

where d is the average particle size; k is a constant dependent on apparatus; D_v and D_a are the diameters of the vessel and agitator, respectively; η_m and η_c are the viscosities of the monomer phase and the continuous phase, respectively; γ is the interfacial tension between the two phases; R is the volume ratio of the monomer phase to the continuous phase; C_s is the concentration of the suspension aid; and N is the agitator speed. The droplet size and the final particle size are determined by the amount and type of suspending agent used and the intensity of agitation. Larger amounts of suspending aid or more vigorous agitation promotes smaller particle size. Conversely, if larger particles are desired, less suspending aid and less agitation are required. However, these factors are detrimental to maintaining a good suspension and preventing agglomeration. Reduced agitation is less efficient in preventing settling of the dispersed phase in the reactor, and the lower particle velocity causes collisions to be longer in duration. This promotes coagulation and ultimately suspension failure (the whole dispersed phase has fused into a more or less polymerized mass of coagulated beads at the bottom or top of the reactor). Typical agitation rates in laboratory-scale reactors are 200–400 rpm.

With this brief description of a generic suspension polymerization process, the next section covers, in more detail, the suspension process as it has been applied to the production of superabsorbents. As for bulk-polymerized superabsorbents, most of the published literature is in the form of patents and patent applications. Much of the chemistry of the process is identical to that covered in other sections of this book. Those features that are truly unique to the suspension process are discussed in sections 3.3.3 and 3.3.4. Scale-up of agitation and the nature of the suspending aid and its effect on particle size are covered in section 3.3.3. In section 3.3.4 the discussion focuses on particle shape. One may say, "but a bead is a bead." Quite true, but even though the droplets begin as spheres, they need not maintain that shape during the polymerization. Crosslinked gels are being made, and whatever shape the polymerizing droplet has at the gel point determines the shape of the final product particle. As is described in section 3.3.4, a variety of shapes have been made. In section 3.3.5 some advantages and disadvantages of a suspension process are discussed and the status of this technology in the commercial sector is briefly reviewed.

3.3.2. Suspension Polymerization of Poly(Sodium Acrylate-Co-Acrylic Acid) Gels

The monomer-containing phase is essentially the same as that used for a bulk polymerization. Acrylic acid in water is neutralized with base, typically sodium hydroxide, to the desired degree of neutralization and total monomer concentration. To help with heat dissipation, the neutralization can be accomplished in a vessel that already contains the continuous phase. However, to ensure that even mixing of the two aqueous phases is achieved, the suspending aid should not yet be present. A crosslinker and, optionally, a chelating agent to prevent metal ion interference with the initiator system are added. If the initiator has poor solubility in the continuous phase it should be added to the monomer solution prior to suspension formation to ensure even distribution in the monomer phase. The crosslinker and initiators used are typically the same as are used in bulk polymerization. One advantage of a suspension system over a bulk system is that a much higher monomer concentration can be used in the polymerizing phase because the intimate contact of the continuous phase with the polymerizing particles facilitates heat transfer from the polymerizing particle. A disadvantage is that suspension stability limits the useful phase ratio (ratio of monomer to continuous phase) to 1:1 or less. Thus, the advantage of higher monomer concentration in the polymerizing phase is essentially negated by the volume dilution caused by the continuous phase.

The continuous phase can be nearly any liquid that, when compared to water, is a poor solvent for the monomer. This solubility requirement limits the degree of neutralization to a minimum of about 50%. Because any hydrocarbon is a good solvent for acrylic acid, too much acrylic acid may partition into the continuous phase so that it becomes impossible to prepare a suspension that is stable throughout the whole polymerization. This is presumably because the poly(acrylic acid) that is being formed in the continuous phase is sticky and serves to glue the polymer particles together. The choice of continuous phase is determined primarily by a combination of cost, ease of removal, recycle, and safety concerns, from both a flammability and a toxicological standpoint. Higher molecular weight aliphatic hydrocarbons (C_{10}–C_{14}) generally satisfy these requirements. Aromatics are less desirable since they are more toxicologically active and are better solvents for acrylic acid. The suspending aid is discussed in detail below. However, in general it should be preferentially soluble or dispersible in the continuous phase, but with enough ampholytic character to concentrate at the interface when the two phases are combined. Suspending aids of greater hydrophilicity promote formation of an oil-in-water suspension as opposed to the desired water-in-oil suspension.

The monomer phase and the continuous phase with suspension aid are combined with agitation. The intensity of agitation, along with the amount and type of suspending aid, determines the particle size. The suspension is purged with nitrogen to remove oxygen. Polymerization is then initiated, typically by application of heat to decompose a thermal initiator. As alluded to above, the addition of initiator to the preformed suspension is not recommended because an even distribution of initiator among the droplets cannot be expected to occur. The gel point occurs at very low conversion in the polymerization. If suspension stability is a problem, more suspending agent can be added and agitation can be increased after the gel point is reached because the particle size and shape are already set.

When the polymerization and an optional step to achieve ultrahigh monomer conversion (residual monomer burnout) are complete, the product can be isolated by any one of several means. An azeotropic removal of water can be carried out in the reactor or a separate vessel followed by filtration and evaporation of residual continuous phase.[121] Alternatively, the product can be filtered while still wet (if the suspending aid is effective at preventing the particles from sticking together). Then, the water and residual continuous phase can be removed by a more conventional drying method. If the suspending aid leaves a hydrophobic surface on the particles, a wetting agent can be added.[122]

The product is now essentially ready for packaging. One clear advantage of a suspension process is that postreactor particle sizing is unnecessary. Drying is also efficient relative to most bulk processes. Because the particles can be produced at the desired final particle size, the surface area of the particles is high (relative to a solution process polymerization) and the diffusion path is short.

3.3.3. Suspension Aids and Agitation

The primary requirement of a suspension aid is to stabilize the droplets against coalescence before, during, and after polymerization. Suspending agents impact particle size (Eq. 3.4) through their interfacial tension with the disperse phase and through the coverage of the surface of the dispersed-phase particles (inverse dependence in Eq. 3.4). A large variety of suspension aids have been used. A common thread in the patent literature is the application-driven need to prepare a larger particle size than is conveniently achieved by a standard surfactant-type system. Emulsions (particle sizes less than 1 μm) are easily prepared by dispersing the phases in a blender with a monomeric surfactant. Preparation of suspensions with particle sizes of hundreds of micrometers requires higher molecular weight surfactants, polymers, and even particulate suspending aids.

As a rule, if an oil-continuous system is desired, the suspension aid must be soluble or dispersible in the oil phase but with enough ampholytic character to reside at the oil–aqueous interface. A simple explanation is that if the molecule is more compatible with the oil phase, then a larger portion of its volume will be on the oil side of the interface. This will then force the interface to curve in a convex fashion with respect to the aqueous phase, resulting in the oil phase being continuous.

The favored lower molecular weight suspension aids are nonionic surfactants with a hydrophil–lipophil balance (HLB) of 3–6, particularly those surfactants that are solid at room temperature,[123] for example sorbitan monostearate or monooleate and polyethyleneglycol 200 distearate or dioleate (Fig. 3.11). They function at 1–2 wt% in aliphatic hydrocarbons. An embellishment that leads to a sharper particle size distribution, because of fewer fine particles, is to carry out the polymerization as a continuous addition of monomer-phase and continuous-phase containing suspension aid.[124] If this is done, surfactants with HLB in the range of 2–12 can be employed. A major disadvantage of these low-molecular-weight suspension aids is that the average particle size is usually small, 200 μm or less.

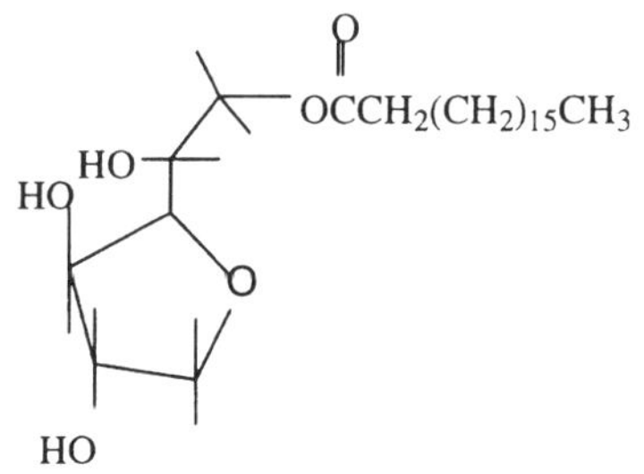

Sorbitan monostearate

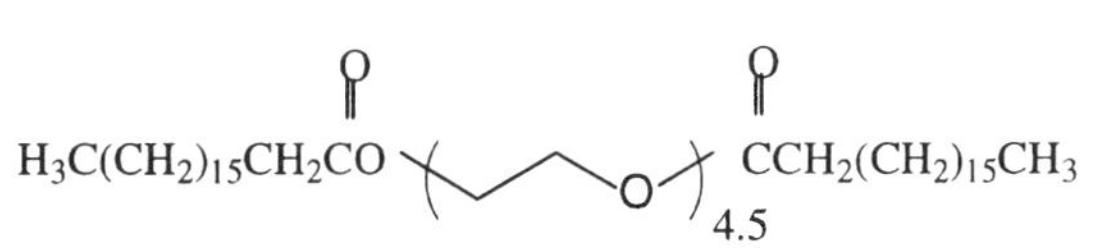

Polyethyleneglycol 200 distearate

Figure 3.11. Chemical structure of two commonly used low-molecular-weight suspending aids for preparing poly(sodium acrylate) by suspension polymerization in aliphatic hydrocarbons.

There are several ways to obtain larger particle size. Higher molecular weight, that is, polymeric, suspension aids may be used, including ethylene-co-maleic anhydride,[125] poly(α-olefin-co-maleic anhydride),[126] ethyl cellulose,[127] poly(laurylmethacrylate-co-acrylic acid),[128] and cellulose esters such as acetates, propionates, and butyrates.[122] Particle sizes in the 200–350 μm range can be obtained with these suspension aids. These polymeric suspension aids are soluble in hydrocarbons, although in some instances (ethyl cellulose in toluene, for example) heat is required to enhance solubility, and the polymerization process is run by slowly adding the monomer mix to the hot oil phase. These suspending aids work by forming a polymeric film around the monomer droplets that protects the individual particles during particle–particle collisions.

Even larger particle sizes (400–1,000 μm) can be made by using particulate suspension aids, such as hydrophobic fumed silica or cationic bentonite clays.[129] These materials are not truly in solution, but are merely dispersed. They function by accumulating at the surface and forming a protective sheath around the droplets. Anyone who has observed the film of dust encasing the first few drops of a rainfall has observed this phenomenon. Figure 3.12 shows the layer of fumed silica surrounding a dried polyacrylate bead.[129]

Both the type of suspending aid and the rate of agitation affect the particle size of the suspended products. In addition to the particle-sizing role, the agitation system must be capable of keeping the particles suspended through all phases of the process. These roles can easily conflict. The energy required to suspend the dense monomer phase (density of monomer phase neutralized to about 80% is about 1.2 g/cm^3) in the less dense organic continuous phase (0.75–0.85 g/cm^3) can easily shear the monomer droplets into too fine a particle size. Scale-up of a suspension process may worsen this conflict. Suspension of the disperse phase is dependent on bulk fluid motion while sizing, in combination with the suspending aid, is dependent on shear forces and microscale turbulence.

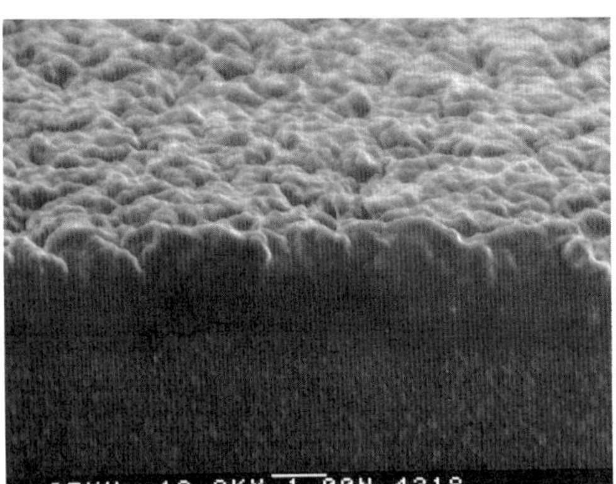

Figure 3.12. Cross section of a superabsorbent polymer bead showing the surface layer of hydrophobic fumed silica suspension aid.

The average droplet size of the dispersed phase obtained in a stirred vessel depends on the mechanism by which the droplets form. Droplet formation may be dominated by dispersive forces so that the droplet size decreases uniformly with agitation intensity. For systems of this type, mean droplet size is often found to be proportional to impeller tip speed (Eq. 3.4). This mechanism of droplet formation is common, and the theory is treated in standard works.[130] The nonparticulate suspending aids cited above would all be expected to exhibit this behavior.

When polymeric suspending aids are used in combination with particulate suspending aids, droplet coalescence at high agitation rates may lead to particle sizes larger than expected based on the results at lower agitation rates.[131] In a system of this type, mixing intensity must only provide small enough droplets so that the droplets may coalesce back to the desired larger size. Then the final droplet size is only weakly dependent on the agitation rate. Gregory et al.,[132] while determining the scale-up rules for a system of this type, found that particle size depended only on agitator speed N to the 0.3 power. As a result, once the droplets are sized, the agitation system need only keep the disperse phase suspended. It may then be possible to independently vary agitation rate and colloid concentration to achieve a desired particle size and thereby reduce the conflict between suspending and sizing droplets.

3.3.4. Particle Shape

Superabsorbents can be prepared in a variety of particle shapes by the suspension polymerization technique. Typically, suspension polymerizations yield spherical particles because this shape yields the minimum single-particle surface energy. The motivation to produce nonspherical shapes is related to certain shortcomings of spherical particles in the diaper application. First, spheres are susceptible to redistribution in the diaper, that is, they roll around.[133] Second, the minimal surface area leads to slow absorption rate when compared with particles of the same sieve cut prepared by grinding the product from a bulk polymerization.

One way to increase the surface area is to create porosity in the particle. One method of doing this is to carry out the polymerization by continuously adding the monomer solution to the preheated continuous phase.[134] The temperature, initiator concentration and identity, monomer concentration, and continuous-phase composition are chosen such that the water from the monomer phase and the oil in the continuous phase azeotropically boil and form vapor bubbles in the monomer beads at the precise moment of gelation. In this way the bubbles leave behind pores in the product, which lead to a faster absorption rate.

An alternate route to porosity is to polymerize an oil-in-water-in-oil (o/w/o) suspension wherein the monomer droplets themselves contain a second water-immiscible dispersed phase, which can be removed after the polymerization is complete.[135] Typical oil phases are hexane and cyclohexane for either or both the inner and outer oil phases. The oil-in-water phase is stabilized with poly(vinyl alcohol) and the o/w phase is stabilized in the outer oil phase by ethyl cellulose. The product is dried under vacuum to remove the inner oil phase. Void volumes up to 0.2 cm^3/g can be produced. Increases in absorption rate by up to fivefold are reported. Figure 3.13 shows the morphology of this product.

Another route to porous particles is described in those patents in which small particles are agglomerated to form larger aggregates. Such a reaction is conveniently carried out in an inert carrier, so a suspension process is ideal. The starting small particles can be prepared *in situ* prior to the aggregation step, or they can be prepared in a separate bulk polymerization. The primary polymerization can be carried out in concert with the aggregation as well in what amounts to controlled suspension failure. In one example the seed particles are slurried in the continuous phase, and more monomer mix is then added and polymerized.[127] In a second example, the agglomeration is carried out by simply adding a solution of a water-soluble polymer to a partially dried suspension polymer still dispersed in the continuous phase and then azeotropically drying the mixture.[136] As the last of the water is removed, the surfaces of the gel particles become sticky due to adsorbed polymer, and agglomeration occurs.

A process that not only results in a greater ratio of surface area to volume, but also produces a nonspherical product, has been described.[133] To understand how this process works it is necessary to consider the details of agitation. The monomer droplets are continually undergoing elongation as they move in and out of shear fields set up by the agitator. Typically, upon leaving areas of high shear, the particles immediately relax to spherical geometry. However, if the viscosity of the particles is high enough, the relaxation is slow enough that the particles enter another region of

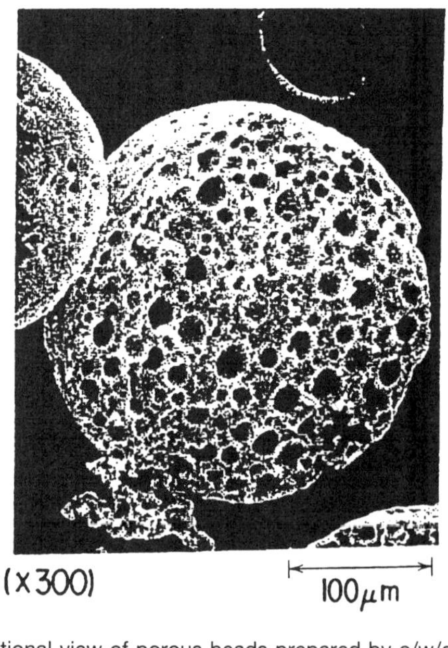

Figure 3.13. Cross-sectional view of porous beads prepared by o/w/o suspension polymerization.

high shear before the spherical shape is restored. In this way the particles can be made to have an elongated "Vienna sausage" shape. If the monomer phase has a viscosity of 15–5,000 mPas, spheres with a diameter of 100–600 μm are produced. If the viscosity is 5,000–1 million mPas, particles with a length of 1,000–10,000 μm and aspect ratios of 1.5–20 are produced. Hydroxyethyl cellulose or methyl cellulose can be used as the thickener for the monomer phase. Examples are shown in Figure 3.14.

As a final variation on the theme of morphology control, highly convoluted particles of superabsorbent polymer can be made by suspension polymerization.[137] These particles are so convoluted as to be almost all surface (Fig. 3.15). The important feature of the process used to produce these shapes is that the polymerization is caused to occur only near the surface of the particles. The initiation is by a redox system of sulfur dioxide and organic hydroperoxide. The sulfur dioxide is continuously added as a gaseous or liquid solution to the oil phase, and the hydroperoxide, typically *tert*-butylhydroperoxide, is homogeneously distributed throughout the monomer phase. The sulfur dioxide reacts with the water to produce bisulfite ion, which is the actual reducing agent of the redox couple. Initiation occurs at the surface of the droplets where the sulfur dioxide first contacts the aqueous phase. Gelation occurs almost immediately, as a gel layer forms on the surface of the droplets while the particle interior has not yet begun to polymerize. Details of the mechanism are not known. However, either hydrostatic pressure induced by the slight shrinkage that accompanies polymerization or simple solubilization of the gel by unreacted monomer solution could cause monomer to move from the interior of the droplet to the surface of the droplet. This monomer then polymerizes to form new surface. The rate of polymerization versus the rate of monomer movement causes the locus of polymerization to remain at the particle surface throughout the entire polymerization.

New surface is continuously produced by the transport of monomer from the interior of the particle to the surface. This results in a continual increase in the ratio of surface to volume as the surface undergoes a deformation reminiscent of that of a

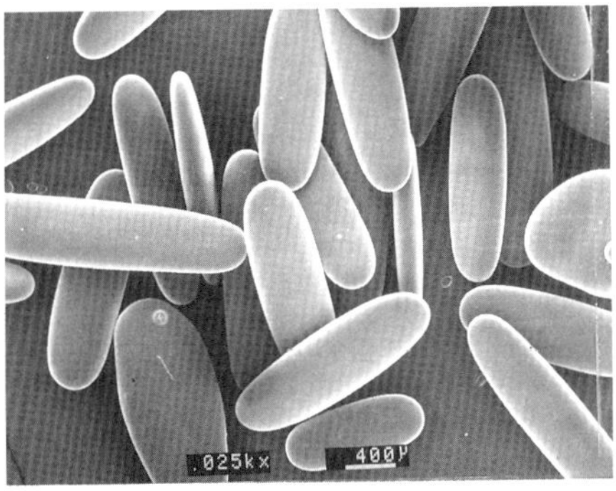

Figure 3.14. Elongated superabsorbent particles prepared by viscosifying the aqueous phase of a suspension polymerization. (Sample courtesy of Nippon Shokubai Co. Ltd.)

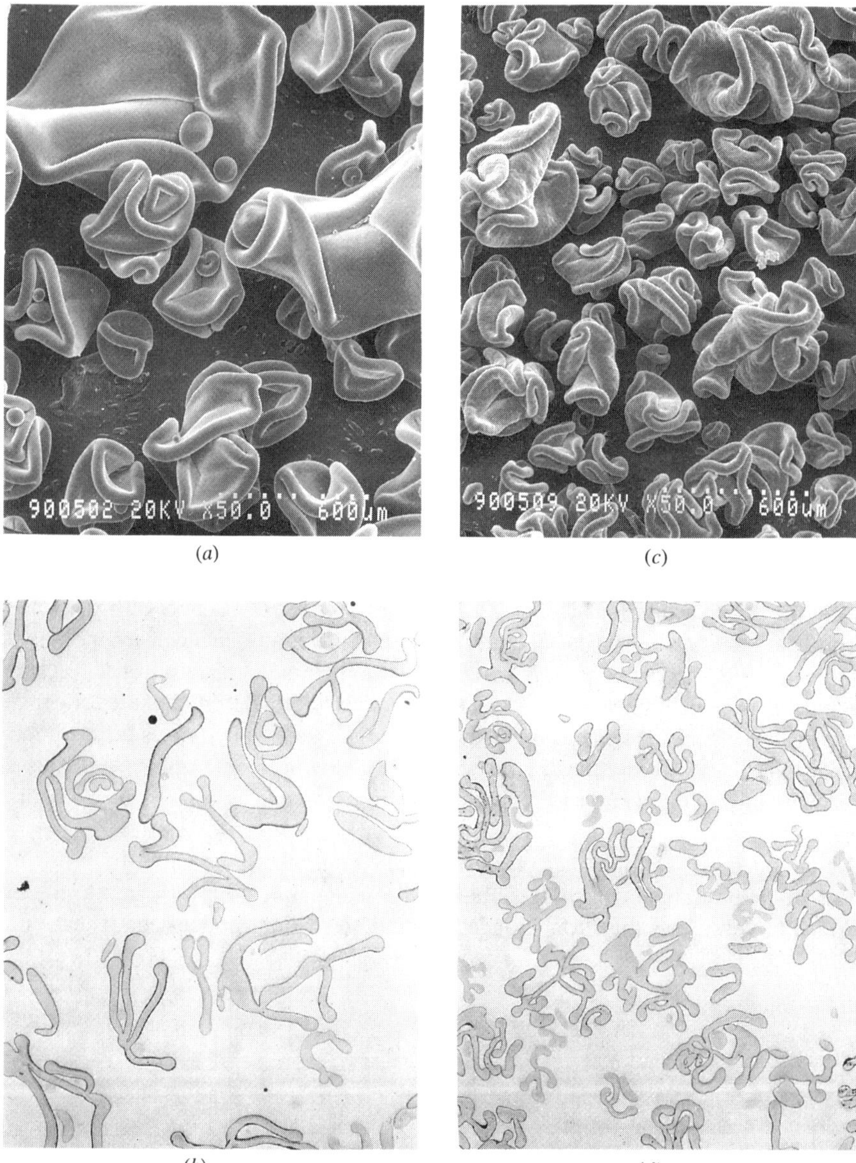

Figure 3.15. **(a–d)** Highly convoluted superabsorbent particles prepared by suspension polymerization: **(b)** and **(d)** are cross-sectional views of **(a)** and **(c)**, respectively.

ball being deflated. At low conversion, early in the polymerization, the particles are observed to be tetrahedral in shape with lobes of unreacted monomer forming the edges of the tetrahedra. This whole process continues to such an extreme that a second and even a third stage of bifurcation can occur in these lobes. The particles are fractal in nature in the sense that the surface area depends on the measurement scale, and there is self-similarity at each stage of bifurcation. After drying, the edges curl up to form the crinolated surfaces observed in Figure 3.15. The particles have such a high surface area that rate increases in fluid absorption of 30-fold over that of a sphere of equivalent cross section are achieved. The degree of convolution is controlled by the initial monomer droplet size, monomer concentration, crosslinker concentration, and hydroperoxide concentration.

3.3.5. Commercial Status of Suspension Technologies

There are three principal commercial superabsorbent processes that use suspension technology, and a description of the processing equipment and chemicals used is not available from public sources. Typically, the primary challenge in scaling-up a suspension process from the laboratory to a commercial scale is to manage agitation so that mass and energy transfer needs are met without adversely affecting the desired ability to control particle size. Due to the larger height of a full-scale reactor, more power input (higher continuous-phase fluid velocity) is required to prevent settling-out of the suspension. However, this additional power must not cause the particle size to decrease.

There are several inherent advantages of a suspension process over a bulk process. Most of these are a result of producing a product that needs no further size reduction. Temperature control during the polymerization reaction is easier as a result of the contact of the low viscosity (relative to the gelatinous polymer) continuous phase with the reactor walls. Surface crosslinking chemistry is easily carried out in the reactor either before or after drying. Drying is faster because the particle size, and therefore the diffusion path, is smaller relative to gel particles derived from the solution process. Azeotropic drying can be conveniently used. Particle size distribution is generally sharp and, in particular, fines production is much reduced compared with processes requiring sizing of dry product.

There are disadvantages to a suspension process as well. The continuous phase does not contribute to product performance and must be recycled for the process to be cost-effective. Even with recycling, the extra equipment required for recycling adds to the capital cost of a suspension process. Furthermore, the continuous phase is typically a volatile and flammable organic with the associated considerations for safety on a commercial scale. The reactor volume yield of product is limited by the operable phase ratio, which in turn is controlled by the stability of the system to suspension failure. Finally, as alluded to above, some diaper manufacturers have determined that the use of spherical particles is problematic in their processes due to the redistribution of the particles in the diaper.

To the best of our knowledge, in 1996 there were three commercial superabsorbent polymer manufacturers using a suspension process (Table 3.1). These were Kao Soap Co. and Sumitomo Seika in Japan and Elf Atochem S.A. in France.

Atochem sold about 11,800 tons into diapers and the medical market using a process licensed from Sumitomo Seika. Sumitomo sold about 10,000 ton into diapers, electrical cabling, and soil conditioning, and Kao produced 10,000 tons for captive use in the Japanese diaper market.

REFERENCES

1. F.L. Buchholz, N.A. Peppas, eds., *Superabsorbent Polymers: Science and Technology*, ACS Symp. Ser., No. 573, American Chemical Society, Washington D.C. (1994).
2. F.L. Buchholz, In *Absorbent Polymer Technology,* L. Brannon-Peppas, R.S. Harland, eds., Elsevier Science Publishers, Amsterdam (1990).
3. P.J. Flory, *Principles of Polymer Chemistry,* Cornell University Press, Ithaca, N.Y. (1953).
4. J.E. Mark, B. Erman, *Rubberlike Elasticity, A Molecular Primer,* John Wiley and Sons, New York (1988).
5. B. Obenski, Superabsorbent Polymers—A Growing Commodity, *Nonwovens Industry,* **25,** 44–46 (1994).
6. *Chemical Marketing Reporter,* February 12,1996, Schnell Publishing Co., New York, pp. 10–11 (1996).
7. *Chemical Marketing Reporter,* March 11, 1991, Schnell Publishing Co., New York, p. 4 (1991).
8. *Chemical Week,* March 24, 1993, p. 49.
9. R. Po, *J.M.S. Rev. Macromol. Chem. Phys.,* **C34,** 607 (1994).
10. K. Nukushina, *Yuki Gosei Kagaku,* **38,** 546 (1980).
11. F. Masuda, *Kinoo Zairyo,* April, p. 36 (1982).
12. F. Masuda, K. Nishida, A. Nakamura, *U.S. Patent 4,076,663* (1978).
13. K.G. McCurdy, K.J. Laidler, *Can. J. Chem.,* **42,** 818 (1964).
14. T. Sutton et al., *U.S. Patent 5,064,582* (1991).
15. G.T. Woodrum, T.H. Majette, *Eur. Patent Appl. 463,388* (1992).
16. Y. Irie et al., *Eur. Patent Appl. 454,497* (1991).
17. A.T. Graham, H. Gartner, *U.S. Patent 5,342,899* (1994).
18. T. Tsubakimoto, T. Shimomura, *U.S. Patent 4,286,082* (1981).
19. Y.F. Irie et al., *Eur. Patent Appl. 574,260* (1993).
20. F.M. Wampler, *Plant/Operations Progress,* **7,** 183 (1988).
21. T. Sato, M. Baba, M. Okane, *U.S. Patent 4,317,926* (1982).
22. N. Kutepow, H. Doerful, *U.S. Patent 3,086,046* (1963).
23. T. Horlenko, A. Fisher, *U.S. Patent 3,868,410* (1975).
24. Anonymous, *RD 37,631 Res. Disclosures*, **376,** 538 (1995).
25. T. Shimomura et al., *Eur. Patent Appl. 372,706* (1990).
26. E.N. Zil'berman et al., Translation of *Zhur. Org. Khim.,* **19,** 33–37 (1983).
27. M. Chmelir, J. Pauen, *U.S. Patent 4,857,610* (1989).
28. J. Siddall, T. Johnson, *U.S. Patent 4,833,222* (1989).

29. H. Brehm, K. Dahmen, R. Mertens, *Ger. Offen. 4,021,847* (1992).
30. F.L. Buchholz, *Chemtech,* **24,** 38–43 (1994)
31. K. Brandt, S. Goldman, T. Inglin, *U.S. Patent Re. 32,649* (1988).
32. T. Tsubakimoto, T. Shimomura, *U.S. Patent 4,625,001* (1986).
33. K Kimura et al., *U.S. Patent 4,985,514* (1991).
34. F. Masuda, K. Nishida, A. Nakamura, *U.S. Patent 4,076,663* (1978).
35. D.R. Chambers et al., *U.S. Patent 5,145,906* (1992).
36. T.A. Kay, F.J. Rodriquez, *J. Appl. Polym. Sci.,* **28,** 663–666 (1983).
37. A.O. Tonojan et al., *Vysokomol. Soedin.,* **A15,** 1847–1851 (1973).
38. G.T. Woodrum, T.H. Majette, *U.S. Patent 5,350,799* (1994); *Eur. Patent Appl. 463,388* (1992).
39. H. Pohleman, H. Spoor, E. Heil, *Ger. Offen., 1,218,157* (1966).
40. A. Echte, *Handbuch der Technischen Polymerchemie,* VCH Verlagsgesellschaft, Weinheim, BRD (1993).
41. F.L. Buchholz, *U.S. Patent 4,861,849* (1989).
42. Anonymous, *RD 38,514 Res. Disclosures*, **385,** 288 (1996).
43. B. Nowakowsky et al., *U.S. Patent 4,873,299* (1989).
44. K. Hennig et al., *U.S. Patent 4,729,877* (1988).
45. G. Nöltner, *Chem.-Ing.-Techn.* **57,** 1005–1013 (1985).
46. Y. Haruna, A. Yano, Y. Irie, *U.S. Patent 5,124,416* (1992).
47. B. Nowakowsky et al., *U.S. Patent 4,769,427* (1988).
48. M. Nakaki et al., *U.S. Patent 4,656,232* (1987).
49. T. Tsubakimoto, T. Shimomura, Y. Irie, *U.S. Patent 4,666,983* (1987).
50. T. Tsubakimoto et al., *U.S. Patent 4,734,478* (1988).
51. T. Irie et al., *U.S. Patent 4,920,202* (1990).
52. J. Svarz, *U.S. Patent 3,732,193* (1973).
53. M. Chmelir, J. Pauen, *U.S. Patent 4,893,999* (1990).
54. W. Alexander, M. Anderson, *U.S. Patent 4,985,518* (1991).
55. T. Sumiya, M. Date, K. Tanka, *Eur. Patent Appl. 496,067* (1992).
56. K. Miyake et al., *U.S. Patent 5,079,034* (1992).
57. H. Takeda, Y. Taniguchi, *U.S. Patent 4,525,527* (1985).
58. W. Alexander, M. Teppo, *U.S. Patent 4,820,742* (1989).
59. R.B. Bird, W.E. Stewart, E.N. Lightfoot, *Transport Phenomena,* John Wiley & Sons, New York, Chapter 18 (1960).
60. Y. Irie et al., *Eur. Patent Appl. 508,810* (1992).
61. R.H. Perry, C.H. Chilton, eds., *Chemical Engineer's Handbook,* 5th Ed., McGraw Hill, New York (1973).
62. I. Yoshio, T. Hatsuda, K. Yonemura, *Eur. Patent Appl. 497,623* (1992); *U.S. Patent 5,275,773* (1994).
63. Anonymous, *RD 37,243 Res. Disclosures*, **372,** 273 (1995).
64. G.A. Krudes, *Extrusion, Encyclopedia of Polymer Science and Engineering,* 2nd Ed., Vol. 6, J.I. Kroschwitz et al., eds. John Wiley & Sons, New York pp. 571–631 (1986).

65. K. Wilezynski, *Evaluating Screw Performance in Extrusion, Handbook of Polymer Science and Technology,* Vol. 3, N.P. Chermisinoff, ed. Marcel Dekker Inc., New York, pp. 421–442 (1989).
66. M. Date et al., *Ger. Offen.* 4,127,814 (1992), *JP 89 62,317* (1989).
67. T.C. Johnson, *U.S. Patent 5,075,344* (1991).
68. A.T. Graham, *U.S. Patent 5,385,983* (1995).
69. K. Takada, O. Sato, *Jpn. Kokai Tokkyo Koho 96–73,518* (1996).
70. W.L. McCabe, J.C. Smith, P. Harriott, *Unit Operations of Chemical Engineering,* 4th Ed., McGraw Hill, New York (1985).
71. Anonymous, *RD 37,327 Res. Disclosures*, **373,** 333 (1995).
72. Anonymous, *RD 37,441 Res. Disclosures*, **374,** 442 (1995).
73. J.R. Thygeson, E.D. Grossman, *Am. Inst. Chem. Eng. J.*, **16,** 749 (1970).
74. E. Tittman, Fortschritte Auf Dem Gebeit Der Einbandtrockner-1. Auslegungsverfahren, *Verfahrenstechnik,* **13,** 15–25 (1979).
75. J. Crank, *The Mathematics of Diffusion,* 2nd Ed., Oxford University Press, New York Chapter 9, (1975).
76. J. McKetta, *Encyclopedia of Chemical Processing and Design,* Vol. 17, Marcel Dekker Inc, New York, pp. 17–19 (1982).
77. C.L. Prasher, *Crushing and Grinding Process Handbook,* John Wiley & Sons, New York, (1987).
78. S. Goldman, J. Horney, H. Retzsch, *U.S. Patent 5,061,259* (1991).
79. L.G. Austin et al., *Int. J. Miner. Process.*, **6,** 321–336 (1980).
80. H. Colijn, *Mechanical Conveyors for Bulk Solids,* Elsevier Science Publishers B.V. Amsterdam (1985).
81. R.D. Marcus et al., *Pneumatic Conveying of Solids,* Chapman and Hall, London (1990).
82. H. Takahashi et al., *U.S. Patent 5,369,148* (1994).
83. H. Brehm, *Ger. Offen. 3,741,157* (1989); M. Chmelir, K. Stukenbrock, *Ger. Offen. 3,741,158* (1989); H. Brehm, M. Chmelir, *Ger. Offen. 3,917,646* (1990).
84. S. Nielson, D. Kim, *U.S. Patent 4,990,541* (1991).
85. S. Nielsen, *U.S. Patent 5,011,864* (1991).
86. Y. Fujiura et al., *U.S. Patent 5,002,986* (1991).
87. W. Lewis, K. Bailey, *U.S. Patent 4,950,692* (1990).
88. K. Bailey, G. Flasch, *U.S. Patent 4,970,267* (1990).
89. K. Bailey, J. Cramm, M. Miner, *U.S. Patent 5,122,544* (1990).
90. T. Hatsuda et al., *U.S. Patent 5,140,076* (1992).
91. T. Hatsuda et al., *Eur. Patent Appl. 450,922* (1991).
92. T. Hatsuda et al., *Eur. Patent Appl. 450,923* (1991).
93. K. Kajikawa et al., *Eur. Patent Appl. 496,594* (1992).
94. Anonymous, *RD 38,326 Res. Disclosures*, **383,** 186 (1996).
95. H. Ziche, *U.S. Patent 3,997,508* (1976).
96. N.Y. Yen, F.D. Osterholtz, *U.S. Patent 3,900,378* (1975).
97. S.H. Ganslaw, H.G. Katz, *U.S. Patent 4,043,952* (1977).
98. S. Oobayashi et al., *Jpn. Kokai Tokkyo Koho 57–44,627* (1982).

99. T. Ito et al., *Jpn. Kokai Tokkyo Koho 58–42,602* (1983).
100. K. Nagasuna et al., *Eur. Patent Appl. 605,150* (1994).
101. K. Dahmen, H. Redlof, *Ger. Offen*. 3,713,601 (1988).
102. S. Ueda, K. Tanaka, *Eur. Patent Appl. 555,692* (1993).
103. F. Engelhardt et al., *U.S. Patent 5,331,059* (1994).
104. K. Dahmen, R. Mertens, *Ger. Offen*. 4,020,780 (1991).
105. T. Sumiya, M. Date, K. Tanaka, *Eur. Patent Appl. 629,411* (1994).
106. H. Klimmek, *Ger. Offen*. 4,244,548 (1994).
107. S.R. Kellenberger, *Eur. Patent Appl. 339,461* (1989).
108. M. Melius et al., *Eur. Patent Appl. 615,736* (1994).
109. M. Date, T. Sumiya, H. Takai, *Eur. Patent Appl. 618,005* (1994).
110. G.T. Woodrum, H.C. Crall, *WIPO Int. Appl*. 95–27,739 (1995).
111. H. Brehm, H.-G. Hartan, *Ger. Offen*. 4,333,056 (1995).
112. J.H. Burgert et al., *WIPO Int. Appl*. 94–20,547 (1994).
113. W. Reeves, *Eur. Patent Appl. 690,077* (1995).
114. F. Englehardt et al., *Ger. Offen*. 4,426,008 (1996).
115. T.L. Staples, S.P. Wood, W.L. Treptow, *WIPO Int. Appl*. 94–22,940 (1994).
116. A.D. James, *U.S. Patent 4,490,281* (1984).
117. P.M. Robinson, *U.S. Patent 4,339,373* (1982).
118. K.L. Krumel, A. B. Savage, *U.S. Patent 4,080,358* (1978).
119. T.L. Staples et al., *WIPO Int. Appl*. 97–42,676 (1997).
120. R. Arshady, *Coll. Polym. Sci*., **270,** 717–732 (1992).
121. M. Nakamura et al., *U.S. Patent 4,683,274* (1987).
122. H. Yamasaki et al., *U.S. Patent 4,459,396* (1984).
123. T. Shibano, K. Ito, *Jpn. Kokai Tokkyo Koho 61–157,513* (1986).
124. T. Yamamoto, K. Ito, *Jpn. Kokai Tokkyo Koho 03–195,708* (1991).
125. M. Nakamura et al., *U.S. Patent 5,180,798* (1993).
126. K. Ito, T. Shibano, *Jpn. Kokai Tokkyo Koho 62–62,807* (1987).
127. S. Hirokawa et al., *Ger. Offen*. 3,617,311 (1987).
128. P. Flesher, D. Farrar, *Eur. Patent Appl. 233,014* (1987).
129. F.W. Stanley, J.C. Lamphere, Y. Chonde, *U.S. Patent 4,708,997* (1987).
130. V. Levich, *Physico-Chemical Hydrodynamics,* Prentice-Hall, New Jersey (1962)
131. R.M. Wiley, *J. Coll. Sci.,* **9,** 427 (1954).
132. T.D. Gregory et al., *Scale-up of a Batch Suspension Polymerization Process for Producing Acrylate Polymer Particles,* AIChE Presentation, Chicago (November 1996).
133. K. Nagasuna et al., *U.S. Patent 4,973,632* (1990).
134. H. Aoyama, T. Moriya, S.Kondo, *U.S. Patent 5,210,159* (1993).
135. K. Masamizu, S. Hirokawa, F. Sanae, *U.S. Patent 4,742,086* (1988).
136. Mitsubishi Petroch. KK, *Jpn. Kokai Tokkyo Koho 63–210108* (1988).
137. F.W. Stanley, J.C. Lamphere, *Eur. Patent Appl. 516,925* (1992).

4

ANALYSIS AND CHARACTERIZATION OF SUPERABSORBENT POLYMERS

Sergio S. Cutié, Patrick B. Smith,
Robert E. Reim, Andrew T. Graham

4.1. INTRODUCTION

Superabsorbents are prepared by polymerizing a mixture of acrylic acid, generally partially neutralized, and a multifunctional crosslinking agent (0.001%–1% by mass) to form a three-dimensional network or gel. The crosslinked nature of superabsorbents makes them insoluble in water, apart from a small amount of non-crosslinked, extractable polymer and traces of monomer. The insoluble, crosslinked polymer network cannot be analyzed by routine chromatographic procedures. Spectroscopic techniques have, however, been used to characterize the polymers. Nuclear magnetic resonance (NMR), infrared (IR), and mass spectrometry coupled with pyrolysis–gas chromatography have been invaluable as tools to positively identify polyacrylate superabsorbents. The morphology of the particles, fibers, and films can be scrutinized by optical and electron microscopy. Destructive methods, for example, iodine tests and hydrolysis of the crosslinker followed by gel permeation chromatography, can add information about manufacturing characteristics.

Because the degree of polymerization of poly(acrylic acid) in superabsorbents

Modern Superabsorbent Polymer Technology. Edited by Fredric L. Buchholz and Andrew T. Graham.
ISBN 0-471-19411-5.

may be on the order of a few tens of thousands, even low-frequency events can have a significant impact on the synthesis and subsequent performance of these materials. Initiators and crosslinkers used in the manufacture of commercial superabsorbent polyacrylate products are typically used in the range of a few parts per million (ppm) to a few parts per thousand by mass. The significant effects of minor impurities, for example, chain transfer agents, inhibitors, or accelerants, on the polymerization and the polymer product require the most pure available grade of acrylic acid to be specified for use in the manufacture of superabsorbents. The quality of the other raw materials, for example, crosslinkers or neutralents, may also influence the process or the final product properties. Thus, careful analysis of raw materials has been of significant interest in understanding the performance properties of commercial acrylate superabsorbents. Gas chromatography, liquid chromatography, NMR, IR, and mass spectrometry have all been used.

Because the superabsorbent products are used in personal care applications, analysis of product purity has also been of significant interest. Procedures of interest include determination of unreacted monomers, initiators, and other trace components by extraction from the polymer followed by analysis, typically by gas or liquid chromatography. Soluble polymer is also determined by analysis of an extract, but we have chosen to include it in the discussion of the polymerization reaction.

The determination of bulk polymer properties may be accomplished in a number of ways. Thermal methods have been used to determine the polymer's glass transition temperature, heat capacity, and degradation at elevated temperatures. Product properties that are specific to the handling of the dry final polymer product are also measured. Product moisture has been measured by a variety of methods. Polymer bulk density and flowability may affect diaper machine settings or the performance of the polymer in a particular diaper machine and are determined, typically, using variations of ASTM methods for plastic powders.

Central to ensuring the performance of superabsorbent polymers in use are measurements of properties characteristic of the network structure and of the product's end-use performance. Network theory relates network structural parameters to swelling and modulus (Chapter 5). Important structural parameters include primary chain molecular weight, molecular weight between crosslinks, polymer gel fraction or its complement, the fraction of soluble polymer, and topological heterogeneity, for instance, amount of branching and concentration of dangling ends. In addition to these primary network characteristics, a number of analysis methods have been proposed to predict polymer performance in the end-use. These methods include absorption under an applied load (AUL) and permeability of the swollen gel mass. These methods are related to, but are not explicitly determined by, more fundamental network characteristics. Absorption rate has also been commonly measured as a fundamental performance property of superabsorbents.

In some cases, a variety of methods are used to measure the same fundamental property. As appropriate for measurement of that product property, various analysis methods and their strengths and weaknesses are discussed. For analysis descriptions, published documents and patent sources are used. Key methods are emphasized to help promote an "industry standard," some of which have been collected by others.[1] Previous reviews may be referred to for the viewpoints of other authors or for end-uses different from the personal care market.[2,3]

4.2. MONOMER CHARACTERIZATION

4.2.1. Acrylic Acid

4.2.1.1. Gas Chromatography. Capillary gas chromatography using free fatty acid phases composed of nitroterephthalic acid–modified polyethylene glycol, with flame ionization detector and oven temperatures from 70°C to 220°C, has been used to determine the quantity of various impurities in acrylic acid monomer.[4] This type of analysis allows for the routine determination of aldehydes, acrylates, alcohols, acids, and inhibitors. Figure 4.1 shows a gas chromatographic trace of acrylic acid with several common impurities indicated.

4.2.1.2. Dimer of Acrylic Acid by Liquid Chromatography and NMR Spectroscopy. As discussed in Chapter 2, section 2.2.2.2, a dimer of acrylic acid forms through a Michael addition reaction during storage. The dimer of acrylic acid has been quantified using liquid chromatography with ultraviolet or mass spectrometry detection and by NMR analysis.[5] To ensure the quality of the monomer during storage and handling, it is inhibited against free radical polymerization by the addition of the monomethyl ether of hydroquinone (MEHQ). However, MEHQ does not inhibit the Michael addition reaction in this system, which yields primarily diacrylic acid (dimer).

The rate of formation of the dimer is a function of temperature, time, and water content.[6] Furthermore, the dimer incorporates into the backbone of the polymer during free-radical polymerization, as demonstrated by ^{13}C-NMR spectroscopy. Figure 4.2 shows a ^{13}C-NMR spectrum of a poly(acrylic acid) co-polymerized with

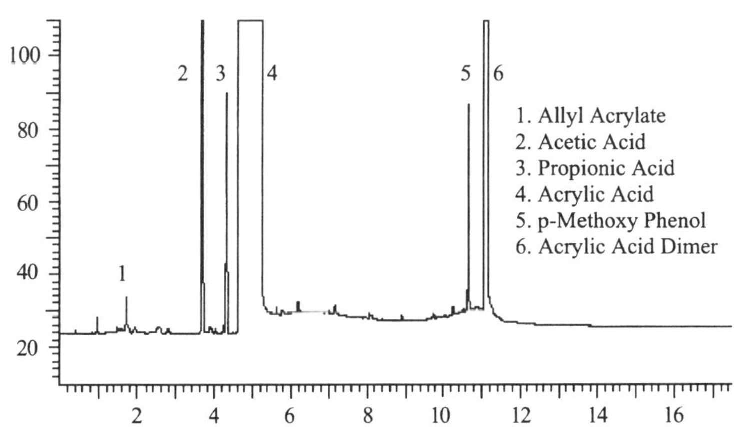

Figure 4.1. A capillary gas chromatographic trace for commercial acrylic acid.

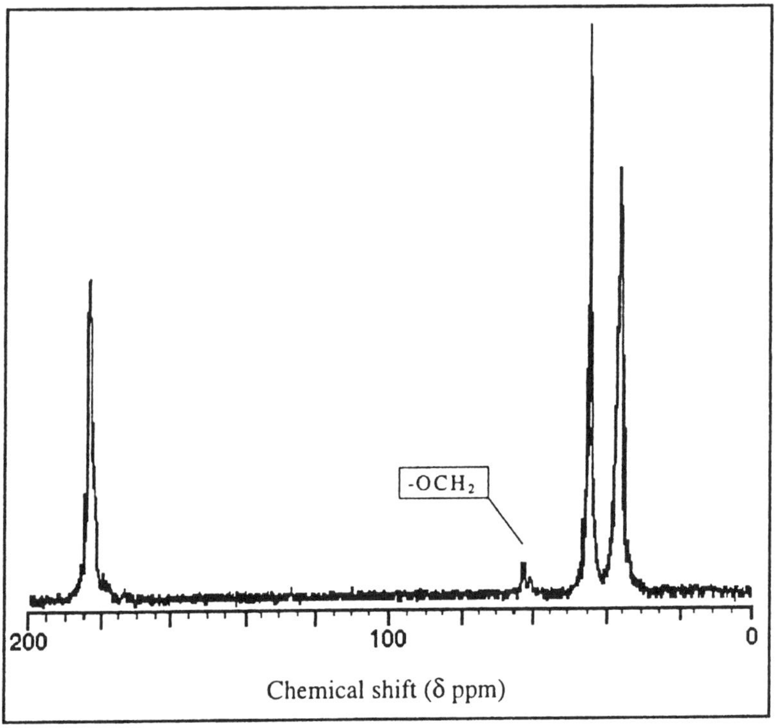

Figure 4.2. The ^{13}C-NMR spectrum of the poly(acrylic acid) made with 6 mol% dimer.

6 mol% dimer. The spectrum clearly shows the incorporation of dimer by the normal free-radical mechanism through the vinyl group. The resonance of the $-OCH_2$ carbon of the incorporated dimer is present at 60–65 ppm and that of the carbonyl is observed at 177–180 ppm. No olefinic resonances are observed. Upon heating this polymer to 125°C, the amount of incorporated dimer was reduced to less than 4%. The 2 mol% dimer lost as a result of heating was observed as acrylic acid monomer according to the reaction in Scheme 4.1. This confirms the ability of the incorporated dimer to undergo the reverse Michael reaction to yield monomer and implicates it as a source of residual acrylic acid in the final polymer.

4.2.1.3. Infrared Analysis. IR spectrophotometry is an important tool for the identification and characterization of chemical compounds, particularly polymers.[7]

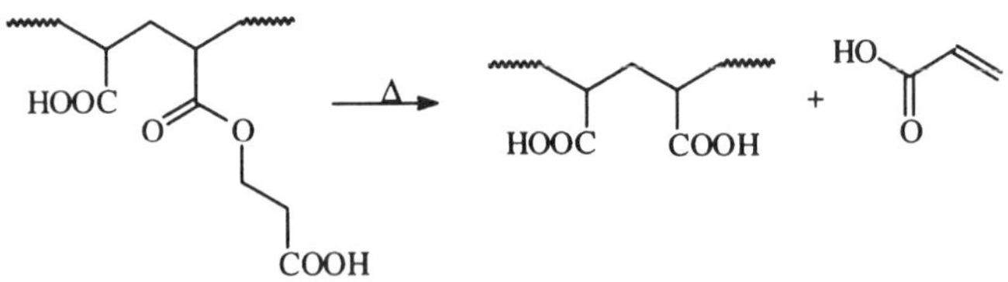

Scheme 4.1

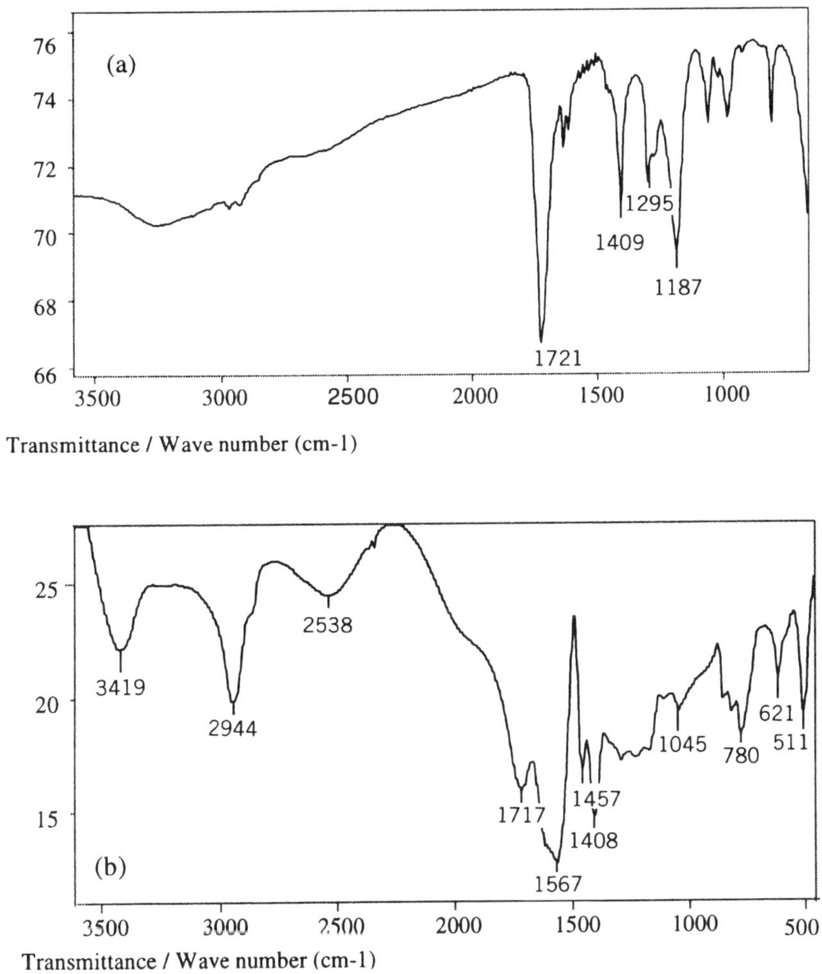

Figure 4.3. Infrared spectra of **(a)** acrylic acid and **(b)** 65% neutralized crosslinked poly(sodium acrylate).

Its main feature is the study of molecular structures that can be determined by the observation of changes in the vibrational motions of atoms within the molecule after interaction with electromagnetic radiation, which is called the *absorption spectrum* of the material. IR spectroscopy is routinely used for the identification of raw materials, intermediates, and finished products in analytical and quality control laboratories. Figure 4.3a,b shows, respectively, an infrared spectrum of acrylic acid and a spectrum of poly(acrylic acid-co-sodium acrylate) made by mixing portions of each material with KBr powder and pressing into pellets. IR spectra were acquired with a Perkin-Elmer Paragon 1000 FT-IR at a nominal resolution of 4 cm^{-1}. IR analysis has also been used to follow the progress of chemical reactions occurring in films of dried polyacrylate (see Chapter 2, section 2.5.4.2).

4.2.1.4. Inhibitor and Residual Crosslinker Quantitation. MEHQ is the inhibitor of choice for prevention, during shipping and storage, of polymerization of acrylic acid or the crosslinkers used to make superabsorbent polymers. Usually about 200 ppm of MEHQ is added by the manufacturers. MEHQ in acrylic acid, 1,1,1-trimethylolpropane triacrylate (TMPTA), and allylmethacrylate can be determined by reverse-phase liquid chromatography using 50% methanol/water as the mobile phase. In the case of ethoxylated TMPTA (HE-TMPTA), the mobile phase must also contain acetonitrile. The samples are prepared by adding about 5 g of sample to 5 g of methanol. Twenty microliters of the mixture are injected into the liquid chromatograph. The MEHQ response is monitored using ultraviolet detection at 280 nm. The TMPTA and HE-TMPTA responses are monitored using ultraviolet detection at 206 nm. Concentration is obtained from a comparison of the peak height of the unknown to the peak height of a standard of known concentration. The limit of detection is about 10 ppm, and a relative precision at the 95% confidence level of ± 7% was found.

MEHQ can also be selectively determined using liquid chromatography with oxidative electrochemical detection at 700 mV versus Ag/AgCl at a glassy carbon electrode.[8] Separations are performed using a Whatman Partisil ODS 10 column of dimensions 4.6 mm × 250 mm. A mobile phase of acetonitrile/water 30/70 (v) containing 0.02 mol/L $HClO_4$ is used at a flow rate of 1.0 mL/min. Samples are prepared for analysis by dilution with mobile phase. MEHQ is quantified by external standard analysis.

4.2.2. Crosslinker Purity

In the manufacture of superabsorbent polymers, one or more of a variety of crosslinkers have been used to form networks through ether or ester linkages. Crosslinker purity is important to the integrity of the network, because heat-reversible Michael adducts are often present in these materials as byproducts of synthesis and storage. The purity of several key crosslinkers has been investigated with a variety of techniques. The separation technique of choice is liquid chromatography because of the reactivity of the crosslinkers. Liquid chromatography may be used with a variety of detectors, from ultraviolet for initial separation and monitoring to mass spectrometry and NMR for final characterization. The crosslinkers may be characterized using IR or NMR spectroscopy.

Ethylene glycol dimethacrylate, TMPTA, and diethyleneglycol diacrylate are examples of crosslinkers made by esterification of multifunctional alcohols. They are typically, as sold commercially, a mixture of several similar compounds. Using liquid chromatography with thermospray mass spectrometry detection, the nominally tri-functional TMPTA was also found to contain di-, tetra-, penta-, and higher functional groups.[9] Figure 4.4 shows a typical ^{13}C-NMR spectrum of this material. At first inspection, it appears from the ^{13}C-NMR spectrum that the sample is greater than 75 mol% pure. However, since TMPTA is a tri-functional molecule, each defect has a multiplicative influence on the overall purity. As a result, the tri-function-

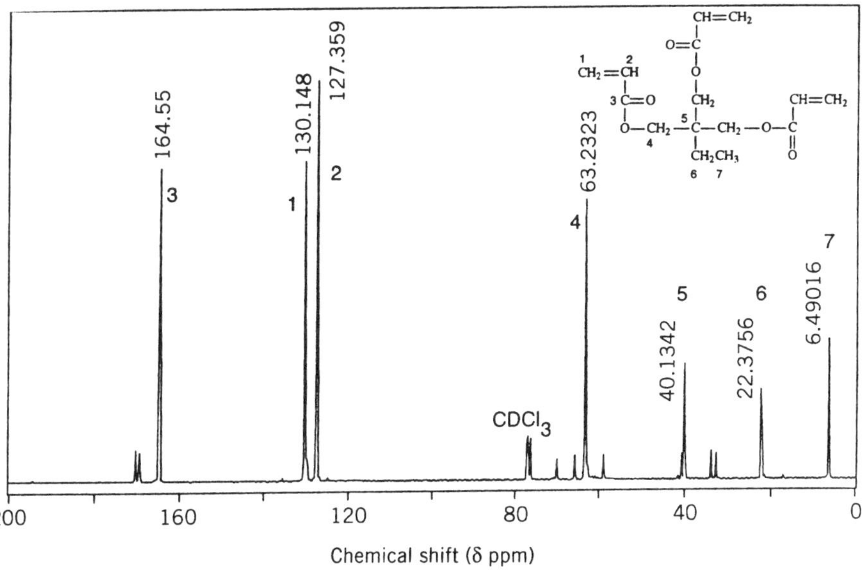

Figure 4.4. The ^{13}C-NMR spectrum of a commercial sample of TMPTA.

al TMPTA molecule constitutes less than about 50 mol% of the sample in agreement with liquid chromatography–ultraviolet results. The structure of TMPTA, and the structures of two of its principal impurities, are shown in Figure 4.5. These impurities, formed by a Michael addition reaction, comprise the majority of the impurities. These structures may be thermally reversible upon heating the gel during drying and may cause the network properties to depend on thermal history.

The number of possible compounds increases when a multifunctional alcohol is ethoxylated. As an example, TMPTA ethoxylated with 15 molar equivalents of ethylene oxide (HE-15-TMPTA) was characterized by liquid chromatography with mass spectrometry detection using electrospray ionization techniques. This material was a mixture of many individual compounds with differing numbers of ethoxy groups and varying numbers of acrylate end caps. Three major component classes were identified, the intended structure and two major impurities, (Fig. 4.6). HE-15-TMPTA consists of a distribution of homologs with 6–22 ethoxy units per molecule. The average number of ethoxy units was 12 per molecule. One of the major impurity classes was an HE-15-TMPTA analog in which one of the methylol units had not reacted with any ethylene oxide. This impurity class contained a distribution of ethylene oxide run lengths on the other two arms. These two major component families also possessed a distribution of acrylate units that varied from 0 to 3, the average being 2.75. The other major impurity class (also a homologous series) was consistent with an oligomer of ethylene oxide with zero, one, or two acrylate units attached.

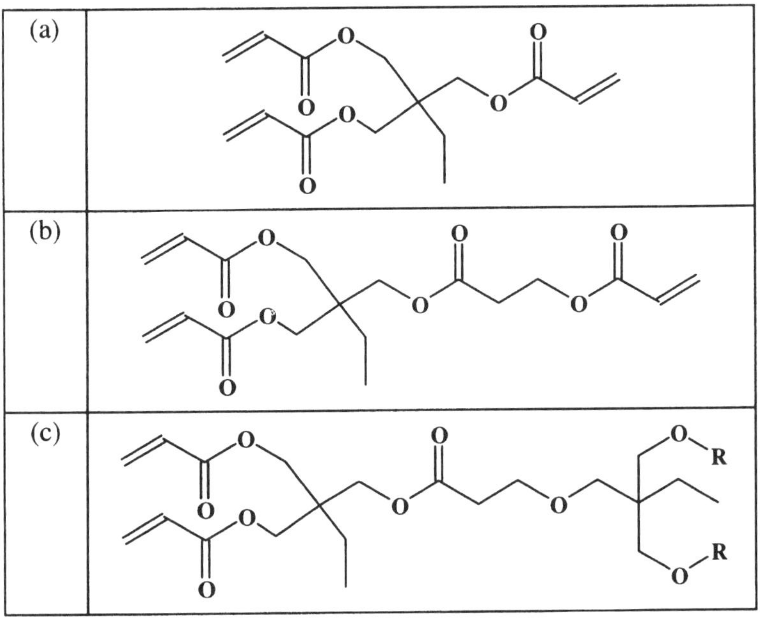

Figure 4.5. The structure of TMPTA and common impurities. **(a)** TMPTA. **(b)** The Michael addition product of TMPTA with one acrylic acid unit. **(c)** The Michael addition product of TMPTA with two TMPTA units.

4.3. CHARACTERIZATION OF THE POLYMERIZATION REACTION

4.3.1. Reaction Kinetics by NMR

NMR provides a convenient method for monitoring the polymerization of acrylic acid since it is capable of distinguishing the olefinic monomer from the polymer in the same sample. Of the different NMR techniques, ^{1}H is preferred over ^{13}C because of the greater sensitivity, faster acquisition rates, and ability to maintain isothermal conditions by using smaller diameter tubes. This makes ^{1}H-NMR useful for monitoring the polymerization kinetics (see Chapter 2) of acrylic acid as a function of temperature, initiator concentration, neutralization, monomer concentration,[9] and inhibitor (MEHQ) concentration,[10] provided the reaction mixture remains homogeneous and there are no extremes of temperature or pressure. For example, the dependence of the polymerization rate on initial monomer concentration was determined at 55°C by measuring the rates of polymerization for feed batches containing partially neutralized acrylic acid concentrations ranging from 37.2% (4.56 mol/L) to 9.3% (1.17 mol/L) at constant total initiator level. The reaction mixture, containing sodium persulfate as the initiator, was loaded into the NMR tube using a glass pipette and deoxygenated for a minimum of 30 min. The tubes were capped and placed into the temperature-equilibrated, nitrogen-purged probe of the

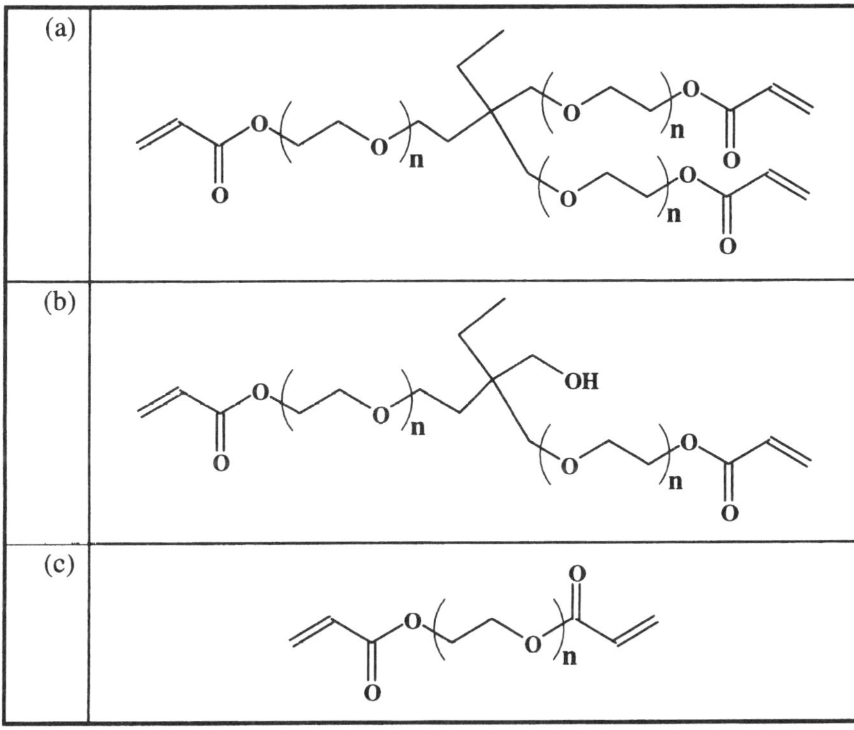

Figure 4.6. The structure of HE-TMPTA and some common impurities. **(a)** HE-TMPTA. **(b)** A representative structure from impurity class I. **(c)** A representative structure from impurity class II.

NMR spectrometer. The temperature of the probe was controlled to within ±1°C. The time of data accumulation for each spectrum was determined by the number of scans and the delay time between scans. Figure 4.7 gives a typical series of ^{1}H-NMR spectra, from low to high conversion. The areas of the peaks at 1.5–2.5 ppm (polymer) and at 5.5–6.5 ppm (monomer) were used to determine the relative concentrations of polymer and monomer as a function of time (see Fig. 2.3, Chapter 2). The rate of polymerization was calculated from the initial slopes of these curves.

4.3.2. Crosslinker Incorporation

Crosslinks are typically incorporated into the network by the use of co-polymerized multi-vinyl crosslinkers. The primary factors influencing the ultimate structure of the polymerized, crosslinked network are the amount, type, and uniformity of incorporation of the multifunctional crosslinker. To understand the distribution of crosslinks in the network, the reactivity ratios of all of the various double bonds in the system with each of the other types of double bonds must be known.

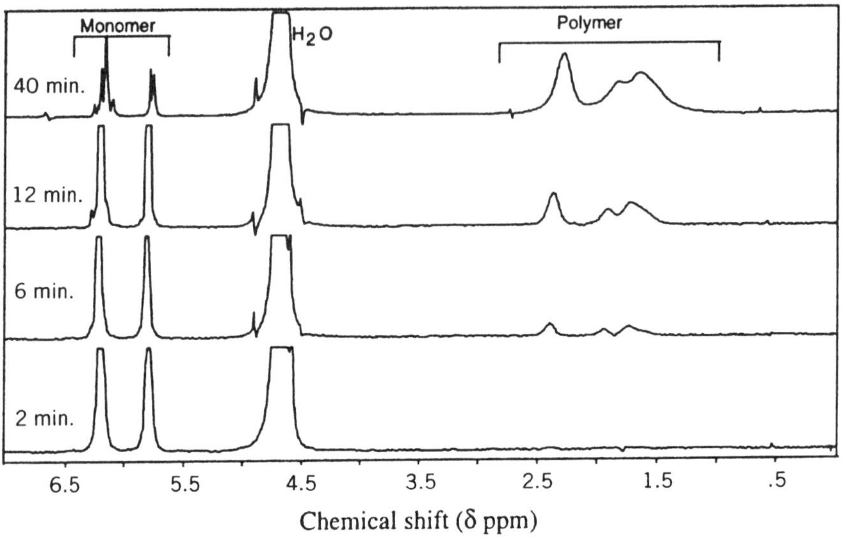

Figure 4.7. The ^{1}H-NMR spectra of the polymerization of acrylic acid as a function of polymerization time.

Arriola et al.[11] studied TMPTA as a model multifunctional acrylate crosslinker for co-polymerization with acrylic acid. The incorporation of TMPTA, through the reaction of its first double bond, was monitored by analysis of unreacted TMPTA using chromatographic methods. However, the incorporation of the second and third double bonds required spectroscopic characterization. Crosslinkers in these systems are generally present at concentrations below about 1%, making them difficult to detect spectroscopically. TMPTA was ^{13}C-labeled at the carbonyl position to increase its ^{13}C-NMR signal 100-fold. The reactivity of the carbonyl moieties was monitored by ^{13}C-NMR methodology similar to that used in the kinetic analysis discussed above. The incorporation of TMPTA was rapid relative to the incorporation of acrylic acid, giving rise to a somewhat nonuniform network.

4.3.3. Primary Chain Molecular Weight

The molecular weight and molecular weight distribution of backbone polymer is important in determining the quality and performance of the polymer. These two properties can affect the quality and performance of a polymer, since even polymers with identical molecular weights and different molecular weight distributions can have different performances. Superabsorbent polymers are manufactured using one of a variety of crosslinkers, including diethylene glycol diacrylate or TMPTA, both of which contain ester bonds in the molecular backbone between the crosslinking sites. Although crosslinked superabsorbent polymers are networks of essentially infinite molecular weight and are, therefore, impossible to analyze directly by chromatography, hydrolysis under heat and high pH will break the ester bonds arising from the crosslinker molecule. This allows traditional size-exclusion chromatography to be used to determine the molecular weight distribution of the primary chains so that various average molecular weights may be calculated. (Note that this technique does not work for polymers crosslinked with multifunctional allyl compounds like triallylamine or ethers, for example, tetraallyloxyethane.)

Poly(acrylic acid) polymers, upon neutralization, are polyelectrolytes and exhibit a number of properties in solution characteristic of their electrostatic nature. The chromatography of water-soluble polymers and the solution properties of polyelectrolytes have been explored.[12–22] Separation in size-exclusion chromatography is determined by the selective permeation of the polymers into and out of the mobile-phase-filled pores of the column packing. The elution of the polymer is governed by the time it spends in the pores. Hence, the larger molecules, which spend less time in the pores, elute first, and the smaller molecules elute later. The principal factor affecting the size-exclusion chromatography of a polymer is, therefore, the hydrodynamic volume of the polymer and not the molecular weight.

The apparent molecular weight of polymer samples is determined by comparison of the chromatograph of a sample to that of standards with known molecular weights. Poly(ethylene oxide) narrow distribution standards obtained from Toso-Haas (Montgomeryville, PA), ranging in molecular weight from 26,000 to 913,000 g/mol, were used as a "primary" standard. Poly(acrylic acid) standards with a polydispersity of about two, obtained from American Polymer Standard Corporation (Mentor, OH), were used to compare with the poly(ethylene oxide) standards since the hydrodynamic volume of the poly(acrylic acid) standards is closer to that of the sodium polyacrylate analyzed and serve as a "secondary" standard. A close correlation between the two sets of standards was observed.[23]

Molecular weight and molecular weight distributions of sodium polyacrylate polymers were determined using TSK-GEL $GMPW_{XL}$- and TSK-GEL 1000-PW-type columns, introduced by Toyo Soda in 1985.[24,25] For analysis of low- to medium-molecular-weight materials,[23] two TSK-GEL $GMPW_{XL}$ columns and one TSK-GEL 1000-PW column with 1 mL/min mobile phase flow were used. For ultrahigh-molecular-weight sodium polyacrylate, one TSK-GEL $GMPW_{XL}$ was used, and the mobile-phase flow was reduced to 0.5 mL/min. TSK GEL PW packing is composed of a methacrylate backbone and carboxylic groups attached to the surface. The particles are rigid, spherical, and porous polymeric beads that have excellent chemical and mechanical stability. The mobile phase was buffered with sodium chloride and sodium phosphate salts to eliminate ionic interactions. Detection was by refractive index. The software used for the calculations of molecular weight and molecular weight distribution was obtained from Polymer Laboratories Inc. (PL Caliber brand GPC/SEC software, Ver. 6.0). Any low-molecular-weight components below 12,600 g/mol were not included in the calculation. Figure 4.8 shows chromatographic traces of a sample of molecular weight about 4 million g/mol. The trace identified as (b) shows a leading edge typical of polymer exclusion, while the trace identified as (a) shows a more Gaussian envelope reflecting more ideal size-exclusion chromatography separation.

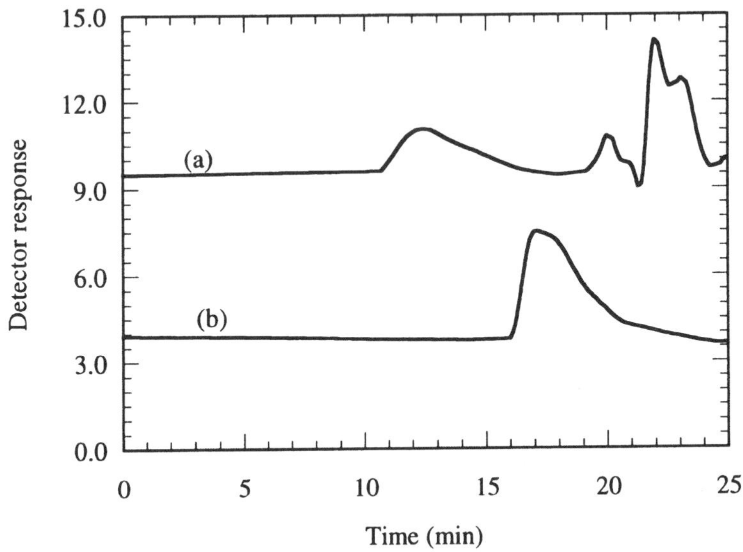

Figure 4.8. SEC traces of samples of molecular weight about 4 million g/mol. **(a)** Analytical conditions for ultrahigh molecular weight. **(b)** Analytical conditions for low to medium molecular weight.

4.3.4. Extractables and Percent Neutralization

A logical result of the free-radical process used to synthesize superabsorbents is a distribution of chain lengths and crosslinker concentrations during the polymerization process as monomer and initiator concentration vary with conversion. Low-molecular-weight chains have a lower probability of being incorporated into the network and, when not incorporated, are the soluble or extractable fraction. These uncrosslinked chains can be solvated and contribute to swelling if they are retained in the gel phase; however, they do not contribute to the modulus of the sample (see Chapter 5, section 5.1.1). As a result of the presence of the soluble polymer in the gel and the consequent lowered chemical potential of the water in the gel, the difference in chemical potential of water between the gel phase and the external phase, which is the driving force for swelling, is increased.

The soluble fraction can be readily extracted from the polymer when the polymer is swollen in excess liquid, hence the name *extractables*. The molecular weight of the soluble polymer, relative to the molecular weight between crosslinks in the network, determines whether these molecules will diffuse from the gel and how rapidly they will be extracted.[26] For commercial superabsorbent samples, the non-crosslinked, polydisperse polymer has molecular weights ranging from 1,000 to about 1,000,000 g/mol and the extraction process may take several days to reach equilibrium (see Fig. 2.15, Chapter 2). In practice, the amount of soluble polymer is determined by extracting the polymer with 0.9% NaCl for 16 h and quantitating a portion of the extract.

High-performance gel permeation chromatography[27] with refractive index detection was very successful in quantifying levels of extractables in the polymer and also in determining soluble poly(acrylic acid) in water at the parts per million level. Poly(acrylic acid) can be identified using pyrolysis–gas chromatography after derivatization of the polymeric acid to the methyl ester.[28] However, for superabsorbent polymers, the soluble fraction actually exists as the sodium salt or sodium polyacrylate. The derivatization of the acid group in the presence of the salt does not proceed as with the polymeric acid. The sodium polyacrylate when pyrolyzed produces several unique products, one of which, cyclopentadiene, can be used to identify sodium polyacrylate. Electron impact mass spectra and matrix isolation IR detectors connected to the pyrolysis–gas chromatography unit were used to positively identify cyclopentadiene.

Characterization of the soluble fraction for commercial superabsorbents for quality control procedures or patent use, of necessity, is simpler than the above procedure. A portion of polymer is extracted in 0.9 wt% NaCl (1 g polymer in 200 g solution works well). After extraction, the sample containing the gel and the extracted polymer is carefully filtered to remove all traces of crosslinked polymer. One procedure specifies a 0.22-μm filter.[29] The resulting filtrate is titrated between a pH of 10.0 and a pH of 2.7. For samples containing about 1%–15% extractables, the standard error for this method was about 6% relative at the 95% confidence level. A dual titration, starting from the pH of the solution, will allow the determination of percent neutralization of the starting polymer in addition to the percent extractables. The percent neutralization of superabsorbent polymers can also be measured after exchanging the sodium present in the polymer with ammonium ions and then determining the amount of sodium present in the exchange solution using a sodium ion-selective electrode.[30]

4.4. ANALYSIS OF COMPONENTS PRESENT IN THE POLYMER AT LOW CONCENTRATIONS

Because of the crosslinked nature of the solid or solvated polymer, direct analysis of minor components or impurities is usually not possible. One technique commonly used to overcome this difficulty is extraction of the minor component into a solvent for that component, followed by quantitation using chromatography. The time required for complete extraction depends on the molecular size of the component (as noted above for extractable polymer) and the degree of swelling of the polymer in the chosen solvent. Better solvents allow the polymer to swell more and the component to be extracted more rapidly. An additional criterion used for choosing the extraction solvent is the choice of mobile phase used for chromatographic separation and the compatibility of the extraction solvent and the mobile phase.

NMR has also been used in several ways to analyze the effects of minor components. In one case, the crosslinker was ^{13}C-labeled to enhance its signal. An example of this approach is discussed in section 4.3.2, where the ^{13}C-labeling of the crosslinker is used to follow its incorporation into the polymer network. A second use of NMR is discussed in section 4.4.3.2 to characterize poly(acrylic acid) network swelling and, thereby, the network structure.

4.4.1. Acrylic Acid and Dimer of Acrylic Acid

Residual acrylic acid monomer in superabsorbent polymers has been determined using gas chromatography[31] and liquid chromatography after extraction from the polymer with 0.9 wt% sodium chloride.[5,32–35] The same liquid chromatography conditions used for the determination of residual monomer in the polymer have been used for the determination of acrylic acid dimer, trimer, and tetramer in acrylic acid monomer (Fig. 4.9).[5] Liquid chromatography is preferred to other methods for the determination of residual acrylic acid in the polymer and residual oligomers in the acid, because of the reactive, water-soluble, polar nature and potential for thermally induced polymerization of acrylic acid. Also the dimer is thermally labile and can decompose under gas chromatography conditions. A typical liquid chromatography method to determine residual acrylic acid is to extract 1 g of polymer in 200 g of 0.9 wt% NaCl. Then 100 μL of filtered extract is injected into a chromatographic column with detection at 200 nm. Several types of liquid chromatography columns that have been used in the past include Bio-Rad's Fast Acid Analysis Column, Hamilton's PRP-1, and Reverse Phase. Quantitation is by external standards and is obtained by comparison of the peak height of the sample to the peak height of a known standard. For analysis of residual acrylic acid from less than 100 ppm to about 3,000 ppm, the precision at the 95% confidence level is about ±6%. A limit of detection in the low ppm range can be attained. The limit of detection is defined as a response equal to three times the background noise.

4.4.2. Determination of β-Hydroxypropionic Acid

The sodium hydroxide neutralization of acrylic acid was explored to reduce the level of acrylic acid dimer which in turn produces residual acrylic acid monomer during the polymerization to make superabsorbent polymer (see Chapter 2, section 2.2.2.2). The hydrolysis of the ester group in the acrylic acid dimer was found to produce species like β-hydroxypropionic acid with no vinyl bond that would remain as a residual monomer. A reverse-phase liquid chromatography method was developed to monitor β-hydroxypropionic acid in acrylic acid monomer and in superabsorbent polymer. The liquid chromatography consisted of a Whatman reverse-phase Partisil 10 ODS-2 column and 0.01 N H_2SO_4 mobile phase with refractive index detection. The method was found linear at the concentration of interest of 60–600 ppm. Quantitative recoveries were obtained at the same concentrations.

4.4.3. Analysis of Crosslinkers

4.4.3.1. Quantitation of Unreacted Crosslinker. Reverse-phase liquid chromatography has been used to determine residual crosslinkers in the superabsorbent polymer after extraction with combinations of methanol–water, acetonitrile–water,

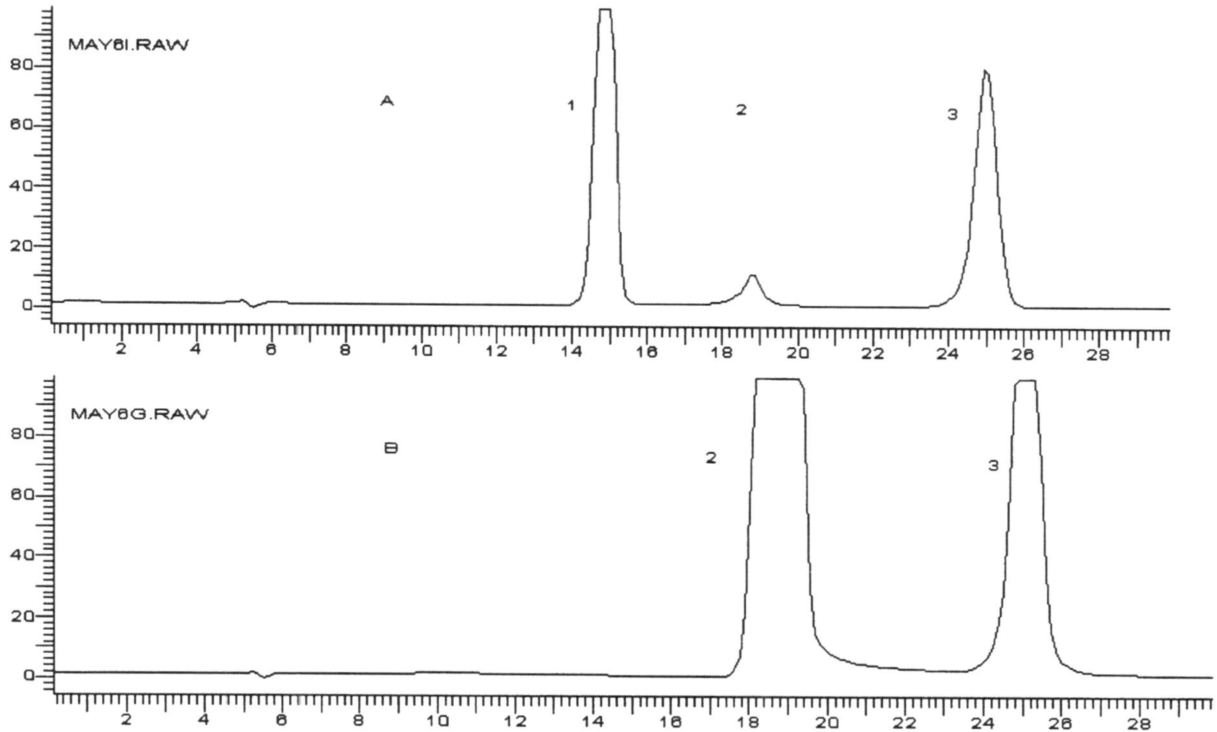

Figure 4.9. Liquid chromatographic traces of **(a)** a 12.5 ppm standard of acrylic acid dimer in acetic acid and **(b)** of acrylic acid. Peak 1 is acetic acid, peak 2 is acrylic acid, and peak 3 is acrylic acid dimer.

or methanol–acetonitrile–water, depending on the mobile phase used for the analysis. Detection is by ultraviolet spectroscopy at 206 nm, and the detection limits obtained are in the low ppm. The separation is performed using Zorbax-ODS or Zorbax Rx-C18, 5 μm, 4.6 × 250 mm. Crosslinkers analyzed under these conditions are typically multifunctional and usually not very water soluble, for example, TMPTA, diethylene glycol diacrylate, allylmethacrylate, and HE-TMPTA. Ether-type crosslinkers, like ethylene glycol diglycidyl ether, are better analyzed using capillary gas chromatography with a DB-1 (methyl silicone) column and flame ionization detection.

4.4.3.2. NMR Characterization of Swelling and Network Structure.

^{13}C-NMR line widths depend on the degree of swelling of superabsorbent poly(sodium acrylates)[36] and, as a result, provide a convenient empirical measure of the extent of crosslinking. The ^{13}C-NMR method is not conducive to determining the crosslinking level quantitatively for two reasons. First, the line widths cannot be fit with a single lorentzian or super-lorentzian line shape. The lines contain a distribution of line widths due to the distribution of crosslinking environments that exist within the gel and are, therefore, very difficult to treat quantitatively. Second, the NMR methods correlate directly with the swelling of the network rather than its crosslink concentration. Swelling is dependent on both the extent of crosslinking and the primary chain molecular weight as discussed in section 4.6 and Chapter 5.

An alternate ^{1}H-NMR method was devised to determine the degree of swelling of poly(acrylic acid) gels in a more quantitative fashion.[37] Figure 4.10 shows the ^{1}H-NMR spectra of three crosslinked, water-swollen poly(acrylic acid) polymers made as homogeneous in crosslinker as possible. Two ^{1}H-NMR resonances were observed for water in this system, one from bulk water outside the gel and the other from absorbed water. The resonance for absorbed water was shifted from that of bulk water because the environment of the network pores was more acidic than that of bulk water. The shift difference between absorbed and bulk water increased as the amount of crosslinking increased, consistent with a simple model based on the swelling of the network. Figure 4.11 gives a plot of the inverse of the shift difference ($1/\Delta\delta$) versus the swelling capacity for the three standard samples. The predicted straight line behavior is observed.

4.4.4. Detection of Graft Substrates—Starch and Poly(Vinyl Alcohol)

The presence in a superabsorbent polymer of a graft substrate such as starch or poly(vinyl alcohol) may be qualitatively determined by the sequential addition of a dilute solution of iodine and boric acid to the polymer.[38] The analysis is based on the blue color of the starch/iodine complex and the green color of the complex of poly(vinyl alcohol) with iodine in the presence of boric acid. A solution of 25 mL of 0.1N aqueous iodine in 500 mL of deionized water is sprayed or dropped onto a few granules of superabsorbent polymer. If a blue to purple color is observed, starch is present. If no color, or only a very light color, is observed, the polymer is treated with a 3.8 wt% of boric acid in deionized water. If a green color is observed,

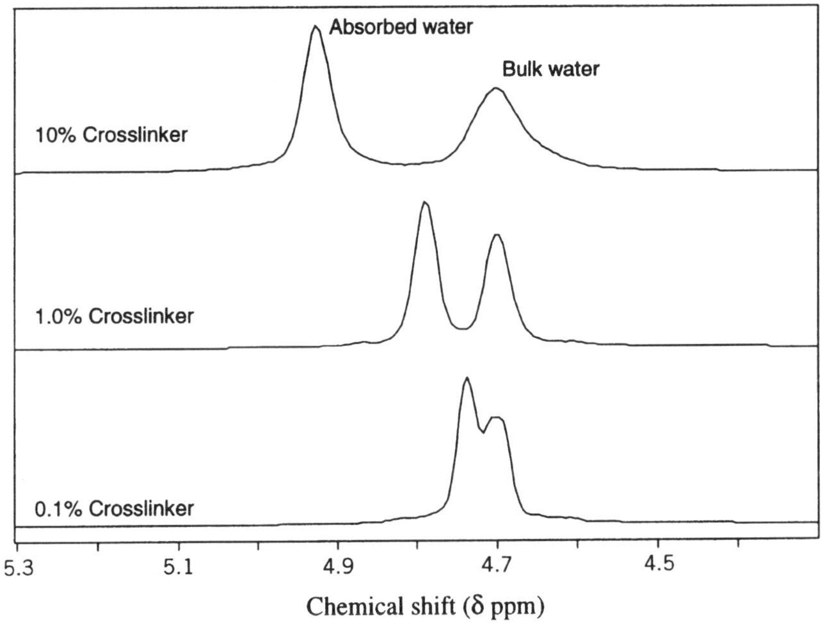

Figure 4.10. The ^{1}H-NMR spectra of water-swollen, homogeneously crosslinked poly(acrylic acid) gel standards, acid form.

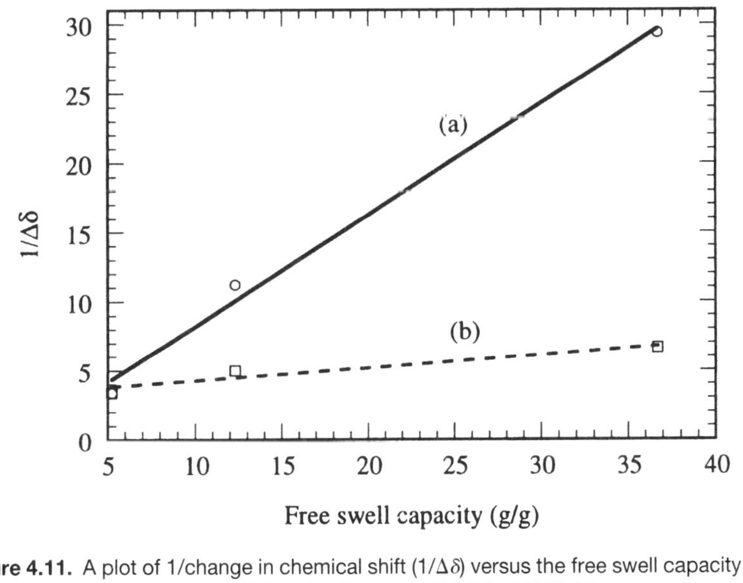

Figure 4.11. A plot of 1/change in chemical shift ($1/\Delta\delta$) versus the free swell capacity for three standard samples swollen in **(a)** deionized water or **(b)** 0.9 wt% NaCl.

poly(vinyl alcohol) is present. Used qualitatively, as is described here, this method is capable of detecting starch or poly(vinyl alcohol) when they are present at greater than about 0.1–0.3 wt%. The complete method of Finley[38] may be adapted for quantitative determination.

Polarized light microscopy has been used to further identify starch in superabsorbent polymers. The microscope used was a Zeiss Axiomat light microscope equipped with transmitted polarized optics (PLM). If starch is present it exhibits a characteristic bright-field transmitted light of a smooth spherical to elliptical shaped morphology. In addition, the characteristic Maltese cross is observed under PLM conditions.

4.4.5. Sodium Persulfate

4.4.5.1. Sodium Persulfate Quantitation. Sodium persulfate is often used as a thermal polymerization initiator to achieve high conversions of monomer to polymer and to control the concentration of unreacted residual acrylic acid in the product. It is often desirable to determine the concentration of residual persulfate in dry polymer or to determine persulfate concentrations in process samples such as gels prior to, during, and after polymerization reactions to solve process problems and to provide data for process kinetics and modeling. The determination of the concentration of residual initiator present after reaction requires a sensitive and selective technique because the initiator is frequently present at very low concentrations and difficult to detect directly. Indirect colorimetric methods based on reaction with the highly colored α,α-diphenyl β-picryl hydrazyl radical have been used.[39] Persulfate radicals generated by the thermal decomposition react with the hydrazyl radical and the decrease in color is directly proportional to the persulfate concentration. Other indirect methods used to quantify persulfate, such as the back-titration of Fe^{3+} produced from the oxidation of added Fe^{2+}, are nonspecific and affected by other oxidizing species and dissolved oxygen.[40]

Sampled-current polarography has been used to directly quantify persulfate in latex media.[41] At a mercury electrode and at potentials negative of +0.50 V, persulfate is reduced according to Eq. 4.1:

$$S_2O_8^{2-} + 2e^- \longrightarrow 2\, SO_4^{2-} \tag{4.1}$$

Species that yield anodic current at the mercury electrode, e.g., halides and chelants, or high concentrations of species that suppress the polarographic reduction current, for example, polyacrylate, can interfere with the polarographic determination. Furthermore, the more sensitive techniques of differential pulse and square wave polarography, often required for trace analysis, cannot be applied to persulfate determination because its positive reduction potential (in aqueous media) is merged with the mercury dissolution process.

The separation of persulfate from mono- and divalent anions using ion chromatography with perchlorate mobile phase and ultraviolet detection was reported.[42] However, attempts to determine persulfate in superabsorbent polymers using this approach are often unsuccessful due to the presence of high concentrations of ultra-violet-absorbing species like acrylic acid and acrylate, which mask the persulfate response. The ion chromatographic method was improved by using electrochemical detection to measure persulfate in superabsorbent polymer media.[43] The chromatograms in Figure 4.12 show that sensitive and selective determination of persulfate was achieved. Anions and the superabsorbent polymer matrix were separated from persulfate on an anion-exchange analytical column prior to quantitation of the persulfate using constant-potential detection at a renewable dropping-mercury electrode. Sulfate, sulfite, thiosulfate, and polythionates do not interfere. The method is applicable to dry and process superabsorbent polymer samples.

Sodium persulfate is extracted from superabsorbent polymer into aqueous 5% (w/v) Na_2SO_4 and then determined by ion chromatography using reductive electrochemical detection. Process samples are analyzed by weighing 5.000 g of gel into a clean 8-oz glass bottle and adding 100 mL of 5% (w/v) Na_2SO_4 solution. The solution is homogenized and shaken before filtration and injection into the chromatograph. The mobile phase is 0.07 mol/L $NaClO_4$, continuously purged with water-saturated prepurified helium to remove dissolved oxygen, which is co-reduced with

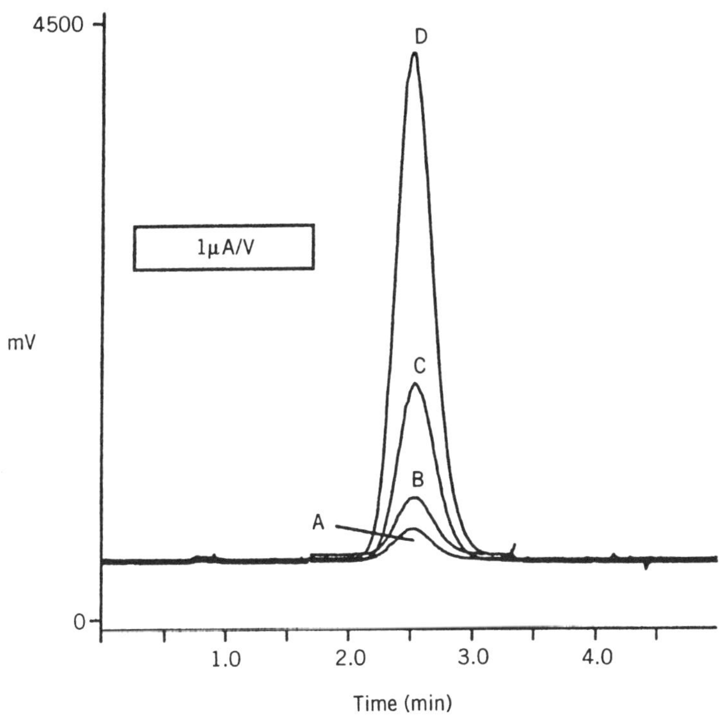

Figure 4.12. Chromatograms showing response for **(A)** 3.72, **(B)** 7.43, **(C)** 22.3, and **(D)** 55.75 ppm sodium persulfate prepared in 5% (w) sodium sulfate.

persulfate. Sodium persulfate is quantified by external standard analysis. Superabsorbent polymer products are analyzed similarly using 2,500 g of dry product.

The analytical column has dimensions of 3.0 mm i.d. × 300 mm o.d. and was filled with a quaternary ammonium latex anion exchanger on 10–20 μm substrate. The mobile phase flow rate was 1.4 mL/min. The injection size was 100 mL. The detector was a static mercury drop electrode (EG&G Princeton Applied Research with a model 174A potentiostat). The drop time was 1.0 s; mode, sampled dc; drop size, large; applied potential, –300 mV versus Ag/AgCl (1 mol/L LiCl). Other details are given by Henton et al.[43] and Theoharides et al.[44]

4.4.5.2. Sodium Persulfate Decomposition. Depending on the pH and temperature of aqueous solutions, persulfate decomposes by two different mechanisms.[45] Below pH 2, the decomposition is acid catalyzed and does not follow a free-radical route. Byproducts include sulfur tetroxide and bisulfate. In dilute acid solution, the sulfur tetroxide decomposes ultimately to sulfuric acid and oxygen. It is a fast reaction and if not buffered is auto-catalytic in that the sulfuric acid that is generated further increases the acidity. At a pH greater than 4, persulfate decomposes by homolytic cleavage, with an activation energy of 140.2 kJ/mol, to generate two sulfate ion radicals. Between pH 4 and pH 13, the thermal decomposition rate varies only by a factor of two, with the maximum rate occurring at about pH about 7.[46]

The rate of decomposition of persulfate is affected by the concentration of reactive vinyl monomers, generally up to a limiting concentration above which no additional effect is observed. In the presence of monomers such as vinyl acetate, the rate of decomposition of potassium persulfate increases up to 50-fold.[47]

The activation energy and rate constants for the decomposition of sodium persulfate in the presence of acrylic acid have been determined for monomer compositions representative of those used to prepare sodium polyacrylate superabsorbent polymers.[43] Figure 4.13 shows that the rate of decomposition of sodium persulfate was first-order. Relative to buffered aqueous conditions, the decomposition rate was increased by a factor of 2–7 in the presence of acrylic acid, low concentrations of neutralized acrylic acid, or high concentrations of acrylic acid at low percent neutralization as a result of an induced decomposition mechanism by the monomer. The activation energy of sodium persulfate decomposition in 65% neutralized acrylic acid at 33.8% solids was found to be 94.6 kJ/mol.

In contrast, the decomposition rate of persulfate in high concentrations of partially neutralized acrylic acid was decreased relative to its decomposition rate in the presence of acrylic acid (not neutralized) (Fig. 4.14). An ionic gel cage mechanism in which persulfate radicals become trapped as the polymerization progresses, was proposed to explain the influence of solids on the rate of decomposition of sodium persulfate.

4.4.6. Inhibitors

Quantitation of MEHQ, present in acrylic acid as a polymerization inhibitor, was discussed in section 4.2.1.4. MEHQ may also be found in the polymer product (see

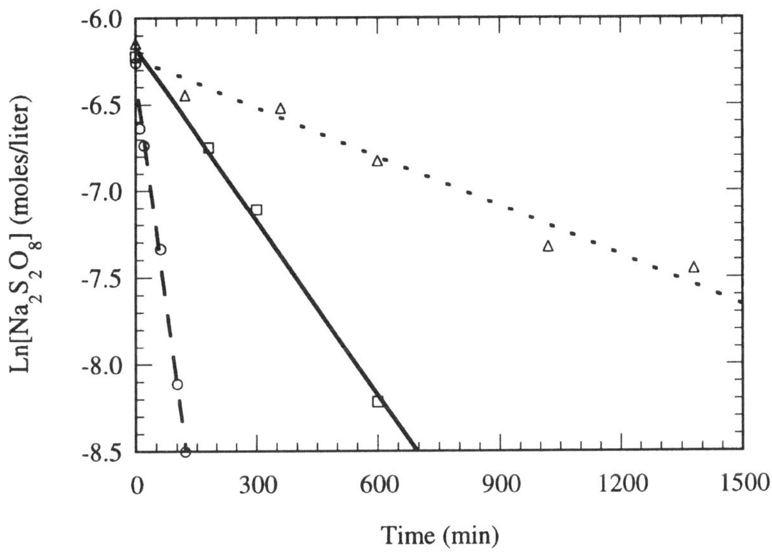

Figure 4.13. First-order decomposition plots for sodium persulfate at 55°C (△), 70°C (□), and 85°C (○) in the presence of 33.8 wt%, 65% neutralized acrylic acid.

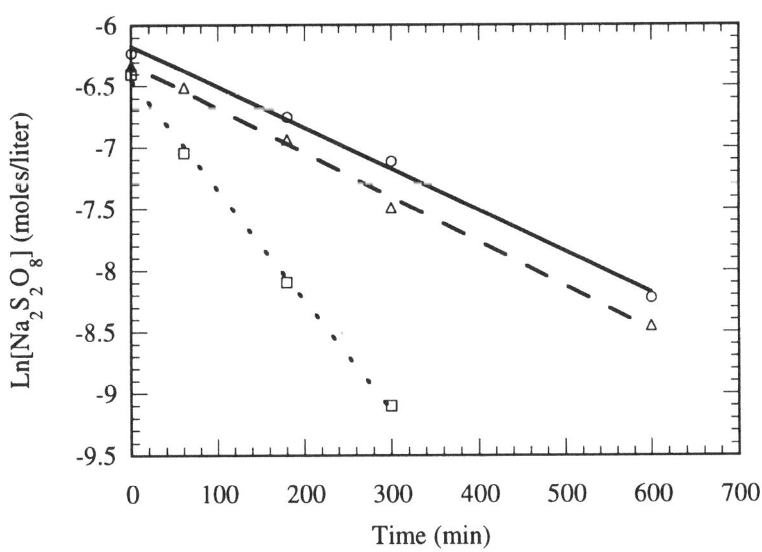

Figure 4.14. Effect of monomer concentration (65% neutralized acrylic acid), 3.4 (○), 16.9 (△), or 33.8 (□) wt%, on the rate of decomposition of sodium persulfate at 70°C.

Chapter 2, section 2.5.3). MEHQ is extracted from polymer product by swelling 0.1–0.2 g of sample in 5.0 mL of 5% (w) Na_2SO_4 and adding to 5.0 mL of mobile phase. The solution is filtered through a 0.45-μm Nylon 66 syringe filter prior to injection and analyzed by liquid chromatography with electrochemical detection as described in section 4.2.1.4. MEHQ is quantitated by standard addition analysis.

4.4.7. Surface Tension

In a diaper core, capillaries exist between fluff pulp fibers, polymer particles, or the two in combination, as discussed in Chapter 1. Distribution of liquid in the diaper core will be affected by the surface tension of the liquid that is flowing in these capillaries. In use, impurities may be extracted from a swollen superabsorbent polymer into the solution external to it. Through the surface tension of the solution, the movement of fluid through the capillaries in a diaper core may be affected by these impurities. Measuring the surface tension of a solution of extract allows an estimate of the effect on capillary flow. Surface tension measurements on a series of samples must be done in a consistent manner, because in the end-use the amount of liquid and of swollen polymer cannot be accurately known. One method for obtaining this information has been outlined for polyacrylate superabsorbents.[48] One gram of polymer is swollen in 200 mL of 0.9% NaCl. The surface tension of the supernatant is measured using a Du Nouy tensiometer.

4.5. BULK POLYMER CHARACTERIZATION

4.5.1. Polymer Identification by IR and ^{13}C-NMR

A number of analytical techniques can be used to identify poly(acrylic acid-co-sodium acrylate). NMR spectroscopy and IR spectroscopy have had wide application. An IR spectrum of poly(acrylic acid-co-sodium acrylate) is shown in Figure 4.3b. ^{13}C-NMR spectroscopy is also a convenient method for the identification of poly(acrylic acid). Figure 4.15 shows a ^{13}C-NMR spectrum of a poly(acrylic acid) network with the chemical shift assignments given on the spectrum. The shift of the carbonyl resonance is pH dependent. At low pH values, it resonates at about 180 ppm and is difficult to differentiate from poly(acrylamide). At elevated pH (salt form), the carbonyl resonance occurs at about 184 ppm.

4.5.2. Thermal Techniques

Thermal techniques have been used in polymer characterization to determine specific heats, melting points, and polymer glass transition temperatures (T_g), and to study thermal degradation and oxidation reactions.[49] Previous studies of (uncrosslinked) poly(acrylic acid) salts have determined the influence of hydration on electrostatic interactions, as well as the influence of the extent of neutralization and the species of alkali metal ions of poly(acrylic acid) salts, on the drying process.[50] It

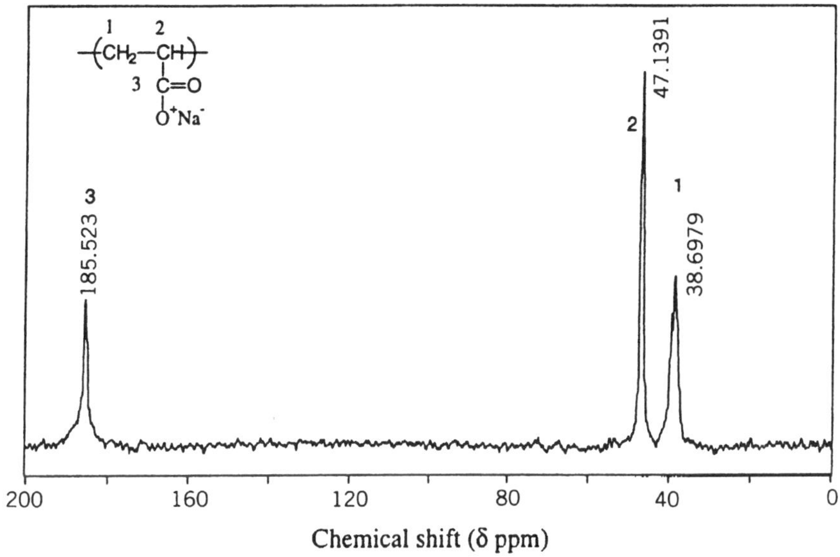

Figure 4.15. The ^{13}C-NMR spectrum of a poly(acrylic acid) gel swollen with 0.9 wt% NaCl.

has been reported that the T_g of poly(acrylic acid) salts increased with increased neutralization and decreasing water content.

Thermal techniques have also been used to estimate the degree of hazard associated with processes and conditions of operation. Adiabatic reaction calorimetry, differential scanning calorimetry, thermogravimetric analysis, and other techniques may be combined to understand the amount and rate of heat that may be evolved if an unplanned event occurs in the process. A review of the general subject is available elsewhere.[51]

4.5.2.1. Glass Transition Temperatures. The glass transition process has been described as the freezing (upon cooling) or the unfreezing (upon heating) of micro-Brownian motion of chain segments of 20–50 atoms in the chain backbone.[52] The transition between the glassy and the rubbery phase can be observed as a step in the heat capacity of the system upon heating or cooling in a differential scanning calorimeter (DSC). Figure 4.16 shows a typical DSC scan for a sample of partially neutralized, crosslinked poly(acrylic acid) that contains about 40% water, based on dry polymer mass. This DSC scan was obtained using a TA Instruments 912 Dual Sample DSC with a 2100 Thermal Analyzer for data reduction. The sample was weighed into a high-volume Perkin-Elmer stainless steel pressure DSC pan. This pan was chosen since it creates an excellent seal and does not allow water loss from the system upon heating, even at 200°C, this sample having lost less than 1% mass upon heating. The downward step, centered around 25°C, is the glass transition. For comparison, Figure 4.17 shows the DSC scan of a sample containing only about 5%

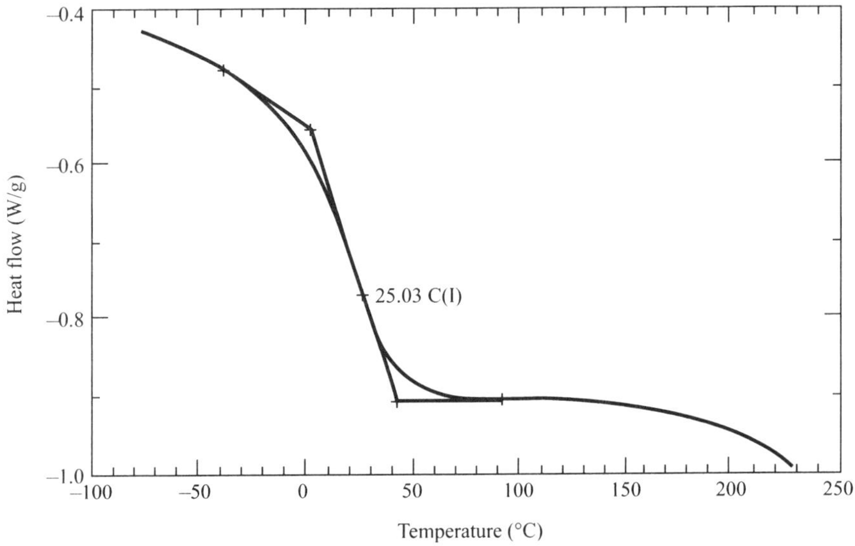

Figure 4.16. DSC scan of a poly(acrylate) containing about 40% moisture.

water. In this case, there are three inflection points, with the one at 139°C being the largest. This suggests that T_g is a function of water concentration and shows some indication of how T_g and water concentration are related. This also points out that in a complex system (partially neutralized, partially crosslinked) a single description of T_g may not be appropriate. Figure 2.8 (see Chapter 2) displays the T_g as a function of polymer moisture for a 65% neutralized polyacrylate crosslinked with trimethylolpropane triacrylate (A.J. Pasztor, The Dow Chemical Co., unpublished results).

In other studies using crosslinked poly(acrylic acids), it has been reported that there is a direct relationship between T_g and the extent of crosslinking of a polymer system[53–56]. It was proposed that for co-polymers T_g can increase or decrease with increasing crosslinking density depending on the chemical nature of the co-monomer,[53] although for a crosslinked homopolymer T_g always increases with increasing crosslinking density. Thus the effect of adding a crosslinker to the system cannot always be predicted and should be studied experimentally.

4.5.2.2. Heat Capacity Measurements. Specific heat measurements have been obtained (S.J. Martinez, The Dow Chemical Co., unpublished results) on superabsorbent polymer made by the process of Siddall and Johnson.[48] The measurements were carried out using a SETARAM differential scanning calorimeter (DSC-111, Serial No. 051035387). Approximately 100 mg of sample was encapsulated into a stainless steel crucible and sealed using an aluminum O-ring. Thermal scans were measured from 40° to 230°C at a rate of 2°C/min. The reference standard was synthetic sapphire (NBS, SRM 720). Corrections to the data consisted of a heat

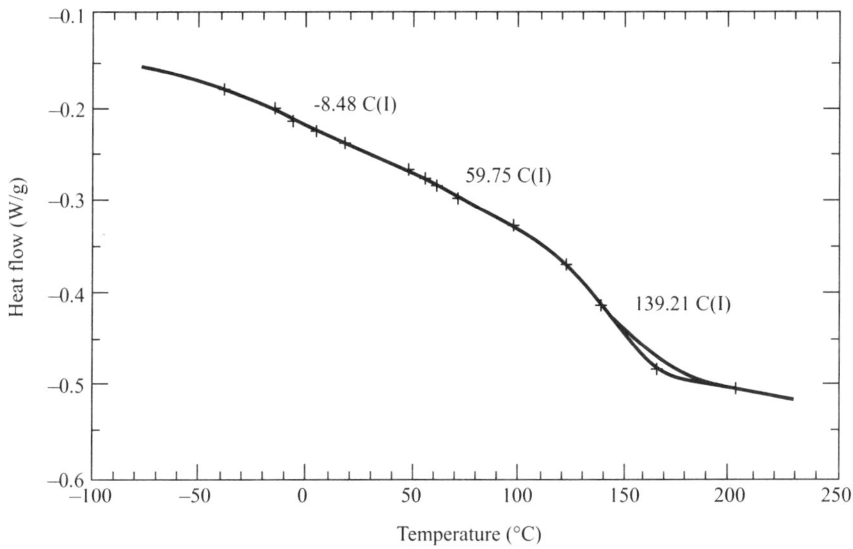

Figure 4.17. DSC scan of a poly(acrylate) containing about 5% moisture.

flow correction based on the reference specific heat of sapphire. The specific heat results for the temperature range from 50° to 150°C are shown in Figure 4.18.

4.5.3. Moisture Analysis

Product moisture does not directly affect product performance properties. However, moisture level is important because of its effect on handling of the product for sale. Products that have been heated for surface crosslinking reactions and the like may be dry enough that problems with the build-up of static electricity may arise. At the other extreme, in high humidity environments, the product may pick up enough moisture to become sticky and difficult to handle.

4.5.3.1. Drying. Traditionally the determination of moisture in superabsorbent polymers has been obtained gravimetrically after heating the sample at 105°C for 3 h. These conditions were recommended by the Technical Association of the Pulp and Paper Industry (TAPPI)[57,58] to determine moisture in paper and paper products. The release of water from superabsorbent polymers, however, does not necessarily follow the same mechanism or the same rate of release as with paper or paper products. Simple experiments show that significant moisture remains in a commercial sample of superabsorbent after 3 h in a 105°C oven. Heating the superabsorbent polymer for various times and at different temperatures yields different percentages of mass loss. Drying to constant mass at a higher temperature, for example, 160° or 170°C, gives moisture values several percentages higher than found in 3 h of heating at 105°C. Temperatures higher than about 200°C are not recommended because

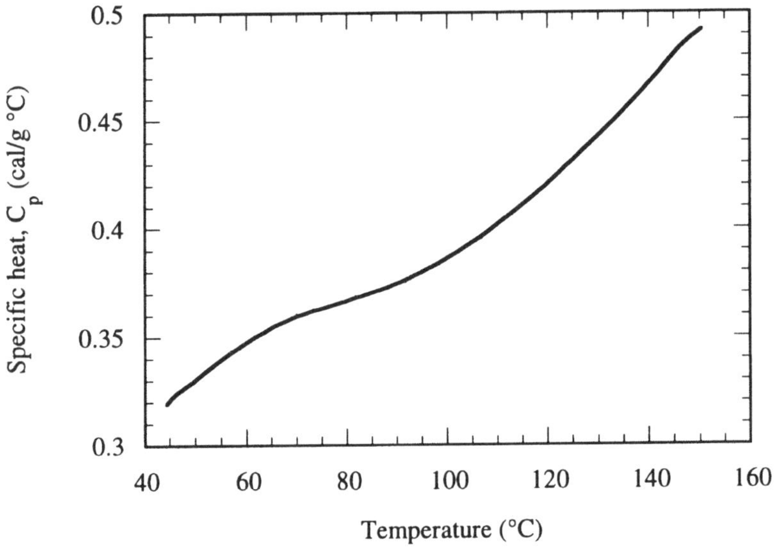

Figure 4.18. The specific heat of a 65% neutralized polyacrylate superabsorbent from 44° to 150°C.

other reactions occur that can lead to mass loss, for example, oxidation of the polymer, anhydride formation, and decarboxylation (see Chapter 2, section 2.5.4.2).

An approach to the drying of superabsorbent polymers that avoids the problem of oxidation and still removes essentially all of the water is as follows. About 1 g of sample is placed in an empty 100 × 7.8 mm liquid chromatographic column. The column is connected to a line with a flowing helium atmosphere inside a temperature-controlled chamber (a gas chromatograph), and the temperature is raised. To remove essentially all of the water, a temperature and time of heating of 180°C for 1 h are recommended. The sample is maintained under a helium atmosphere until cooled. The column is capped to avoid reabsorption of water. Following the heat treatment the mass loss is determined gravimetrically using a Mettler AE160 balance.

4.5.3.2. Anhydride Formation and Decarboxylation. Under aerobic and anaerobic conditions thermogravimetric/mass spectrometry (TG/MS) and TG/gas chromatography/MS (TG/GC/MS) have been used to determine that the main products evolved when heating superabsorbent polymers are water and carbon dioxide.[59–61] When the superabsorbent polymer is heated, the mass loss depends on the time and temperature of heating. IR spectroscopy has been used to follow the time dependence of this process and to characterize changes in the polymer, for example, in the amount of anhydride functionality present. The formation of anhydride was reported when slightly crosslinked poly(acrylic acid) was heated at 250°C.[60] In a separate study, superabsorbent polymer at various percentages of neutralization were heated at 250°C for 10 min and IR spectra were obtained (Fig. 4.19). (M.L. McKelvey and S.S. Cutié, The Dow Chemical Co., unpublished results). The intensity of the anhydride band at 1,805 cm^{-1} decreases as neutralization increases. These techniques have been combined to understand the chemical processes, including anhydride formation and decarboxylation, occurring during heating of poly(acrylic acid) polymers and their salts (see Chapter 2, section 2.5.4.2).

4.5.4. Microscopy

4.5.4.1. Optical Microscopy. Optical microscopy, as an analytical tool, has the advantage of being a non-destructive technique. Microscopy has been extensively used to characterize absorbent materials (Chapter 1, Fig. 1.1), superabsorbent polymers (Chapter 1, Fig. 1.2), and composite absorbent structures (Chapter 1, Fig. 1.6). In any particular application, image enhancement may be obtained by using a variety of lighting techniques, including polarized light. Optical microscopy, combined with a videocamera, may be used to obtain a continuous record of time-dependent phenomena, for example, swelling.

4.5.4.2. Electron Microscopy. Scanning electron microscopy (SEM) is a method capable of examining surface detail with a great depth of field and high res-

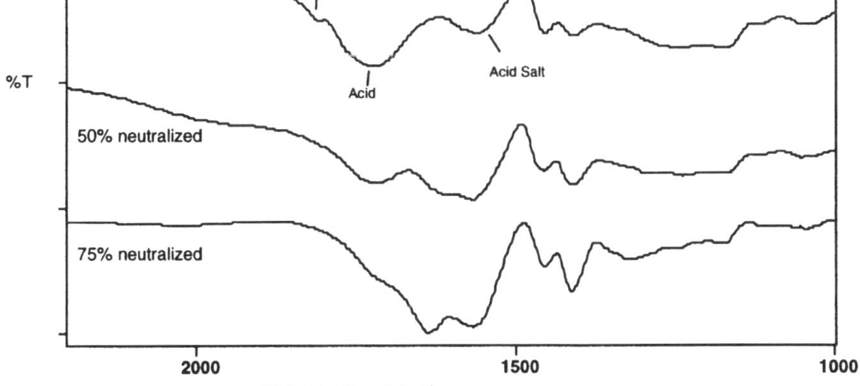

Figure 4.19. Infrared spectra of superabsorbent polyacrylate, at various levels of neutralization, heated at 250°C for 10 min.

olution. Feature size as small as 0.1 nm can be imaged. Magnifications ranging from ×15 to ×150,000 are usually obtained. Most polymer characterization work involves imaging the contrast generated through the dependence of secondary electron emission on surface topography, sample composition, and crystallinity. Morphologic features such as step height, aspect ratio, size distribution, and porosity can be determined. It is common to obtain a photographic record of a sample over a range of magnifications to provide a complete understanding of the microstructure. An example of the great depth of field available is shown in Chapter 3, Figure 3.15, a highly convoluted suspension polymerized polyacrylate superabsorbent. Figure 3.12 shows a cross section of a superabsorbent bead with a clearly visible surface layer of silica.

Environmental scanning electron microscopy (ESEM). Sometimes called a wet-SEM, is a relatively new technique that makes possible, for the first time, the unrestricted, undisturbed, direct observation of samples in their native state. With the ESEM, wet specimens, even hydrated superabsorbent particles, can be examined. A unique vacuum system permits continuous observation of the hydrogel under fully saturated water vapor conditions or other gas environments. Wet specimens stay wet indefinitely or may be slowly dried by controlled evaporation and then rehydrated to reveal the dynamics of adsorption. Magnification from ×50 to ×50,000 is possible with a resolution of 100 Å. Elemental composition contrast, density differences, and crystal orientation are visible using ESEM. Still images or real-time videotapes of wetting, drying, adsorption, melting, and crystallization are obtained using ESEM's fully integrated, TV-rate frame store system and VHS-compatible video signal output.

Analytical scanning electron microscopy (ASEM) is SEM capable of elemental analysis. Elemental analysis at a resolution of approximately 1 μm is accomplished by energy-dispersive x-ray spectroscopy and may be used to determine neutralization (i.e., sodium, potassium), the distribution of elements such as aluminum (Chapter 2, Fig. 2.17), as well as the presence of coatings, for instance, those used on superabsorbents to prevent dusting or caking. A semiquantitative elemental analysis of elements heavier than boron is possible. The detection limit is typically in the 0.2–0.5 wt% range. Compositional contrast imaging on polished polymer samples is possible through back-scattered electron detection or element-specific x-ray maps.

4.5.5. Particle Size Determination

The determination of particle size and particle size distribution of superabsorbent polymers creates very interesting challenges. Superabsorbent polymers are hygroscopic and adsorb moisture from the atmosphere. The amount of moisture present in the particles and the rate of absorption of moisture by the particles are not the same for all superabsorbent polymers. In the presence of moisture finer particles may become attached to larger particles, which can produce misleading results. Also, these materials are polyelectrolytes and may attach themselves to any surface, affecting particle size distribution measurements. Several techniques have been used to determine particle size and particle size distribution in superabsorbent polymers.

The Rotap sieving, similar to the ASTM method used for polymer powders,[62] has been the technique of choice within the superabsorbent polymer industry. The process consists of placing a powder sample in a nest of sieves that are shaken horizontally while the column is being hit at the top with a hammer.[63] The idea is to reproduce the type of motion that has been successful when sieving shallow beds of powder by hand, combining a horizontal swirling action with a vertical jarring action. Care should be taken to eliminate blinding of the sieves.

The aerodynamic particle size analyzer,[64,65] based on the principle of particle lag in a flow field, has been used to quantify the relative concentration of particles of less than 10 μm. However, it has not been fully evaluated for general superabsorbent polymer particle sizing. The particles to be measured are first suspended in air using a disperser, and then the suspension is expanded through a nozzle into a partial vacuum. The air velocity remains constant in the measurement region and the particles accelerate at different rates according to their size. Two laser beams determine the particle velocity. As the particles pass through each laser beam, light is scattered and detected by two photomultipliers. The time between the measurements at the two photomultipliers is converted into particle size.

4.5.6. Bulk Density and Flowability

In commercial usage, polymer bulk density[66] and polymer flowability[67] affect equipment design in the polymer manufacturing plants and machine settings on diaper manufacturing lines. These characteristics are determined by timing the flow of a sample of superabsorbent granules through a funnel into a cylinder of known volume. The bulk density is the mass/volume ratio of polymer in the cylinder. The flowability is the time for the sample (typically 100 g) to flow through the funnel.

4.6. POLYMER ANALYSIS AND NETWORK STRUCTURE

4.6.1. Swelling Capacity: Theory and Practice

Certainly the first property, and arguably the most important, measured in a commercial superabsorbent used in the personal care market is the extent of swelling. This is true not only because swelling is related to the properties of the network, but also because the principal performance criterion for diapers is the amount of liquid contained per unit cost of diaper. For a complete discussion of these relationships, see Chapter 5. The following discussion is intended as a brief summary relating the methods of measurement to the equations describing swelling, modulus, and swelling rate and the assumptions that have entered into those equations.

Using the "phantom network approximation," and following the approach of Mark and Erman,[68] an expression (Eq. 5.10) for equilibrium swelling of polymers, q_m, is derived in Chapter 5 and is shown below as Eq. 4.2. Equilibrium swelling depends on the polymer concentration υ_{20} at the time of preparation (e.g., during crosslinking), the solvent (swelling liquid) interaction parameter, χ, the molar volume V_l of the swelling liquid, the dry polymer density ρ_0, the backbone molecular weight M_n of the polymer, and the average molecular weight between junctions, M_c.

$$q_m^{5/3} = \frac{(1/2 - \chi)2M_c}{V_1 \rho_0 \upsilon_{20}^{2/3}(1 - 3M_c/M_n)} \tag{4.2}$$

q_m is defined as the swelling ratio from the reference state volume V_0, which is the volume at which the polymer was prepared, to the final volume, V_{eq}, so $q_m = V_{eq}/V_0$.

Restating Eq. 4.2 in terms of q_m yields Equation 4.3:

$$q_m = \left(\frac{2kM_c}{\rho_0 \upsilon_{20}^{2/3}(1 - 3M_c/M_n)} \right)^{3/5} \tag{4.3}$$

where

$$k = \frac{(1/2 - \chi)}{V_l}$$

The derivation of this equation assumes a "low concentration" of crosslinks and a "good" solvent, which means that chain deformation during swelling is small relative to the length of the extended chain. This derivation also does not specifically take into account the ionic character of the typical superabsorbent polymer backbone, but this effect can be accommodated by using an effective value of the interaction parameter χ.

To relate an experimentally determined swelling capacity to the derived theoretical capacity, the manner in which swelling measurements are made must be taken into account. First, the theoretically derived swelling equations assume a volumetric measurement of swelling in that the relevant quantities are volume fractions of polymer and solvent. However, swelling measurements are typically done gravimetrically. It is assumed implicitly that the density of the swollen gel is the sum of the densities of the components with the fraction of each component as a weighting factor. The experimental gravimetric swelling ratio, Q, is converted to a volumetric swelling ratio, q_m, using Eq. 4.4, assuming a polymer density, ρ_0, of 1.66 g/cm^3 and a density of swelling liquid of 1.00 g/cm^3.

$$q_m = \frac{\frac{1}{\rho_0} + Q}{\frac{1}{\rho_0}} = \frac{0.6024}{0.6024 + Q} \tag{4.4}$$

Second, the commonly measured swelling capacity of industry is taken from a minimum dried volume, V_{dry}, rather than from the volume of preparation, V_0. This is not a measurement of q_m, but rather of a different swelling, ratio q_{dry}, which is related to q_m through Eq. 4.5:

$$q_{dry} = q_m \frac{V_0}{V_{dry}} \tag{4.5}$$

For measurements of swelling and modulus, this difference is not particularly significant. However, if the relationship between swelling and the volume of preparation is important, this difference becomes significant. Substituting into Eq. 4.2 for q_m, and treating $(1/2 - \chi)/V_1$ as a constant, gives Eq. 4.6:

$$(q_{dry})^{5/3}(V_{dry}/V_0)^{5/3} = \frac{2kM_c}{\rho_0 \upsilon_{20}^{2/3}(1 - 3M_c/M_n)} \tag{4.6}$$

In Eq. 4.6, the term υ_{20} is the volume fraction of polymer at the time of crosslinking and is the volume of the polymer divided by the volume at the time of preparation, V_0. Rearranging Equation 4.6, collecting the terms in V_0, and solving for q_{dry} gives Eq. 4.7:

$$q_{dry} = \frac{(V_0)^{7/5}}{V_{dry}}\left(\frac{2kM_c}{\rho_0(1 - 3M_c/M_n)}\right)^{3/5} \tag{4.7}$$

Swelling from the dry state (neglecting any changes in the network structure caused by the processing steps) does not change the dependence of swelling on crosslink density and network characteristics. This also means that swelling from the dry state does not affect the correlation between measurements of modulus and swelling. However, the exponent on V_0 is 7/5 instead of 3/5. Also, q_{dry} is volume fraction based, as is q_m, and polymer and swelling liquid densities must still be taken into account for Eq. 4.7 to be used for gravimetric measurements.

4.6.1.1. Variables in the Swelling Analysis. A number of variables may affect the measured value of the swelling capacity. The discussion above described some of these explicitly. The swelling liquid has a strong impact on the equilibrium swelling. The size of the polymer particles before swelling affects the length of time required to reach equilibrium swelling, hence the absorption or swelling rate. Swelling liquid held between the particles by capillary forces must be accounted for, or excluded, when comparing the experimental results to theory. In relating swelling measurements to network characteristics, the theory assumes the volumetric measurement of an equilibrium swelling capacity. Either volumetric or gravimetric techniques may be used. However, in the latter case, it is implicitly assumed that the densities of the polymer and swelling liquid are additive in the swollen gel.

For the most direct comparison of experiment to theory, the volume of a polymer sample with regular geometry is measured in both the dry state and final swollen state. Yin[69] has adapted this technique to determine the swelling extent of rectangular slabs of polyacrylate superabsorbents. It has also been used to determine the swelling of poly(acrylamide) networks in acetone–water solvent systems.[70] Ricka and Tanaka,[71] using a microscopic technique, measured the swelling of cylinders polymerized in micropipets. We have also used the microscopic technique to measure the swelling of spheres produced in inverse emulsion polymerizations. These methods work well for swelling determinations over a range of swelling liquids and may be preferred for swelling of polyelectrolytes at low ionic strength.

Commercial superabsorbents, however, are sold as irregular granules. In addition, superabsorbents used in personal care products are used to absorb and retain body fluids of relatively high ionic strength (about 0.9 wt% NaCl), and the fluid is sensibly constant in composition. Thus, the issues surrounding swelling a polyelectrolyte gel in a low ionic strength swelling liquid are not of great concern, and differences in swelling that result from differences in the ionic strength of the liquid are very small (for instance, Donnan equilibrium and migration of counter ions from the gel to the outer solution, although important, are relatively constant factors in the personal care products). As a result, alternate swelling methods were developed that were suitable for the morphology of commercial superabsorbents.

4.6.1.1.1. Particle Size and Swelling Time. Swelling capacity methods described in the early patent literature did not specify a particle size distribution for the analysis. It seems likely that the interest of the inventors was the relative performance of polymers in the end-use, rather than in swelling as a method of characterizing the network structure. Later, the particle size used for analysis was limited to particle size distributions useful in diapers, covering the range of particle diameters from 100 to 850 μm.[72] More recent versions of industrial swelling methods have narrowed the particle diameter distribution specified for the analysis still further to 300–600 μm.[73,74]

Industrial swelling methods have not always considered swelling to equilibrium. In several cases swelling times of 10 min or less are described.[75–77] Gradually, as the industry has developed, the swelling times in the patent literature have lengthened to 30 min[73,74] or more.[78] Figure 4.20 shows the relationship between mean particle diameter, swelling time, and the approach of the measured swelling extent to an equilibrium value. As the particle size increases (smaller mesh number), the time required to reach equilibrium increases. Thus, for 40–45 mesh particles (385 μm average particle diameter), equilibrium swelling is achieved in about 20 min, while for 20–25 mesh particles (780 μm average particle diameter), only about 90% of equilibrium swelling is reached at 30 min. Depending on the size of the particles before swelling, the swelling time, and the amount of interparticle liquid, the swelling capacity result may be greater than or less than an equilibrium swelling value.

4.6.1.1.2. Interparticle Liquid. In early, industrially used analysis methods for determining the swelling extent of collections of small particles, the additional mass of interparticle, unabsorbed liquid was included in the swelling ratios. The developers had not attempted to meet the criteria of theory, just as theory had been neglected with respect to swelling time and particle size distribution. In one free swell capacity analysis,[79] the sample first was swollen in 0.9 wt% NaCl and then the gel particles were separated from the liquid by gravity filtration. Values from this analysis are typically larger than for methods where interparticle moisture is excluded. The particle size distribution of the sample and the modulus of the swollen

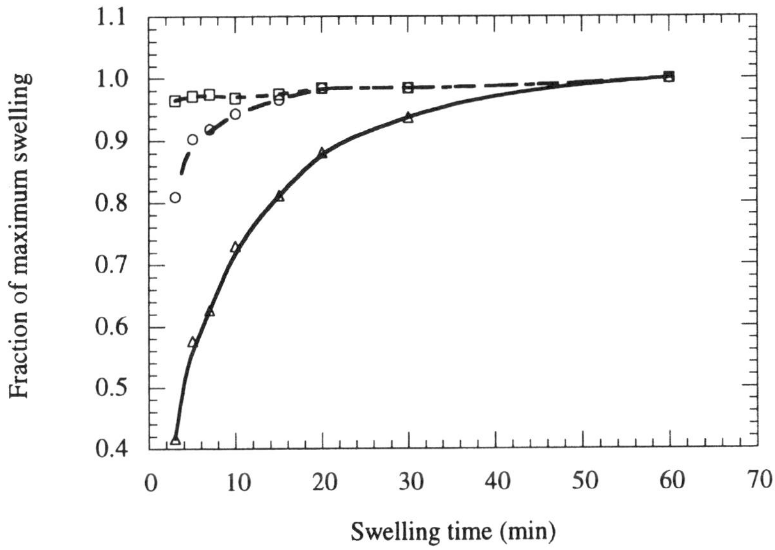

Figure 4.20. Fraction of equilibrium swelling capacity as a function of swelling time for narrow particle size fractions of a 30 g/g capacity sample of uniformly crosslinked poly(acrylate) superabsorbent 20–25 mesh (△), 40–45 mesh (○), 70–100 mesh (□).

polymer also affect the way in which the liquid drains from the gel. Relative error on multiple analysis was about 1%.

Another swelling capacity measurement is based on the analysis long used to measure absorption of liquid into a fiber pad or sheet. Chatterjee and Nguyen[3] have reviewed methods of this type. This method is sometimes referred to as *demand absorbency* and may be obtained using a "Gravimetric Absorbency Testing System" or an adaptation of it. A portion of polymer is placed on a screen, and swelling liquid is absorbed either from zero pressure head or from some prescribed pressure head. The mass gain of the sample or mass loss of the liquid reservoir is recorded. An alternate approach, which does not require much equipment and is suitable for granular dry superabsorbent, is to place the dried granules in a small porous bag (tea bag), immerse the sealed bag in liquid, and allow the polymer to swell. The swollen gel in the bag is allowed to drain for several minutes in order to remove excess liquid.[73,74] The mass is recorded and, accounting for the mass of the dry and wet bag, divided by the dry mass of the polymer to yield a swelling capacity.

Soon, however, improved swelling capacity methods were developed in which the excess liquid is accounted for or removed from the swollen gel. A colorimetric analysis described by Brandt et al.[29] has solved the problem of unabsorbed, interparticle liquid. The concentration of a blue macromolecular dye is determined spectrophotometrically in a known amount of swelling liquid before and after swelling. This technique assumes that the dye is rigorously excluded from the gel and concentrates in the free liquid phase as the polymer swells. This method is referred to as a *gel volume determination*, although a ratio of mass of swelling liquid per mass of superabsorbent is reported. Although the procedure is straightforward, it is time-consuming and cumbersome to use. Poor redistribution of the dye into the free liquid as the polymer swells is probably the principal source of analysis error. The dye tends to concentrate at the polymer–liquid interface. Analysis results correlate well to swelling capacities measured using the centrifuge capacity technique described below. Relative error on duplicate analysis using this method was about 2%.

A second method for excluding interparticle liquid is based on the technique of swelling the polymer in a small porous bag. Good descriptions for methods of this type are available in several patents.[74,80] A small quantity of superabsorbent polymer is placed in a porous heat-sealable bag, a "tea-bag." The sealed bag is then immersed in 0.9% NaCl solution, and the polymer is allowed to swell for a specified period of time. In general, the polymer attains its equilibrium swelling value in 30–90 min, depending on particle size. After removing the bag from the swelling liquid, the interparticle liquid is removed by centrifugation (see below). The correlation between this "centrifuge capacity" analysis and the "free swell capacity" analysis described above is surprisingly good (Fig. 4.21).

The centrifuge capacity method, with the assumption that the density of the swollen polymer is a simple additive density of the polymer and the swelling liquid, as in Eq. 4.4, meets the assumed criteria for the derivation of swelling capacity as a function of network structure. One drawback of this analysis, however, is the high velocity (55 km/h) used during centrifugation. This may lead to water evaporation from the gel if the centrifugation step proceeds for too long a time or is performed

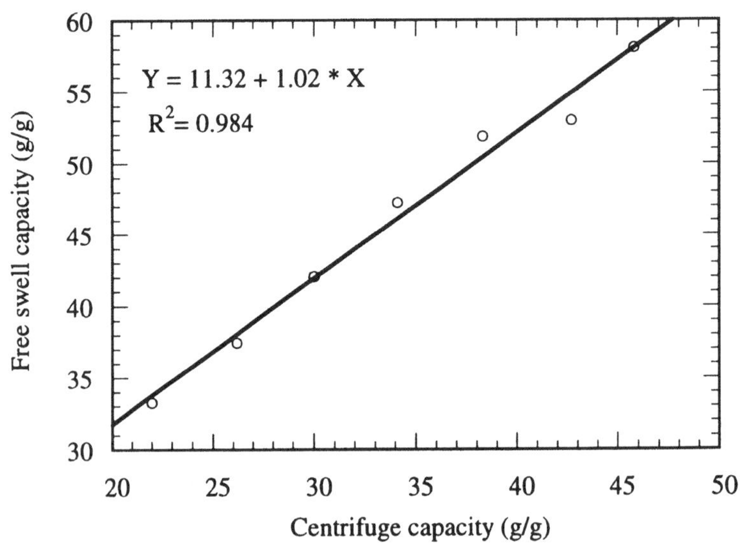

Figure 4.21. A correlation of free swell capacity to centrifuge capacity; 30 min swelling of 30–50 mesh cut.

in a very dry laboratory atmosphere. The relative precision for the analysis summarized below is about ±1.45% at the 95% confidence level.

4.6.1.2. Absorbent Capacity Method. This method is adapted from that of Byerly et al.[74] The swelling liquid, 0.9 wt% NaCl solution, is prepared from deionized water and reagent grade NaCl and poured into a stainless steel pan of about 20 × 30 cm. A glass beaker will suffice if the tea bags can be held submerged for the time of the analysis. A 2.5 × 3 inch rectangular pouch is made by folding a 5 × 3 inch piece of tea bag material and heat-sealing two edges of the formed pouch (tea bag fabric, heat-sealable, Grade 542, commercially available from Kimberly-Clark Corp., Neenah, WI). (Note: we have also successfully used commercial diaper liner fabric; however, the heat-sealing operation to make the bags is much more difficult.) The sealing may be accomplished using a hand-held iron. The tea bag is weighed empty, to at least the nearest 0.001 g, to give mass, M_1. A sample of polymer is screened to give the desired particle size cut for analysis. In many cases, a 300–600 μm (30–50 mesh) cut is desired.

A 0.200 g sample of polymer is added to the tea bag. The bag then is heat-sealed on the open edge. The bag is weighed and the mass recorded as M_2. Before placing the filled bags in the salt solution, the polymer must be evenly distributed over the 2.5 × 3 inch area in the bag. Duplicate measurements are recommended for each sample. The sealed bags are placed flat between two pieces of fiberglass screen (Teflon-coated fiberglass screens having 0.635 cm openings; Taconic Plastics Inc., Petersburg, NY), which serves to keep the bags submerged, and then put into the stainless steel pan containing the 0.9% NaCl. After wetting, the samples remain submerged for the desired time, typically 30 min.

The wet bags are then placed in the basket of the centrifuge (Clay Adams Dynac II, Model 0103, with water collection basket, digital speed gauge and drainage basket). (A description of a suitable basket is given by Lahrman et al.[80]). The tea bags are arranged to keep the centrifuge basket balanced. The centrifuge is operated to give a G-force of about 350. For the suggested centrifuge and basket, this is about 1,600 rpm. The desired rotational speed is held for 3 min and should be controlled to ±300 rpm. The bags are removed and weighed, with empty blank bags being weighed first. The mass of the wet tea bag containing superabsorbent polymer is recorded as M_3, and the mass of the wet, empty bag is recorded as M_4. The absorbent capacity, S, is calculated using Eq. 4.8:

$$S = \frac{M_3 - M_2 - (M_4 - M_1)}{M_2 - M_1} \tag{4.8}$$

The absorbent capacity, S, obtained from the centrifuge capacity method, differs from the gravimetric swelling capacity, Q, in Eq. 4.4. Q is defined as the ratio of the mass of gel to the mass of dry polymer, whereas S is the ratio of the absorbed liquid to the mass of dry polymer. The two are related by Eq. 4.9:

$$Q = 1 + S \tag{4.9}$$

4.6.2. Kinetics of Swelling

Absorption rate or swelling rate involves diffusion of a liquid into the particle. As such, it is governed by Fick's law of diffusion (Eq. 4.10),[81] where C is concentration of the diffusing species and D is the diffusion coefficient of the diffusing species in the polymer. For a spherical particle, where D is a function of C, the diffusion equation reduces to

$$\frac{\partial C}{\partial t} = \frac{1}{r}\frac{\partial}{\partial r}\left(rD\frac{\partial C}{\partial r}\right) \tag{4.10}$$

Thus, the swelling rate is a function of diffusion path length r, and the diffusion coefficient. Through the diffusion coefficient, the swelling rate is expected to depend on temperature (Chapter 5, section 5.2.4) and salt concentration (Chapter 5, section 5.2.5).

4.6.2.1. Swelling Kinetics Methods. Several methods have been used to obtain swelling kinetics for polymer networks. Simple gravimetry versus time is common.[82] Several optical methods have been described in which the dimensional change of the sample is measured by means of the attenuation of transmitted light,[83] microscopy,[84] or profile projection.[85] These latter two methods work well when a regular geometric shape is used as the sample. A viscometric technique has also been developed,[86] based on the dependence of suspension viscosity on the volume fraction of the suspended particles.

The swelling rate is often measured by determining the swelling extent as a function of time, for example, by means of the centrifuge capacity analysis (Fig. 4.20). (Gravimetric swelling ratios are measured because diffusion, the driving force for swelling, is a concentration-driven process, not merely a volume-driven process.) An absorption rate may also be obtained from data of swelling versus time using the Gravimetric Absorbency Testing apparatus or the apparatus described for the absorbency under load (AUL) measurement (Fig. 4.22). Data of absorbency versus time, at any load, may be fit to a simple exponential. The swelling may then be described as a simple process wherein the rate is first-order in the amount of capacity remaining unswollen:

$$\text{Rate of swelling} = \frac{dQ}{dt} = k(Q_{max} - Q[t]) \tag{4.11}$$

where Q_{max} and Q are the swelling capacities at equilibrium and at any time, t, under the conditions of the experiment. The first-order rate constant, k, depends on the diffusion coefficient and the particle radius. Equation 4.12 shows the inverse of the rate constant as a characteristic time, τ:

$$\tau = \frac{r^2}{\pi^2 D} \tag{4.12}$$

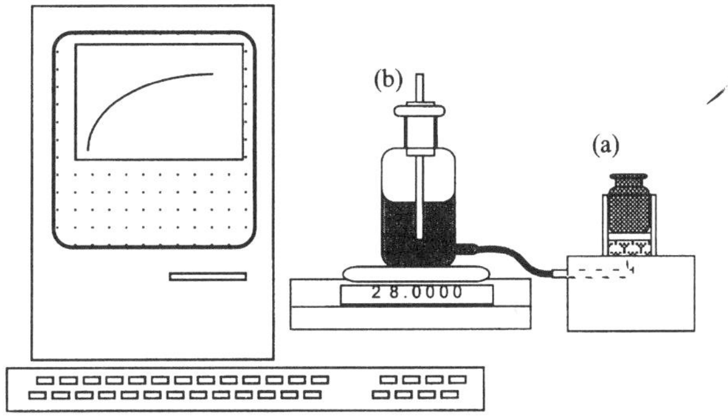

Figure 4.22. An automated device for measuring absorbency of a polymer or pad sample, with or without an applied load, as a function of time. The loss of mass of the liquid reservoir is monitored by the computer and converted by the program into a graph of swelling versus time. The apparatus may be modified for a variety of methods, including permeability.

4.6.2.2. Characteristic Swelling Time by the "Vortex Time" Method. A simple, single-point measurement of absorption kinetics, which takes advantage of the idea of a characteristic time, has been used in the patent art,[87–89] and its characteristics have been described by Buchholz.[86] This method has been referred to as the *vortex time* analysis. It is based on a rearrangement of the integrated form of Eq. 4.11 to obtain an expression for time in terms of the values of Q, Q_{max}, and the rate constant k.

$$t_v = -\frac{1}{k}\ln(1 - Q/Q_{max}) \tag{4.13}$$

If the relative swelling of the sample Q/Q_{max} is known or is held constant, then t_v is a measure of the rate constant, or characteristic time of the swelling process. The swelling time using this method is proportional to the square of the particle diameter, as predicted by kinetics based on Fickian diffusion. The dependence of swelling time on equilibrium swelling arises because the mass of liquid absorbed during the swelling rate measurement depends on the equilibrium swelling capacity of the polymer.

In the original method,[87] 2 g of polymer sample is mixed with 50 mL of the desired test fluid, which is stirred by means of a magnetic stirring bar in a small beaker. As the fluid absorption proceeds, the viscosity of the suspension increases until the stirring vortex disappears at the "sorption time," t_v. The swelling extent of the polymer at time t_v was determined by measuring the sorption times and equilibrium swelling capacities for a single polyacrylate polymer sample in several salt solutions of varying concentration up to 9% NaCl. In this way the volume swelling just at the time of disappearance of the vortex was determined to be equal to the equilibrium swelling of the sample at one particular ionic strength. The sorption time for the sample in 7% NaCl solution was 839 s, but for the 8% NaCl solution the sorption time was greater than 90 min. The swelling capacity was therefore estimated to be 13 g/g when using a 7.5% NaCl solution by interpolation of the values at 7% and 8% NaCl. The volume fraction of swollen polymer in the suspension when the vortex disappears then is estimated to be 2(13)/ 50 = 0.52. For nonpolyelectrolytes, the ratio of polymer to test fluid can be changed until the vortex just disappears at full swelling of the particular sample.

When using the above experimental ratio of 2 g polymer to 50 mL of liquid, if the equilibrium capacity (swelling capacity) of a particular polymer sample is less than 13 g/g, the vortex will never disappear, and the sorption time will appear infinite, clearly a misleading result. Therefore, the equilibrium swelling capacity of the sample must be taken into account to obtain the best results. It is simply accounted for by adjusting the amount of polymer added to the liquid, based on the value of the equilibrium swelling capacity of the sample, in the following way.

Previously, we determined that the vortex disappeared when the ratio of volume V_{abs} of liquid absorbed to initial liquid volume V_{init} equaled 0.52. The value of V_{abs} is equal to the mass of polymer m_p used in the test multiplied by its swelling extent Q_t at the instant the vortex disappears. These relationships may be combined mathematically as follows:

$$\frac{V_{abs}}{V_{init}} = \frac{m_p Q_t}{V_{init}} = 0.52 \tag{4.14}$$

The value of Q_t may be taken as any fixed fraction f_Q of the equilibrium swelling capacity Q_{max} of the sample. Choosing a value of $f_Q = 0.67$ yields the following expression:

$$Q_t = f_Q Q_{max} = 0.67 Q_{max} \tag{4.15}$$

Combining Eqs. 4.14–4.15 through the variable Q_t, and rearranging in terms of m_p, yields Eq. 4.16 by which the amount of polymer needed (per 50 mL of liquid) for the vortex test may be specified:

$$m_p = \frac{0.521 V_{init}}{0.67 Q_{max}} \tag{4.16}$$

As the equilibrium swelling capacity increases, less polymer is used in the test so that each sample reaches the same relative swelling extent at time t_v.

4.6.2.3. Vortex Time Method. The speed of absorption of a 0.9% NaCl solution by the polymer is determined at a given temperature, typically 20°–22°C. Fifty grams of solution is weighed into a 100-mL beaker. The solution is stirred on a magnetic mixer so that a significant vortex is present; 600 rpm with a 25-mm stir bar is satisfactory. Alternately, the magnetic stirrer may be turned on fast enough so that the vortex is just above the stir bar. Two grams of polymer is added to the center of the vortex, taking care to distribute the polymer sample evenly, but quickly, in the liquid. As the polymer swells, the viscosity of the slurry increases until the stirring vortex disappears. The time for the vortex to just disappear is taken as the vortex time. The ratio of polymer to solution may be adjusted, by means of the equation given above, to yield characteristic swelling times that can be related to the swelling rate constant. Without adjusting the ratio of polymer to solution, a relative swelling speed only is obtained. Typically, the granular polymer is analyzed as received. However, a particle size range may be analyzed by obtaining a narrow screen cut as for the equilibrium swelling analysis. The measured time is sensitive to temperature through the viscosity of the swelling liquid and the diffusion coefficient. The analyst's perception of the exact end-point also leads to some variability of results between different analysts.

4.6.3. Elastic Modulus

The elastic modulus of superabsorbent polyacrylate polymer is believed to be important to their performance in personal care articles,[29] especially for advanced diaper designs. In combination with swelling capacity data, modulus measurements can yield much useful information about the efficacy of a particular polymer synthesis and, in combination with end-use performance data, guide the chemist or engineer to improved materials. As with measurements of swelling capacity, the most desirable morphology on which to make a direct rheological measurement of modulus would be a monolithic piece of known geometry, typically a slab. Modulus measurements of this type have been reported on superabsorbents.[90] However, just as with swelling capacity measurements, the most industrially useful measurement techniques measure modulus on commercially available particle morphologies and at the same time yield information about the network structure and end-use performance.

A swelling pressure measurement at an arbitrary set of conditions is one simple technique that has been used to obtain modulus information about the polymer. Equation 4.17 relates the swelling pressure, Π, to the modulus, G, and any applied load, P, on the polymer:

$$\Pi = P + G \tag{4.17}$$

When the applied load is essentially zero, and the polymer concentration and the swelling liquid concentration are constant, the modulus is directly obtainable from the swelling pressure. If the applied load is nonzero, measurements of the swelling pressure at this arbitrary set of conditions will permit a set of polymers to be given a relative order according to their modulus values. This method has been used to characterize and differentiate advanced superabsorbents.[91]

In several other cases, the modulus has been measured rheometrically. Barry and Meyer[92,93] report rheological measurements on a mass of swollen crosslinked polyacrylate gel particles used as thickeners. Their error analysis suggests that the values of shear modulus obtained are reasonable for the materials analyzed. The following is a summary of a method[29] to determine shear modulus on a collection of swollen superabsorbent gel particles. It has been our experience that the relative shear modulus values obtained by this method give the correct relationship between swelling and modulus as predicted by Flory and Erman.[94]

A variety of oscillatory rheometers may be used. Instruments that have been reported in the superabsorbent patent literature include the Rheometrics Fluids Spectrometer,[73] Sangamo Visco Elastic Analyser, Bohlin Instruments Rheometer, and the Yamaden Co. Ltd. Creep Meter.[95] The rheometer comprises a circular lower plate on which the hydrogel is placed. An upper conical plate is positioned at the proper sample gap, at the point at which an untruncated cone would contact the lower plate. An oscillating torque, at a frequency of 100 Hz, is applied to the upper plate, and the resulting angular displacement of the cone is determined. The shear modulus of the hydrogel is calculated from the ratio of the applied stress to the appropriate strain. For a more complete discussion of the experimental and theoretical details of the measurement of modulus, the reader is referred to standard texts on the subject.[96]

The sample to be analyzed is screened to the desired particle size cut such that the particle size after swelling is smaller than the desired gap setting. The sample is swollen to its equilibrium swelling in the desired liquid. Synthetic urine or 0.9 wt% NaCl are typically used. Excess liquid is removed from the swollen hydrogel by blotting. If too little liquid is removed, the particles will move in the excess liquid, and the measured modulus will be too small as a result of interparticle slippage. If too much liquid is removed, the polymer will not be at its equilibrium swelling. This is a critical part of the method and may lead to discrepancies in measurements between analysts.

A useful modification to this procedure is to first measure the swelling capacity by a method that removes the interparticle water. Then, a quantity of polymer is swollen to 90%–95% of its measured equilibrium swelling capacity. No blotting is required. The measured modulus is corrected for the extent of swelling using Eq. 5.7 (see Chapter 5) to calculate the modulus at equilibrium swelling.

In either case, approximately 1.5 mL of the swollen hydrogel is placed on the bottom plate, and the upper conical plate is lowered to the desired gap. Excess hydrogel is carefully trimmed from around the edges of the plate and cone. Stress and strain measurements are taken. The temperature of the rheometer is typically held constant at about room temperature or 21.2°C. The strain amplitude is small, less than about 2.5%. Measurements can be taken at one strain or at a range of strains and extrapolated to the limit of zero strain.

4.6.4. Combining the Concepts: Swelling Under Load

4.6.4.1. Swelling Under an Applied Compressive Load. Superabsorbents in personal care articles must resist deformation and deswelling under an external load because this ability has been correlated to improved performance in pad analysis[91] and in diapers.[89,97,98] When a uniaxial compression is applied to the swelling polymer particle, it is forced to change shape. The magnitude of the deformation depends on the modulus, which in turn depends on the initial crosslink density and extent of swelling. The change in shape will tend to close off channels in the gel mass that allow the convective flow of free liquids. Thus, the shutting off of liquid flow by compression of the gel mass depends on the elastic modulus and the porosity of the sample bed and on variables that affect them. Areal loading of the polymer in the test device and the particle size distribution in the sample are two of these variables. (This effect is discussed more fully in Chapter 5, section 5.3.5.5.) Therefore, for two reasons, more rigid particles of gel maintain their swelling capacity better under load than do the same size distribution of particles of a softer, smaller modulus, more deformable gel.

Water-absorbent crosslinked polymers have been analyzed for swelling capacity and wicking action (the movement of fluid through a bed of polymer particles) under an applied load since at least 1977.[99] The ability to control the liquid pressure head of the swelling liquid was primitive, and the areal loading was not clearly defined in this early method. (A 1-g sample is placed on a fritted-glass funnel of undefined dimensions.) However, the intent to swell polymer samples under loads ranging from 0.68 to 2.8 kPa (0.1–0.4 lb/in^2) is clearly stated. By 1985, measurements of swelling under load for fluff–polymer composites or polymer alone were reported.[100,101] The applied load in this analysis simulates the effect on the polymer of applied load during use. A 0.1-g sample of polymer is spread evenly over a circle of 4.88 cm diameter. This is a polymer loading of about 50 g/m^2 and is said to represent the loading of superabsorbent polymers in products of that time.

Turk and Werner[101] (as discussed by Painter[100]) state that "all superabsorbents absorb less fluid when the quantity of superabsorbent is increased on the same area." (Nagorski[91] shows this effect for two types of polymers.) Therefore, to predict performance of the superabsorbent in the diaper, the analysis should mimic the superabsorbent area density in the polymer–fluff composite product. Because an average of about 6 g of polymer was used in baby diapers with about 60 cm^2 of area, a polymer area density about twice that reported previously,[102] an analysis was developed that confined the polymer to the analysis cell and used a higher polymer loading.[98] The details of this procedure are discussed in section 4.6.5.2. This procedure and slight modifications of it have become an industry standard.

The earliest descriptions of methods of AUL indicate the use of the analysis for measuring absorption rate.[91,99,100] Considerations of polymer area density and particle size distribution affect the determination of swelling rate, just as they do the determination of equilibrium swelling under load. These variables affect the rate at which the bed porosity and permeability change, (see Chapter 5, sections 5.3.5.1 and 5.3.5.2). This prediction of the rate of swelling under pressure qualitatively, but not quantitatively, matches the absorption rate under pressure determined by these methods. The slowing of diffusion in gels of different shapes and, therefore, the decrease in the rate constant of swelling under a compressive load are discussed in Chapter 5, section 5.3.3, and in the following sections.

4.6.4.2. Measurement of Swelling Under a Compressive Load: Absorbency Under Load. Both the equilibrium swelling under a compressive load,

Q_p, and the rate of that swelling may be determined using the same analysis apparatus. A demand absorbency apparatus described by Chatterjee and Nguyen[3] (available from M/K Systems, Danvers, MA) as well as the automated data acquisition system shown in Figure 4.22 may be used for this procedure.

A defined quantity of superabsorbent polymer is weighed into a cylindrical cell. The desired load, typically up to about 7 kPa, is applied, and the polymer is allowed to imbibe liquid, at zero liquid pressure head relative to the bottom of the gel layer, through the screen on the bottom of the cell. A 2.0-kPa load has been equated to the compressive load applied by an average diaper-wearing child while the child is sitting.[73] The results are reported in grams of liquid absorbed per gram of polymer. The relative precision depends on the nature of the polymer and the applied load. A precision of about 1 g/g at a 2.0-kPa load or 1.5 g/g at a 3.9-kPa load is acceptable for replicates. At small compressions, the procedure tends to leave interstitial liquid so that measured swelling values can be higher than the equilibrium swelling. An excellent description of this method is available elsewhere.[89] The following is an abbreviated procedure.

The test apparatus consists of a plastic cylinder, 2.54 cm inside diameter, with 100-mesh wire cloth (150-μm openings) fastened across the bottom of the cylinder, a liquid reservoir, and a cell holder equipped with a porous plate. The liquid reservoir is filled with the appropriate liquid, typically 0.9 wt% NaCl. The apparatus is adjusted so that the height of the liquid menisci in the reservoir and porous plate are equalized. The swelling liquid just produces concave droplets with respect to the base height of the porous plate. A piece of filter paper (GF/A glass filter paper circles of 2.4 cm diameter) is placed on the porous plate to ensure good contact of the liquid with the polymer while minimizing evaporative losses.

A sample of polymer is screened to the desired particle size, typically 30–50 mesh. A 0.160 ± 0.005–g sample of polymer is loaded into the plastic cell so that it is uniformly distributed over the screen. Failure to achieve uniform distribution can reduce the measured swelling value. Also, different sample sizes (areal polymer distribution; see Chapter 5, section 5.3.5.5) can yield different measured values.[91] The piston (4.4 g polycarbonate machined to fit the cell) and the appropriate mass (typically 100, 200, and 317 g for 2.0-, 3.9-, and 6.2-kPa loads, respectively) are placed on top of the piston. The grouping of cell and mass is placed on the porous plate, which is covered with the filter paper. The mass loss of the reservoir or the mass gain of the sample group is recorded to determine the swelling as a function of time and applied pressure. The swelling capacity under the applied load, Q_p, is reported as the mass of liquid gained per mass of polymer. When the balance on which the reservoir is placed is connected to an automated data recording device, as is shown in Figure 4.22, a rate constant, in addition to the equilibrium swelling, may be obtained from the absorbency versus time data.

4.6.4.3. Permeability of Masses of Swollen Gels. Gel permeability is a property of the mass of particles as a whole and is related to particle size distribution, particle shape, the connectedness of the open pores, and compressibility or modulus of the swollen gel. These factors are accommodated by considering the swelling mass of gel as a viscoelastic medium undergoing a creep process. The application of these models to deformable superabsorbent gels is discussed in Chapter 5, section 5.3.4.

The permeability, κ, of the gel layer may be calculated from the measurements using a steady-state flow method and Darcy's law:

$$\kappa = \frac{qL\mu}{A\rho gh} \tag{4.18}$$

where q is the volumetric flow rate, L is the thickness of the gel layer, μ is the viscosity of the liquid, A is the cross-sectional area of the gel layer, ρ is the density of the liquid, g is the acceleration due to gravity, and h is the average height of the liquid head above the gel. For a given cell geometry, the permeability is given by a cell or analysis constant times the ratio of the gel layer thickness to the measured flow time.

Permeability of a layer of swollen superabsorbent has been related to diaper performance using a simple permeability measurement[103] and has been used to characterize prospective new superabsorbent polymers.[104] A modified version of these methods, used for the measurements in Table 5.8 (Chapter 5), is described below.

Using a modified version of the apparatus of Figure 4.22, a sample of polymer is pre-swelled under load, and then the flow rate of saline solution through the compressed mass of gel particles is measured. A 15 cm tall, 2.5 cm diameter cell was used for these measurements. The dry polymer was loaded onto the stainless steel screen of the cell. Then a fritted-glass disk (coarse grade, 3 mm thick) was added on top of the polymer, followed by a hollow, machined-brass mass that supplied the desired load. The hollow mass was necessary to eliminate capillary flow of liquid between the walls of the cell and the brass mass: Flow was restricted before the masses were modified. The polymer sample was allowed to absorb liquid under load for 60 min, using the cell unit and reservoir (absorption rate can be measured on the same sample before permeability is measured). The cell was removed from the base unit and suspended in air by clamping it on a ring stand. Saline solution was carefully poured into the top of the long cell and the time for a given volume of liquid to pass through the column was measured. A plastic ruler was attached to the side of the cell to ensure that the same liquid height was achieved for each sample. Time was recorded for a drop of liquid head in the cylindrical cell of 1 cm. The height of the layer of gel at the bottom of the cell was also measured.

REFERENCES

1. *EDANA Recommended Test Methods for Superabsorbent Materials,* European Disposable and Non-Wovens Assoc., Brussels, Belgium (February, 1996).
2. K.S. Kazanskii, S.A. Dubrovskii, *Adv. Polym. Sci.,* **104,** 97–133 (1995).
3. P.K. Chatterjee, H.V. Nguyen, In *Absorbency*, P.K. Chatterjee, ed., Elsevier Science Publishers, Amsterdam Ch. 2, pp. 29–84 (1985).

4. J&W Scientific Catalog and Technical Reference, 1996–97.
5. S.S. Cutié, G.J. Kallos, P.B. Smith, *J. Chromatogr.*, **408,** 349 (1987).
6. F.M. Wampler, *Formation of Diacrylic Acid During Acrylic Acid Storage,* AIChE Presentation, Houston (March 1987); F.M. Wampler, *Plant/Operations Prog.*, 7, 183–189 (1988).
7. H.W. Thompson, P. Torkington, *Trans. Faraday Soc.* **41,** 246 (1945).
8. S.S. Cutié, D.E. Henton, C. Powell, R.E. Reim, P.B. Smith, T.L. Staples, *J. Appl. Polym. Sci.*, **64,** 577–589 (1997).
9. R.H. Hall, F.P.B. Van der Maeden, A.C.C.M. Willemsen, *Spec. Chem.,* **7,** 56–64 (1987).
10. S.S. Cutié, P.B. Smith, D.E. Henton, T.L. Staples, C. Powell, *J. Polym. Sci., Part B, Polym. Phys.* (in press).
11. D.J. Arriola, S.S. Cutié, D.E. Henton, P.B. Smith, C. Powell, *J. Appl. Polym. Sci.*, **63,** 439–451 (1997).
12. W.W. Yau, J.J. Kirkland, D.D. Bly, *Modern Size-Exclusion Liquid Chromatography,* John Wiley and Sons, New York, pp 97–122, 242, 397–398 (1979).
13. C. Abad, L. Braco, *Br. Polym. J.*, **19,** 489 (1987).
14. C. Abad, L. Braco, *Br. Polym. J.*, **19,** 501 (1987).
15. V. Soria et al., *Br. Polym. J.*, **20,** 115 (1988).
16. E. Perez-Paya, L. Braco, *J. Chromatogr.*, **461,** 229 (1989).
17. V. Soria et al., *J. Liquid Chromatogr.*, **13**, 1785 (1990).
18. E. Perez-Paya, L. Braco, C. Abad., *J. Chromatogr.*, **548,** 93 (1991).
19. M. Potschka, *J. Chromatogr.*, **441,** 239 (1988).
20. S. Kadokura, T. Miyamoto, H. Inagaki, *Polym. J.*, **14,** 993 (1982).
21. H.G. Barth, *J. Chromatogr. Sci.,* **18,** 409–429 (1980).
22. R.J. Bruessau, *Macromol. Chem. Macromol. Symp.* **61,** 199–218 (1992).
23. S.S. Cutié, S.J. Martin, *J. Appl. Polym. Sci.*, **55,** 605–609 (1995).
24. Y. Kato, T. Matsuda, T. Hashimoto, *J. Chromatogr.*, **332,** 39 (1985).
25. H. Sasaki et al., *Sci. Rep. Toyo Soda*, **29** (1985).
26. M. Zrinyi, F. Horkay, *J. Polym. Sci. Polym Phys. Ed.,* **20,** 815–823 (1982).
27. S.S. Cutié, W.C. Buzanowski, J.A Berdasco, *J. Chromatogr.*, **513,** 93–105 (1990).
28. W.C. Buzanowski et al., *J. Chromatogr.* A, **677,** 355 (1994).
29. K.E. Brandt, S.A. Goldman, T.A. Inglin, *U.S. Patent Re. 32,649* (1988).
30. S.S. Cutié et al., *Anal. Chim. Acta,* **260,** 13–17 (1992).
31. A. Vairavamurthy, M.O. Andreae, J.M. Brooks, *Anal. Chem.*, **58,** 2684 (1986).
32. L. Brown, *Analyst (London)*, **104,** 1165 (1979).
33. M. Richards, *J. Chromatogr.*, **115,** 259 (1975).
34. J.M. Neyer et al., *Chromatographia,* **25,** (1988).
35. A.R. Blanchette et al., *Liquid Chromatographic Methods for Determining Extractable Monomeric Acrylate in Polyacrylate Absorbents,* Analytical subcommittee Institute for Polyacrylate Absorbents, Washington D.C., (October 20, 1987).
36. A. Neppel et al., *Polymer,* **29,** 1338–1342 (1988).
37. P.J. O'Connor et al., *Macromolecules,* **29,** 7872–7884 (1996).

38. J.H. Finley, *Anal. Chem.*, **33,** 1925–1927 (1961).
39. C.E.H. Bawn, S.F., Mellish, *Trans. Faraday Soc.*, **47,** 1216 (1951).
40. *Reagent Chemicals,* 8th Ed., *American Chemical Society Specifications*, American Chemical Society, Washington D. C. (1993).
41. J.D. McLean, R.E. Reim, *Am. Lab.,* March, 122–130 (1983).
42. V.T. Turkelson, R.P. Himes, *Determination of Inorganic Anions by Ion Chromatography with Ultraviolet Detection,* 24th Rocky Mountain Conference, Denver, CO (1982).
43. D.E. Henton, C. Powell, R.E. Reim, *J. Appl. Polym. Sci.*, **64,** 591–600 (1997).
44. A.D. Theoharides et al., *LCGC,* **7,** 925 (1992).
45. I.M. Kolthoff, I.K. Miller, *J. Am. Chem. Soc.*, **73,** 3055 (1951).
46. E.J. Behrman, J.O. Edwards, *Rev. Inorganic Chem.*, **2,** 179 (1980).
47. C.E.M. Morris, A.G. Parts, *Die Makromol. Chem.*, **119,** 212 (1968).
48. J.H. Siddal, T.C. Johnson, *U.S. Patent 4,833,222* (1989).
49. E.A. Turi, *Thermal Characterization of Polymeric Materials*, Academic Press, New York (1981).
50. T. Yokoyama, K. Hiraoka. *Polym. Prepr.,* **20,** 511–513 (1979).
51. T. Grewer, *Thermal Hazards of Chemical Reactions, Industrial Safety Series,* Vol. 4, Elsevier Science Publishers, Amsterdam, (1994).
52. R.F. Boyer, "Transitions and Relaxations" in N.M. Bikales, ed., *Encyclopedia of Polymer Science and Technology Suppl. Vol. 2,* Wiley-Interscience, NY, p. 753 (1977).
53. A.R. Greenberg, R. P. Kusy, *J. Appl. Polym. Sci.,* **25,** 1785–1788 (1980).
54. K. Ueberreiter, G. Kanig, *J. Chem. Phys.*, **18,** 399 (1950).
55. T. G. Fox, S. Loshek, *J. Polym. Sci.,* **15,** 371 (1955).
56. F. Rietsch, D. Daveloose, D. Froelich, *Polymer,* **17,** 859 (1976).
57. TAPPI, T 412 om-90, In *Moisture in Paper and Paperboard*, Technical Association of the Pulp and Paper Industry, Tappi Press, Atlanta GA. (1990).
58. TAPPI, T 550 om-90, In *Determination of Equilibrium Moisture in Paper and Paperboard for Chemical Analysis*, Technical Association of the Pulp and Paper Industry, Tappi Press, Atlanta GA. (1990).
59. J. Lleras, S. Combet, *J. Chim. Phys. Physiochem. Biol.,* **69,** 1620 (1972).
60. J.W. Nicholson, A.D. Wilson, *Br. Polym. J.,* **19,** 67 (1987).
61. A. Frank, K. Weunscher, Int. Congr. Automat. Instrum. Pap., Rubber Plast. Ind. Pap., Antwerp, October 1966; *Chem. Abst.,* **69,** 19737 (1968).
62. *Particle Size (Sieve Analysis) of Plastic Materials*, ASTM D1921-89, American Society of Testing and Materials, West Conshohocken, PA, (1996).
63. B.H. Kaye, In *Direct Characterization of Fine Particles,* P.J. Elving, J.D. Winefordner, eds., John Wiley & Sons, New York, p. 69 (1981).
64. T. Allen, In *Particle Size Measurement,* B. Scarlett, ed., Chapman and Hall, London, p. 112 (1990).
65. M.H. Schwartz, R.P. Andres, *J. Aerosol Sci.,* **7,** 4 (1976)
66. *Apparent Density of Free Flowing Metal Powders*, ASTM B212-89, American Society for Testing and Materials, West Conshohocken, PA (1996).
67. *Flow Rate of Metal Powders,* ASTM B213-96, American Society for Testing and Materials, West Conshohocken, PA (1996).

68. J.E. Mark, B.Erman, *Rubberlike Elasticity, A Molecular Primer,* John Wiley & Sons, New York (1988).
69. Y-L. Yin, *Swelling Behavior of Polyelectrolyte Gels,* Ph.D. Thesis, Chapter 6, Princeton, NJ, Princeton University (June 1993).
70. M. Ilavsky, *Macromolecules,* **15,** 782–788 (1982).
71. J. Ricka, T. Tanaka, *Macromolecules,* **17,** 2916–2921 (1984).
72. R. Mertens, K. Dahmen, H. Brehm, *Eur. Patent Appl. 527,867* (1991); K. Dahmen, R. Mertens, H. Brehm, *U.S. Patent 5,455,284* (1995).
73. W.G.-J. Chiang, J. L. McCrickard, *Eur. Patent Appl. 582,062* (1994).
74. S. K. Byerly et al. *Eur. Patent Appl. 532,002* (1993).
75. T. Tsubakimoto et al., *U.S. Patent 4,286,082* (1981).
76. T. Tsubakimoto et al., *U.S. Patent 4,734,478* (1988).
77. T. Irie et al., *U.S. Patent 4,920,202* (1990).
78. Y. Irie, T. Hatsuda, K. Yonemura, *U.S. Patent 5,275,773* (1994).
79. F. Masuda, K. Nishida, A. Nakamura, *U.S. Patent 4,076,663* (1978).
80. F.H. Lahrman, C.J. Berg, D.C. Roe, *U.S. Patent 5,149,334* (1992).
81. J. Crank, *The Mathematics of Diffusion,* 2nd ed., Oxford Science Publishers, New York, pp 1–10 (1993).
82. E. Southern, A.G. Thomas, *Trans. Faraday Soc.,* **63,** 1913 (1967).
83. D.J. Buckley, M. Berger, D. Poller, *J. Polym. Sci.,* **56,** 163 (1962).
84. T. Tanaka, D.J. Fillmore, *J. Chem. Phys.,* **70,** 1214 (1979).
85. A. Peters, S.J. Candau, *Macromolecules,* **21,** 2278 (1988).
86. F.L. Buchholz, in *Absorbent Polymer Technology,* L. Brannon-Peppas, R. Harland, eds., Elsevier Science Publishers, New York, (1990), pp. 233–247.
87. M. Makita, S. Tanioku, *U.S. Patent 4,587,308* (1986).
88. P. Mallo, M. Moreau, J. Cabestany, *U.S. Patent 5,053,460* (1991).
89. M.K. Melius et al., *Eur. Patent Appl. 615,736* (1994).
90. Y.-L. Yin, *Swelling Behavior of Polyelectrolyte Gels,* Ph.D. Thesis, Chapter 7, Princeton, NJ, Princeton University (June 1993).
91. H. Nagorski, in *Superabsorbent Polymers: Science and Technology,* ACS Symposium Series 573, F.L. Buchholz, N.A. Peppas, eds., ACS Books, Washington D.C. (1994).
92. B.W. Barry, M.C. Meyer, *Int. J. Pharmaceutics,* **2,** 1–25 (1979).
93. B.W. Barry, M.C. Meyer, *Int. J. Pharmaceutics,* **2,** 27–40 (1979).
94. P.J. Flory, B. Erman, *Macromolecules,* **15,** 800–806 (1982); P.J. Flory, B. Erman, *Macromolecules,* **15,** 806–811 (1982).
95. D.R. Chambers et al., *Eur. Patent Appl. 530,438* (1993).
96. J.D. Ferry, *Viscoelastic Properties of Polymers,* 2nd ed., John Wiley & Sons, New York (1970).
97. F. Masuda, *Proc. Products and Technologies of Air-Formed Webs Conference*, Pira Fibramerics Programme, Leatherhead, UK (December, 1987).
98. S.R. Kellenberger, *Eur. Patent Appl. 339,461* (1989).
99. A.R. Reid, *U.S. Patent 4,051,086* (1977).
100. E.V. Painter, *Tappi J.,* December, pp. 54–57 (1985).

101. W. Turk, G. Werner, *Proc. Absorbent Prod. Conf., Sec. IV,* M/TS, Kalamazoo, MI (1982).

102. R.J. Koproski, *Proc. Products and Technologies of Air-Formed Webs Conference*, Pira Fibramerics Programme, Leatherhead, UK (December, 1987).

103. M. Plischke, M. Schmidt, C.P. Bewick-Sonntag, *Eur. Patent Appl. 640,330* (1995).

104. K. Kajikawa et al., *U.S. Patent 5,478,879* (1995).

5

THE STRUCTURE AND PROPERTIES OF SUPERABSORBENT POLYACRYLATES

Fredric L. Buchholz

5.1. NETWORK STRUCTURE AND EQUILIBRIUM SWELLING PROPERTIES

5.1.1. Basic Network Theory

Polyacrylate superabsorbents are used principally to absorb liquids in personal care sanitary products. The polymer properties of principal commercial importance are the amount of liquid that can be absorbed, the amount of liquid retained in the gel under various compressions, and the rate at which the liquid is absorbed. The challenge for industrial chemists and engineers is to continually produce new materials with improved properties. To meet this challenge, it is necessary to have a thorough understanding of the polymer and its applications. Only through this understanding of the structure and property relationships is continuous progress assured. Unfortunately, the precise structure of these polyelectrolyte gels is unknown. They have been studied and modeled for over 50 years, and yet no mathematical description can model and predict all the intricacies of these materials. All is not hopeless, however, because much of the study has resulted in some very good partial descriptions of the structure and properties of these highly swellable, soft, and rubbery substances.

Modern Superabsorbent Polymer Technology. Edited by Fredric L. Buchholz and Andrew T. Graham.
ISBN 0-471-19411-5.

Many models of swellable polymer networks have emerged, and a challenge is to choose model(s) of network structure and behavior that allow progress to be made in synthesizing materials that are more useful than preceding materials. Two types of models have emerged from the decades of study: the molecular models of rubber elasticity with ionic contributions added[1–4] and the scaling models of networks.[5–7] Many variants of these basic models have evolved as theory and experimental observations advance in lock-step. As chemists and engineers involved with the synthesis and manufacture of superabsorbent polymers, we find molecular models most useful in helping to solve product development problems because they most directly connect the synthetic conditions to the final properties. Nevertheless, the advances brought to the field by the scaling theories are very important, and they are used often to solve processing problems of a simplified nature. It is well beyond the scope of this book to develop new models of the structure and properties of polyelectrolyte gels, and the reader is directed to numerous other sources for the basic derivations of the models.[8] However we can show how the use of these models can help in the development of new and improved superabsorbent materials. Even though the models are not perfect, they are quite useful in showing how variations in the major structural features can affect the principal properties. The basic Flory–Rehner model, and its modern incarnation the Flory–Erman model, especially as modified to account for ionic effects, are frequently used to describe the behavior of superabsorbent polymers.

5.1.1.1. Network Free Energies. The basic model of ionic networks as described by Hooper et al.[4] incorporates the principles of the Donnan membrane equilibrium into the network theory of Flory and Erman[9] and relates the structure of nonionic networks to material quantities such as elastic modulus over a wide range of conditions. The Flory–Erman theory includes effects of polymer chain entanglements in the swelling and elasticity equation. The inclusion of the Donnan membrane equilibrium allows for a better understanding of the effects of the macro-ions of the polyacrylate network and how they interact with the ions of simple salts that are present in swelling liquids such as urine. Sophisticated modeling[10] is increasingly precise but more complex to use as well, and simplifications are frequently employed.

The principal structural factors contributing to the properties of superabsorbent polymers are the crosslink density and the charge density. Numerous structural features of the polymer system and the magnitude of their interaction, however, govern how the crosslink density and charge density affect the macroscopic properties. Key parameters that are involved in how the crosslink density and charge density interact to yield macroscopic properties are

1. The molecular weight of the polymer chain between crosslinks, M_c
2. The molecular weight of polymer chain if no crosslinks were present, M_n
3. The amount of uncrosslinked polymer present (the extractable fraction of polymer)
4. The extent of entangling of the chains

5. The volume of the polymer system during preparation
6. The spacing of the charges along the chains
7. The thermodynamic interaction of the solvent with the polymer chains
8. The number of polymer strands that meet at a crosslink point, f

The nonrandom aspects of structure related to the method of synthesis are also important.[11] For example, the reactivity ratios of the monomers affect both how and when the crosslinker is incorporated into the network during preparation and how much soluble and extractable polymer is present in the network.

Network behavior may be described by manipulating equations that describe the thermodynamic interaction of the free energies of the components of the network and how they change over time. The vast majority of models assume that the free energy of the network can be taken as a sum of free energies of (1) the mixing of solvent with the polymer segments, (2) the mixing and interaction of the ions of the network and the swelling liquid, and (3) the entropic changes of the connected network chains as their conformations change with swelling or deformation. Of course, an equilibrium state will be attained when the sum of the free energy changes is zero.

Because the polymer networks are used primarily to absorb liquids and salt solutions, the concentration of the polymer and other components in the gel is of primary interest. The free energies of the polymer network may be described in terms of the chemical potentials μ_i of the system, and the chemical potentials may be described in terms of osmotic pressures according to the following definitions:

$$dG_{T,P} \sum_i \left(\frac{\partial G}{\partial n_i} \right) = \sum_i \mu_i \, dn_i \tag{5.1}$$

where $dG_{T,P}$ is the change in Gibb's free energy at constant temperature and pressure, and μ_i is the chemical potential of component i in the system, which is composed of n_i moles of each component. The osmotic pressure π is given by

$$\pi = -(\mu_i - \mu_i^0)/V_i \tag{5.2}$$

where μ_i^0 is the chemical potential of component i in its reference state and V_i is the molar volume of component i. In terms of osmotic pressures, the additive contributions of the free energies can be written as

$$\pi = \pi_{mix} + \pi_{net} + \pi_{ion} \tag{5.3}$$

where the contributions of mixing, network conformation, and ions may be given additionally in terms of the structural features of the network, according to the particular theory.

The Flory–Erman description of how these factors affect the swelling and modulus, with the ionic contribution added according to Hooper et al.,[4] is given in Eq. 5.4, which describes polymers swollen to their maximum with a solvent:

$$\pi = -\frac{RT}{V_1}[\ln(1-\upsilon_2)+\upsilon_2+\chi\upsilon_2^2] - RT\upsilon_2^{1/3}\left(\frac{\xi}{V_0}\right)[1+(\mu_e/\xi)K(\upsilon_2)] + RT\sum_j (c_j^{gel}-c_j^{ext}) \tag{5.4}$$

The three principal contributions are shown in the same order as in the general equation for the total osmotic pressure (Eq. 5.3). The first contribution, due to mixing of the polymer segments with solvent molecules, is taken as the Flory–Huggins description. The second contribution, due to the elastic response of the network, is taken as the Flory–Erman description. The third contribution is taken as the ideal Donnan description of ionic effects in polyelectrolyte systems. The mixing and ionic contributions tend to increase the solvent content, while the elastic contribution tends to decrease the solvent content. Once again, equilibrium is achieved when the sum of the contributions equals zero.

In Eq. 5.4, υ_2 is the volume fraction of polymer in the gel at equilibrium swelling and is commonly taken as the inverse of the equilibrium swelling capacity. The polymer–solvent interaction parameter is χ, the molar volume of solvent is V_1, the number (moles) of crosslinks is μ, the cycle rank of the network is ξ, and $K(\upsilon_2)$ is a complex function of the entanglements or constraints on junctions.[12] The measure of the magnitude of the constraints is given by an entanglement parameter, κ, which is used to calculate the value of the function $K(\upsilon_2)$. The ionic contribution contains two terms; c_j^{gel} is the concentration of all mobile ions in the gel phase, and c_j^{ext} is the concentration of all mobile ions in the external liquid phase, which is in equilibrium with the gel. The ions in the gel phase include the counterions of the polyelectrolyte and any ions that permeate into the gel from the external liquid. The number of ions that permeate into the gel is controlled by the polyelectrolyte concentration in the gel phase (the Donnan equilibrium). The partitioning of ions into the gel contributes to the swelling by increasing the value of c_j^{gel} in Eq. 5.4, but also is controlled by the crosslink density and the charge density (degree of neutralization) of the gel.

In the Flory–Erman treatment, polymer networks are characterized by their "cycle rank," ξ. The cycle rank incorporates the crosslink density and all other structural features of the crosslinked polymer (except for ionic effects). That is convenient mathematically, but would be synthetically more useful if broken out into conventional structural parameters such as M_c and M_n. For a random network such as is obtained approximately in co-polymerizations of vinyl and divinyl monomers, with crosslink functionality of four, the cycle rank density, ξ/V_0, is

$$\frac{\xi}{V_0} = \frac{\rho N_a}{2M_c}\left(1 - \frac{3M_c}{M_n}\right) \tag{5.5}$$

where ρ is the density of the polymer as prepared, V_0 is the volume of the sample at preparation, and N_a is Avogadro's number.

5.1.1.2. *Crosslink Density.* The number density of effective crosslinks, μ_e/V_0, is

$$\frac{\mu_e}{V_0} = \frac{\rho N_a}{2M_c}\left(1 - \frac{M_c}{M_n}\right) \tag{5.6}$$

and the shear modulus of a swollen gel, G, is[13]

$$G = \xi RT \backslash N_a V_0 (\upsilon_2/\upsilon_{20})^{1/3} \tag{5.7}$$

where R is the gas constant, T is absolute temperature, υ_{20} is the volume fraction of polymer during network formation, and υ_2 is the volume fraction of polymer at which the measurement of shear modulus is made. Equations 5.4–5.7 provide the fundamental relationships between swelling capacity, modulus, and the crosslink density for polymer networks.

5.1.1.3. *Modulus of Elasticity.* When an elastic material is subjected to mechanical stress, it deforms or strains. In the most elementary form, the modulus of the material is the ratio of the applied stress to the resulting strain. The elastic modulus refers to the recoverable strain, or the strain that is recovered when the stress is removed. For crosslinked rubbery polymers such as swollen superabsorbent polymers, the magnitude of the elastic shear modulus depends on the molecular structure. Of course the structure is controlled by exactly how the material is made.

The molecular weight of polymer between crosslinks M_c in superabsorbent polymers made by a crosslinking co-polymerization depends on the ratio of the concentration of crosslinker to the concentration of monomer (acrylic acid). The value of M_c may be approximated by the molecular weight of the extractable polymer fraction, because most of the soluble polymer is made in the polymerization when the molecular weight of the forming chains drops below the molecular weight of chain between crosslinks, as estimated from the stoichiometric crosslink density. This occurs near the completion of the polymerization reaction when the monomer concentration is low, and, as a result, the kinetic chain length is also low. When combined with the molecular weight of the backbone polymer chain (if all the crosslinks are somehow cut), the mole fraction of "dangling chains" is accounted for in Eq. 5.5 by a factor of $2M_c/M_n$ inside the parentheses. The dependence of the modulus on crosslinker concentration is shown in Figure 5.1.

Expressing the shear modulus in terms of polymerization parameters, by inserting the expression for ξ/V_0 obtained from Eq. 5.5 into Eq. 5.7, yields

$$G = \frac{\rho_0 RT}{2M_c}\left(1 - \frac{3M_c}{M_n}\right)\upsilon_{20}^{2/3}\upsilon_2^{1/3} \tag{5.8}$$

The swollen modulus therefore depends on the molecular weight between crosslinks, M_c, the backbone molecular weight, M_n, the concentration during

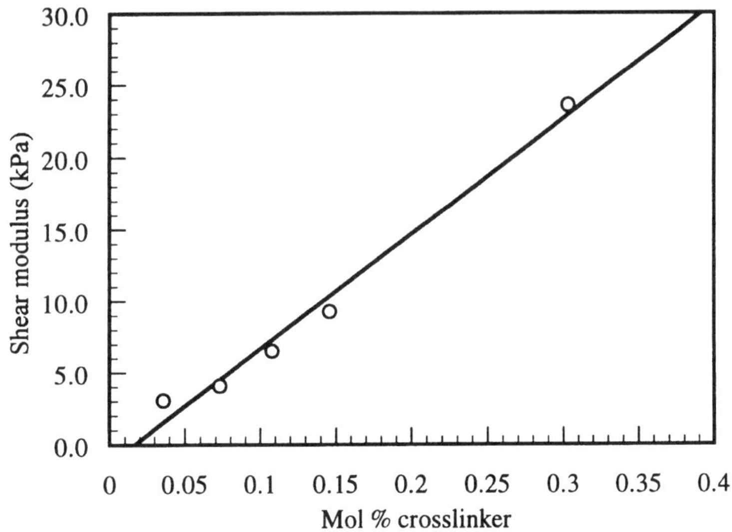

Figure 5.1. The shear modulus of sodium polyacrylate gels as a function of the stoichiometric ratio of crosslinker used in the polymerization, per mole of acrylic acid. The moduli have been normalized to the preparation volume according to Eq. 5.7.

crosslinking, $\upsilon_{2}0$, and the extent of swelling during measurement as expressed through υ_2. A slightly different form of this equation, and also of Eq. 5.10, is given in some of the original literature that describes the properties of rubber. These different forms result from the assumption that the networks are formed in the absence of any solvent, in which case $\upsilon_{2}0 = 1$, and V_0 contains no contribution from the volume of solvent.

The molecular weight between crosslinks and the molecular weight of the polymer backbone together define the amount of "dangling chains" in the network. The fraction of dangling chains, which are network chains having one end attached at a crosslink point and the other unattached to the network, is accounted for in Eq. 5.5 by a factor of $2M_c/M_n$ in the term $(1 - 3M_c/M_n)$. The number of elastic chains meeting at a crosslink that also is connected to a dangling chain is one fewer than if all the chains could support stress. This leads to a lower modulus and is accounted for by the extra factor of M_c/M_n in Eq. 5.5. Even though the dangling chains cannot support stress and therefore cannot contribute to the modulus, they do of course contribute to the swelling of the polymer by solvent (controlled by the thermodynamic solvation of the polymer units and the ions). Branching reactions during synthesis are an additional source of dangling chains that is not accounted for in Eq. 5.8. These reactions occur by radical chain-transfer to polymer coupled with chain termination to a radical not connected to the network.

The polymer concentration during crosslinking $\upsilon_{2}0$ defines the extent of entanglement of the network chains and therefore the number of configurations available to a chain in space. The extent of entanglement is determined by the distance between chains when the crosslinks are formed. The effect of entanglements on properties depends on the ratio of the distance between covalent crosslinks to the distance between entanglements. The concentration of monomer approximately fixes the concentration of chains at crosslinking, although strictly speaking the concentration of elastic chains continuously increases as a free-radical crosslinking co-polymerization proceeds. For example, suppose that a certain polymer at some concentration has an entanglement every 300 monomer units. The entanglements will have less effect on properties if the polymer is crosslinked every 250 monomer units than if the crosslinks are every 500 units.

The straightforward effects of concentration, as suggested by theory, are complicated in crosslinking co-polymerizations because many things change in concert with a change in monomer concentration. The solubility of crosslinkers may increase or decrease as the concentration of undissociated acrylic acid increases in the monomer solution, thereby changing the effective concentration of crosslinker. In addition, the molecular weight of the backbone polymer should increase, in general, as the monomer concentration that is loaded initially into the reactor increases, but decrease as the monomer is depleted during polymerization. Also, free-radical degradation of the polymer chains in the presence of sodium persulfate initiator occurs after the main polymerization is complete, when the polymer gel is still at elevated temperature. For chains attached to the network at only one place (dangling chains), only a single chain scission is required to cleave it from the network and form a soluble chain. As a result, the amount of extractable polymer increases as reaction time is extended. But for each extractable polymer chain formed, there must be many new dangling chains formed. Because only a small fraction of the network chains are initially dangling chains (most are elastic) and because the degradation process is random, the chains that will be most affected are those chains in largest number. The cleavage of chains during this side reaction affects the magnitude of the entanglement effect and hence the modulus. The general effects of concentration on the elastic shear modulus in the fully swollen state are shown in Table 5.1. In these polymers, the amount of crosslinker added to the polymerization, the amount of soluble polymer, and the equilibrium swelling capacity are all very similar, but the modulus increases as the acrylic acid concentration increases.

In deriving Eq. 5.5, the functionality of the crosslinks is taken to be four, as this

Table 5.1. The effect of monomer concentration on the modulus of swollen superabsorbent polymer.[a]

Monomer Concentration (wt%)	Swollen Modulus (kPa)
24	2.65
25	3.04
26	3.26
27	3.46

[a]The networks are all 65 mol% neutralized poly(acrylic acid), swollen in 0.9% NaCl solution to their maximum swelling extent.

is the crosslinker functionality for common divinyl crosslinking agents. Nevertheless, crosslinkers of higher functionality may be used, and this will also affect the modulus through the functionality *f*. Crosslinks of larger functionality will increase the modulus. The factor *A* used to account for the difference in modulus when different crosslinker functionalities are used is given in Eq. 5.9.

$$A = \left(1 - \frac{2}{f}\right) \tag{5.9}$$

N,N′-methylenebisacrylamide and ethyleneglycol diacrylate are both examples of crosslinkers that yield crosslinks of functionality four if both vinyl groups are reacted. 1,1,1-Trimethylolpropane triacrylate (TMPTA) and triallylamine have a theoretical functionality of six. However, if only two of the vinyl or allyl groups are actually used, perhaps due to steric problems in propagating a polymer chain through the crosslink as it forms,[14] the functionality of the crosslink would be four. We have determined by means of enhanced nuclear magnetic resonance spectroscopy (see Chapter 2, section 2.6.3) that residual vinyl groups are not found in superabsorbent polyacrylates made with acrylate ester crosslinkers. Inefficient crosslinking in these cases is most likely due to cyclo-polymerization of the crosslinker into the network structure and to clustering of crosslinks.[15] An example of the different efficiencies of crosslinker incorporation is shown in Figure 5.2. Over a range of crosslinker stoichiometries, the resulting equilibrium swelling capacities fall on separate curves.

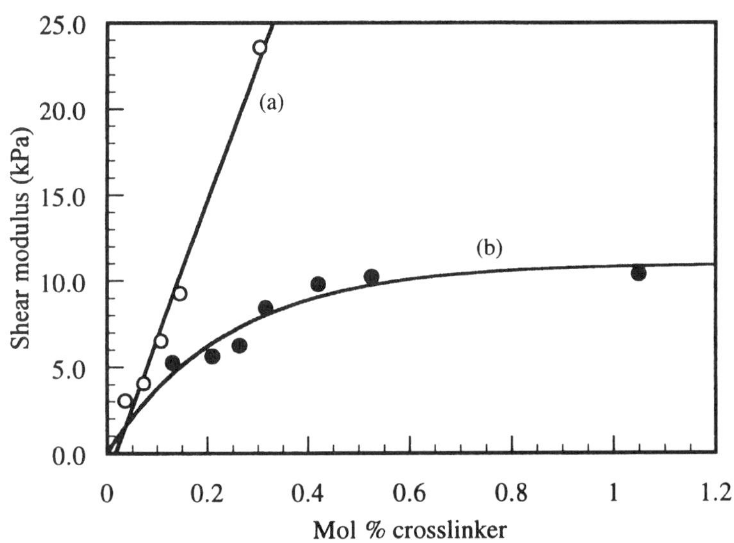

Figure 5.2. Shear modulus as a function of the amount of crosslinker used per mole of acrylic acid, for two structurally different crosslinkers, 1,3-butyleneglycol diacrylate **(a)** and triallylamine **(b)**.

This observation indicates a different efficiency of crosslinker incorporation into the final network polymer.

The extent of swelling of a crosslinked polymer affects the modulus. An increase in the swollen volume decreases the concentration of elastic chains and therefore decreases the modulus. In Eq. 5.8 the modulus of a swollen (non-ionic) polymer is proportional to the cube root of the ratio of the swelling change. The situation is more complex with polyelectrolyte networks. Over a wide range of charge density and ionic strength, the modulus decreases and then increases as swelling progresses. This is thought to be due to the very extended chain configurations that are possible at high charge density and low ionic strength, when the swelling extent is very large. However, for superabsorbent polymers used as liquid absorbents in personal care, the ionic strength of the absorbed fluid usually is sufficiently high that the chain configurations are not greatly extended.[16] The modulus behavior observed for a commercial polyacrylate absorbent polymer with increasing swelling is shown in Figure 5.3. The data were collected by the author by swelling a sample of polymer in a 0.9% NaCl solution to various amounts less than maximum swelling and measuring the modulus according to the methods described in Chapter 4. A fit of the data to a power law shows that the modulus (swollen) decreases as the 0.37 power of the change in swelling, very close to the theoretical value of 1/3 shown in Eq. 5.8. However, changing the concentration of salt in the swelling liquid does affect the magnitude of the modulus, as shown in Figure 5.4, even though an upturn is not observed over this range of swelling ratios. Ionic effects on the modulus of rubbers have been discussed numerous times.[17–20] These effects are not too important, how-

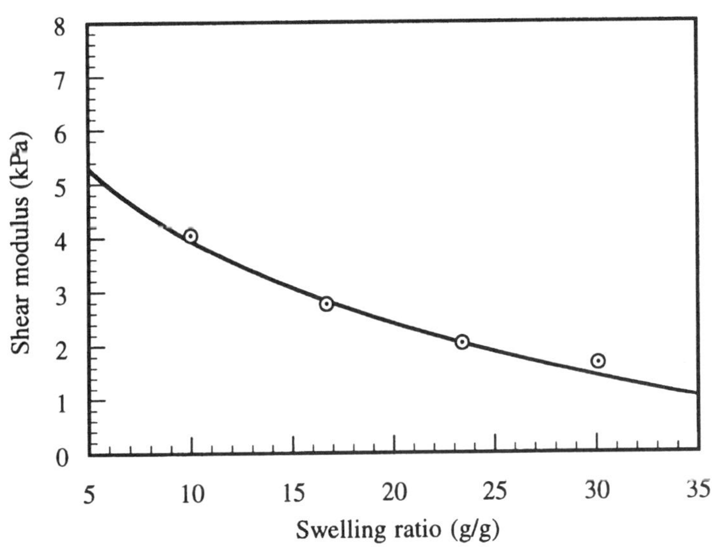

Figure 5.3. The shear modulus dependence on the swelling ratio of the sample. A single sample of superabsorbent polymer was swollen to less than equilibrium by limiting the fluid available. The modulus scales with the –0.37 power of the swelling ratio.

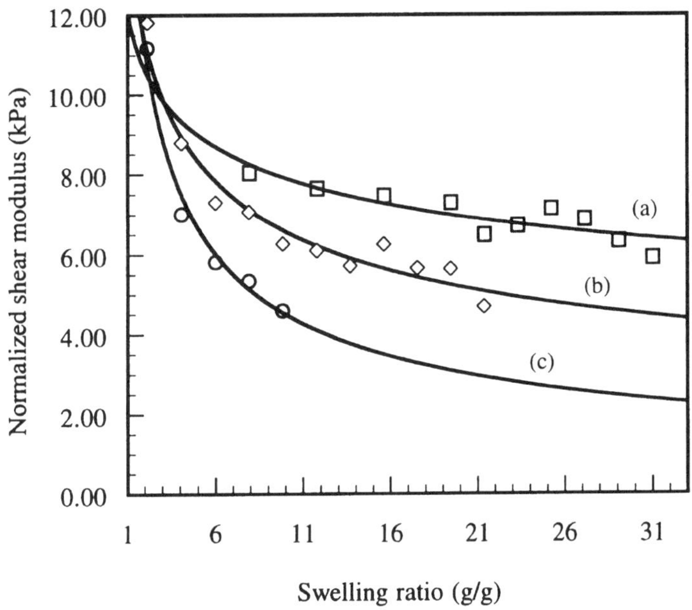

Figure 5.4. Shear modulus of partially neutralized poly(acrylic acid) gel, normalized to the volume of preparation, as a function of the swelling ratio in three different concentrations of sodium chloride, 0.0325% **(a)**, 0.125% **(b)**, and 0.9% **(c)**.

ever, for polyelectrolyte gels swollen in salt solutions of moderate electrolyte concentration (which is the case for superabsorbent polymer used in diapers).

Soluble polymer is that portion of the polymer chains that is unattached to the rest of the infinite network. The effect that this polymer has on modulus is difficult to treat and is usually neglected in the theories. These chains also do not contribute to the modulus but can be solvated and contribute to swelling. However, any soluble polymer may be extracted from the gel during the swelling process to some extent. The extent of extraction depends on the molecular weight of the soluble polymer chains, the extent of swelling of the gel network, the relative kinetics of the swelling and extraction processes, and whether the gel is bathed in excess liquid.

5.1.1.4. Swelling Capacity. From Eqs. 5.4–5.7, and assuming a randomly crosslinked, four-functional network (and neglecting the ionic effects), an equation for the swelling capacity in terms of the key polymerization parameters may be derived. The theory provides that networks prepared in the presence of diluent and then swollen greatly behave according to the so-called phantom network model. In this model the random, thermal fluctuations of the crosslinks are constrained only through the connectivity of the network and not by the presence of other chains in the same region of space. Therefore, the chains, when subjected to a stress, act only on the crosslink points to which they are directly connected. The chains do not affect each other, except through the crosslinks. In a practical sense, the chain deformation of a phantom network during swelling is small relative to the length of the extended chain, and chain entanglements do not restrain the movements of chains or crosslinks (mathematically, the entanglements are explicitly excluded). The phantom network model is obtained from Eq. 5.4 by setting the $K(\upsilon_2)$ term (the term providing for the constraints on the crosslinks) to zero. The logarithmic term may be expanded in a series, keeping only the first two terms as an approximation at small υ_2 (large swelling). Then the terms in υ_2 are collected, and the swelling ratio $q_m = 1/\upsilon_2$ is substituted for the polymer volume fraction at maximum swelling. The resulting approximation for the swelling capacity in terms of polymerization parameters is given below.

$$q_m^{5/3} = \frac{(1/2 - \chi)2M_c}{V_1 \rho_0 \upsilon_{20}^{2/3}(1 - 3M_c/M_n)} \tag{5.10}$$

According to Eq. 5.10, the swelling ratio depends on the length of polymeric chain between crosslinks through M_c and on the kinetic chain length of the polymerization through M_n. In addition, the concentration of monomer at preparation affects the ultimate swelling through the term $\upsilon_{20}^{2/3}$. By comparing Eqs. 5.8 and 5.10, the swelling capacity and the elastic modulus are seen to be related inversely to the same structural parameters.

5.1.1.5. Monomer Concentration Effects. Equations 5.8 and 5.10 predict the swelling and elastic properties of networks as a function of the volume fraction of polymer at which the crosslinks are formed, υ_{20} and molecular weight of the primary chains M_n. In a crosslinking co-polymerization, υ_{20} is usually assumed to be inversely proportional to the monomer concentration. The value of the primary chain molecular weight M_n also depends on the monomer concentration through the kinetic chain length of the polymerization as discussed in Chapter 2. In this section the effect of monomer concentration on the change in primary molecular weight, the magnitude of the soluble fraction, the number of dangling chains in the network, and the entanglement of the chains are discussed in more detail. A simple modification to the theory for calculating the Flory–Erman entanglement parameter κ will also be developed for crosslinking co-polymerizations. The effect of these structural features on the swelling capacity and modulus are also discussed.

The molecular weight of the uncrosslinked chains can be measured after first extracting them from the network. The effect of the monomer concentration on the molecular weight of the primary chains in crosslinked polymers can be determined if the crosslinks are first selectively removed so that the molecular weights can be measured by standard techniques, which were described in Chapter 4, section 4.3.3. For polymers made with acrylate ester crosslinkers (commonly used in superabsorbent polymers), crosslink cleavage may be accomplished by basic hydrolysis followed by size-exclusion chromatography of the hydrolyzed sample. The molecular weight and quantity of the extracted, uncrosslinked chains and the molecular weight of the primary chains after hydrolysis are shown in Table 5.2 for several superab-

Table 5.2. Number average primary chain molecular weights, extractable chain molecular weights, and amounts as a function of the crosslink density of various samples

Sample	Mole % Crosslinker	Swelling Capacity (g/g)	Soluble Polymer (wt%)	MW Primary Chains × 10^{-6}	MW Extractable Chains × 10^{-6}	MW_{extr}/MW_{prim}
a	0.0486	39.5	19.0	0.25	0.08	0.32
b	0.1346	33.1	6.3	0.33	0.05	0.15
c	0.1703	32.9	9.5	0.28	0.05	0.18
d	0.2919	28.5	4.9	0.30	0.04	0.13

sorbent polymers prepared with different crosslink densities but at the same monomer concentration (29 wt% of 65 mol% neutralized acrylic acid).

At this particular constant monomer concentration, the primary chain lengths vary slightly around an average value of 2.9×10^5 g/mol. The swelling capacity decreases, and the amount of extractable polymer tends to decrease with increasing concentration of crosslinker as expected. As noted in section 5.1.1.3, the molecular weight of the extractable polymer is an estimate of the molecular weight between crosslinks and also of the molecular weight of any network dangling chains. The ratio of the molecular weight of the extracted polymer to the molecular weight of the primary chains is then an estimate of the ratio M_c/M_n in Eq. 5.10, which indicates the degree of imperfection of the network structure. The soluble polymer and the dangling chains are usually treated as a diluent of the network modulus. These chains do not contribute to the modulus but can be solvated and contribute to swelling. From the values of the molecular weight, and using the dangling end correction from Eq. 5.8 or 5.10, the modulus of actual networks is calculated to be 30%–90% less than if the networks were "perfectly" connected by the crosslinks. This presents the industrial chemist with a significant opportunity for improvement of the mechanical properties of superabsorbent polymers. Maximizing the modulus requires eliminating or minimizing the content of extractable chains and dangling chains[21] by increasing the primary chain length of the polymer and minimizing side reactions such as chain transfer. The effect of the monomer concentration on the modulus at constant crosslink density is shown in Table 5.1.

The extent of entanglement of the network chains, which is responsible for much of the increase in modulus with increasing monomer concentration, may be accounted for by the parameter κ of the Flory–Erman theory.[12] This parameter measures the intensity of the constraints of the chains on the junctions or crosslinks of the network. The magnitude of the κ parameter depends on the ratio of the molecular mass between crosslinks M_c to the molecular mass of the primary chains M_n and on the concentration at which the crosslinks are formed.

In crosslinking co-polymerizations, the molecular weight of the primary chains decreases continuously during the polymerization as the concentration of monomer decreases. At the same time the molecular weight between the crosslinks may change depending on the relative reactivity of the crosslinker with the acrylic acid.

How then will the value of κ vary with the conversion of the polymerization reaction and on the reactivity ratio of the crosslinker with acrylic acid?

The first crosslinks are introduced into the network when the volume fraction of polymer is small. At these conditions, the value of κ is small and the network behaves more like a phantom network. If in addition the crosslinker is quite reactive, the crosslinks are introduced early in the formation of the network, the molecular mass between crosslinks is small, and κ is smaller as a result.

The dangling chain ends dilute the elastic network chains, much like adding more solvent; this reduces κ. Early in the co-polymerization, when the primary molecular weight is high (relatively high concentration of monomer), the fraction of dangling chains depends on the amount and relative reactivity of the crosslinker. When the amount of crosslinker is high, or when the relative reactivity of the crosslinker is high, the fraction of dangling chains is low, and the value of κ is relatively unaffected by the "dilution" from the dangling chains. However, if the amount of crosslinker is small or if the reactivity is low, which delays the incorporation of the crosslinker into the network, the fraction of dangling ends is higher and the dilution effect is more pronounced. This makes κ smaller, and the network will have properties closer to those of the phantom network in which the chains do not constrain the junctions at all.

As the co-polymerization proceeds, the molecular weight of the primary chains decreases, as a result of the reduction in monomer concentration. For a given value of the crosslinker reactivity, shorter primary chains have the effect of increasing the fraction of dangling chains. Therefore, as monomer conversion progresses, the dilution effect of dangling chains increases and the value of κ decreases. If the relative reactivity of the crosslinker is lower, the incorporation of the crosslinker is delayed until higher conversion. Then, as the primary chain molecular weight decreases with conversion, the number of incorporated crosslinks per primary chain increases. The overall effect under these conditions will be to moderate the change in κ as the reaction proceeds.

What follows is a simple model of the conversion dependence of κ, which has been developed by the author to compare the value of κ for a network formed by crosslinking existing polymer chains to its value during a crosslinking copolymerization. The entanglement parameter κ is defined as

$$\kappa = PM_c^{1/2}\upsilon_2 0 \tag{5.11}$$

where M_c is a function of the ratio of the concentration of acrylic acid to the concentration of crosslinker, and the relative reactivities of the acrylic acid and crosslinker and $\upsilon_2 0$ is a function of the relative reactivities and the fraction of dangling chains, which dilute the network. To account for the total polymer concentration and the dangling chains separately, a Flory–Huggins type correction factor for the dangling chains is used:

$$\kappa = PM_c^{1/2}\upsilon_2 0(1 - 2M_c/M_n) \tag{5.12}$$

In Eq. 5.12, M_n varies with both conversion and total monomer concentration,

$$M_n = [AA]M_{mon}k_p/(fk_pk_t[I])^{1/2} \tag{5.13}$$

where $[AA]$ is the concentration of acrylic acid; M_{mon} is the molecular mass of the monomer; $[I]$ is the concentration of free-radical initiator; k_p, k_t, and k_d are the propagation rate constant, the termination rate constant, and the decomposition rate constant of the initiator, respectively; and f is the efficiency of the free-radical system toward production of chains from the radicals formed. Substituting Eq. 5.13 into Eq. 5.11 yields

$$\kappa = P\left(\frac{[AA]}{[XL]}\right)^{1/2}[AA]\overline{V}_2\left(1 - \frac{2[AA]}{[XL][AA]_0P'}\right) \tag{5.14}$$

where M_c has been replaced by the ratio of acrylic acid to crosslinker, $\overline{V}_2$ is the molar volume of the polymer, $[AA]_0$ is the initial concentration of acrylic acid, and P' is a constant that collects together the various rate constants of the polymerization and the initiator concentration.

The values of κ calculated with Eq. 5.14 are compared with those calculated with no accounting of the change in polymer concentration during polymerization in Figure 5.5. The values of κ are calculated as a function of conversion for a polymerization wherein the reactivities of monomer and crosslinker are equal. The values obtained by assuming that only the concentration of polymer changes throughout the polymerization are plotted as the standard case. The example curves are

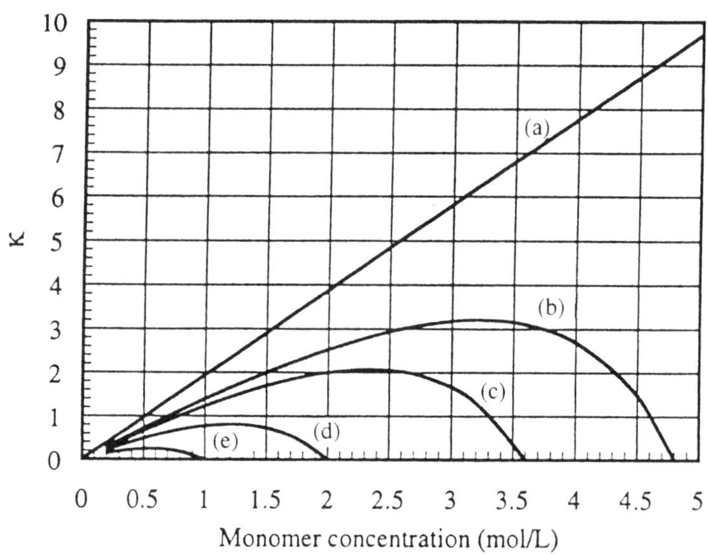

Figure 5.5. Value of the Flory–Erman κ as a function of monomer concentration during polymerization and modified for the effect of dangling chains. The curves are for different initial monomer concentrations: **(a)** standard calculation, **(b)** 4.8 mol/L, **(c)** 3.6 mol/L, **(d)** 2.0 mol/L, and **(e)** 1.0 mol/L.

calculated with different initial concentrations of monomer. This analysis shows that the value of κ does not reach a large value in a typical crosslinking co-polymerization, justifying the use of the phantom network approximation for describing the mechanical and swelling behavior.

5.1.1.6. Ionic Networks. The carboxylic acid groups in the superabsorbent polymer are neutralized with base during synthesis to increase the extent of ionization of the gel. The extent of ionization depends also on the swelling extent of the polymer and on the ionic strength of the swelling liquid. Neutralization increases the swelling capacity by increasing the concentration of ions in the gel. As the gel is increasingly neutralized, salt that is present in the swelling liquid is more excluded from the gel phase during swelling. This so-called Donnan exclusion from polyelectrolyte solutions and networks is described briefly in section 5.1.1.1. The ideal Donnan distribution of ions between an external solution of salt and a polyelectrolyte gel is given by the following equation[8]:

$$\frac{c_j^{gel}}{c_j^{ext}} = \frac{c_j^{ext}}{c_j^{gel} + i\upsilon_2/V_u} \tag{5.15}$$

In this equation, c_j^{gel} and c_j^{ext} and υ_2 have the same meanings as earlier; i is the degree of ionization of the polymer chain, and V_u is the molar volume of a monomer unit in the polymer. As applied to swelling networks, the existence of the Donnan potential means that ions present in the liquid phase external to the gel will be excluded from the gel, more or less depending on the relative concentrations of salt in the external solution and the fixed charges in the polyelectrolyte phase. Because the gel may be viewed as acting as both a polymer solution and a membrane in an osmotic pressure experiment, the gel behaves as if a nonelastic gel substance is contained by an elastic, ionized membrane.[22] Because the magnitude of the salt exclusion or rejection depends on the concentration of the fixed charges of the polyelectrolyte in the gel, salt exclusion depends on the extent of swelling of the gel. As the gel swells, the concentration of the fixed charges decreases, and the Donnan potential decreases. This means that the rejection of salt by the swelling gel decreases as the gel swells, and the maximum extent of exclusion depends on the maximum swelling of the gel.

Figure 5.6 shows the results of calculations for ideal Donnan exclusion from homogeneously crosslinked polyelectrolyte gels in various concentrations of sodium chloride solution. The ratio of salt (NaCl) concentration inside the gel to its concentration outside the gel was determined by solving Eq. 5.15 for its roots and calculating the positive root as a function of υ_2 and the external salt concentration. The result is plotted as a function of the swelling capacity of the superabsorbent polymer (65 mol% neutralized poly[acrylic acid]). The exclusion effect given by the ratio c_j^{gel}/c_j^{ext} depends on the crosslink density as represented by the equilibrium swelling capacity of the polymer. The ratio approaches the value of unity as the swelling capacity increases (polymer concentration decreases). The extent of exclusion also decreases as the salt concentration increases. For a homogeneously crosslinked gel

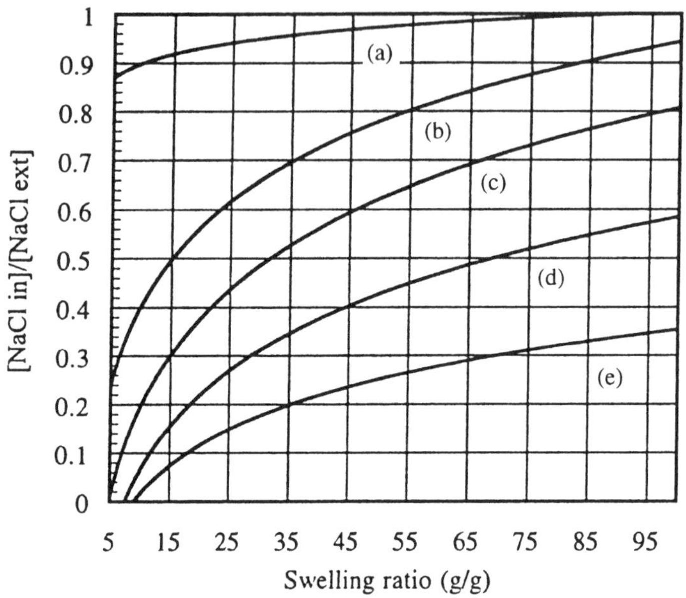

Figure 5.6. The ratio of salt concentration inside the gel to the salt concentration outside the gel, calculated according to the ideal Donnan equilibrium with the external salt concentration equal to 2.85 **(a)**, 0.308 **(b)**, 0.154 **(c)**, 0.077 **(d)**, and 0.0385 mol/L **(e)**.

particle, swollen to 30 times its mass in 0.9% sodium chloride solution, Figure 5.6b shows that the concentration of sodium chloride in the gel phase is about one-half its concentration in the external salt solution. Increasing the extent of neutralization of a superabsorbent polymer over about 30 mol% results in increased absorption of salt solutions, at least partly as a result of increased Donnan exclusion. The extent of ionization of the gel does not increase much past 30 mol% neutralization as a result of counterion condensation,[3] but the gel excludes more salt as its extent of neutralization increases and can absorb additional water.

The dependence of the equilibrium properties of superabsorbents on the extent of neutralization of the network and on the concentration of salts in the swelling liquid has been demonstrated numerous times.[23–27] The equilibrium swelling behavior of representative superabsorbent polymers is shown in Figures 5.7 and 5.8. Over the range of use of superabsorbents in personal care applications, the swelling is smoothly decreased by decreasing the extent of neutralization of the network or by increasing the concentration of salt in the swelling liquid.

5.1.2. Network Compressibility

The principal benefit of using superabsorbent polymers in personal care sanitary products is that they absorb liquid and retain the liquid in a solid phase even under the action of moderate pressure. Compared with the ability of a batt of cellulose

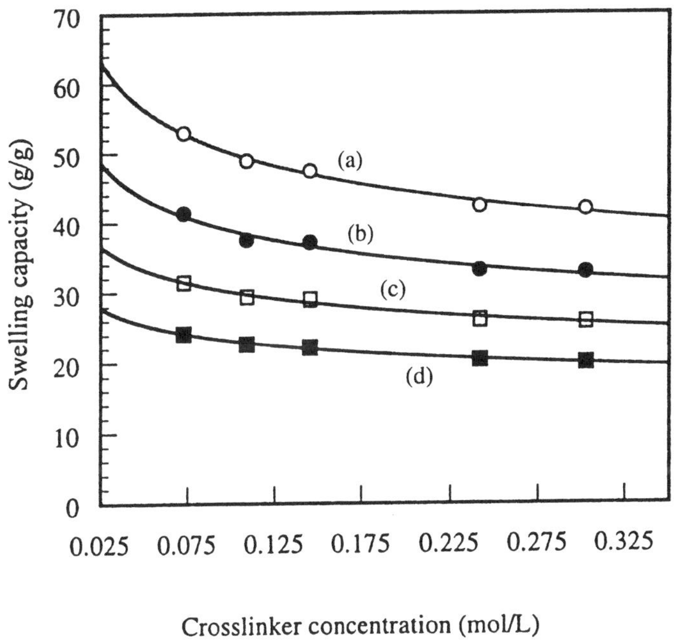

Figure 5.7. The dependence of the swelling capacity on the concentration of crosslinks during preparation of the gel and on the concentration of NaCl in the swelling liquid: **(a)** 0.225%, **(b)** 0.45%, **(c)** 0.90%, **(d)** 1.80%. The gels were polymerized at identical concentrations of monomer, using 1,3-butyleneglycol diacrylate as the crosslinker and with 65% of the carboxylic acid groups neutralized.

fiber fluff to retain water, the high retention ability of superabsorbent polymers is quite amazing. However, just as for fiber batts, liquid can be removed from a swollen superabsorbent gel by application of pressure, but the required pressures are much larger. This deswelling occurs because the application of pressure to the system causes an increase in its free energy and in the chemical potential of any species that can diffuse into or out of the system. The increase in chemical potential serves as a driving force to expel water from the gel into the surroundings when the gel is compressed.

The effect of adding pressure to the gel is similar to adding additional crosslinks to the gel. However, applying pressure to the gel does not change the network structure (the cycle rank) as does changing the crosslink density. The limitation of swelling under pressure and the related deswelling behavior are thus related to the osmotic pressure of the gel, which acts as a polymer solution confined inside a semipermeable membrane. In the following section, the effects of compression on the swelling of gels are described in more detail.

5.1.2.1. Osmotic Pressure of Networks. The osmotic pressure is defined by

$$\mu_l - \mu_1^0 = RT \ln a_1 = -\pi \bar{v}_1 \tag{5.16}$$

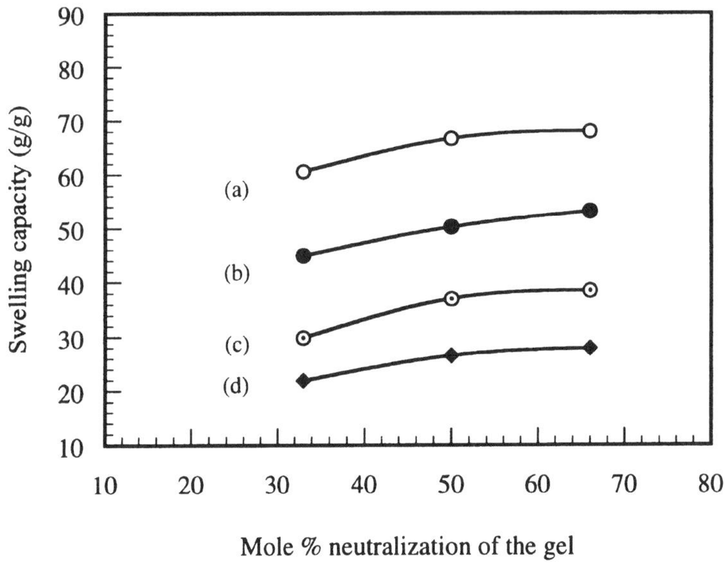

Figure 5.8. The dependence of swelling capacity on the extent of neutralization of the carboxylic acid groups and on the NaCl concentration (by weight) of the swelling liquid: **(a)** 0.225%, **(b)** 0.45%, **(c)** 0.9%, **(d)** 1.8%. The data are for one prepared gel that was neutralized to different extents after network formation.

where μ_1 and μ_1^0 are the chemical potentials of the solvent in the gel phase and in the pure solvent phase, respectively, R is the gas constant, T is the absolute temperature, a_1 is the activity of water in the gel phase, π is the osmotic pressure, and v_1 is the partial specific volume of the solvent. When the phases are not in equilibrium, the differences in the chemical potential of water in the two phases is a driving force for water to enter or leave the gel. The magnitude of this driving force is given by the activity of water in the gel phase and can be alternatively expressed as a pressure, the osmotic pressure π. When the phases are in equilibrium, there is no longer a difference in the chemical potential of water between the two phases and the osmotic pressure becomes zero. In gel swelling experiments, the osmotic pressure is often measured on a gel at constant volume in the presence of excess water. The measured pressure is frequently called the *swelling pressure*, which measures the force required to maintain the volume of the gel constant. The magnitude of this pressure varies with the concentration of polymer in the gel and varies from sample to sample depending on the structural characteristics of the gel and the nature of the swelling liquid.

5.1.2.2. Effect of Network Structure. The structural characteristics affecting the osmotic pressure of a gel are the same characteristics that affect the magnitude of the equilibrium swelling capacity and modulus of the gel. But in an application where limited swelling or deswelling is important, it is the sensitivity of the structural characteristics to a change in external pressure p that is important in determining the magnitude of the effect. The effect of pressure p on a gel has been generally understood for some time, and the following equation gives the expected behavior of a network behaving according to the phantom network model with an added term to account for the ionic effects:

$$p = -\frac{RT}{V_1}\left[\ln(1-v_2) + v_2 + \chi v_2^2 + \frac{\rho V_0}{M_c v_2^{1/3}}\right] + RT\sum_j (c_j^{gel} - c_j^{ext}) \qquad (5.17)$$

In Eq. 5.17 the first three terms inside the brackets are the Flory terms for mixing of the solvent with the segments of the polymer chains, and the last term is the restriction of swelling due to the interconnectedness of the polymer chains through the action of the crosslinks. For ionic networks we must add the term outside of the brackets (from Eq. 5.4) to account for the effects of ions on the swelling pressure. Another simple approach is to ignore the ionic effects specifically and lump the effects of these interactions into the χ parameter. This approach is certainly less satisfying if one is interested in understanding how the extent of neutralization or the concentration of salt in the swelling liquid affects the osmotic pressure of the gel. Equation 5.17 may be used to calculate the effects of added pressure on gel swelling, using estimates of the network structure parameters and the effects of the Donnan equilibrium on ion permeation. A plot of the swelling capacity (inversely proportional to the volume fraction of polymer in the gel) versus the swelling pressure is shown in Figure 5.9. The curve shown is for M_c = 100,000 g/mol, with the

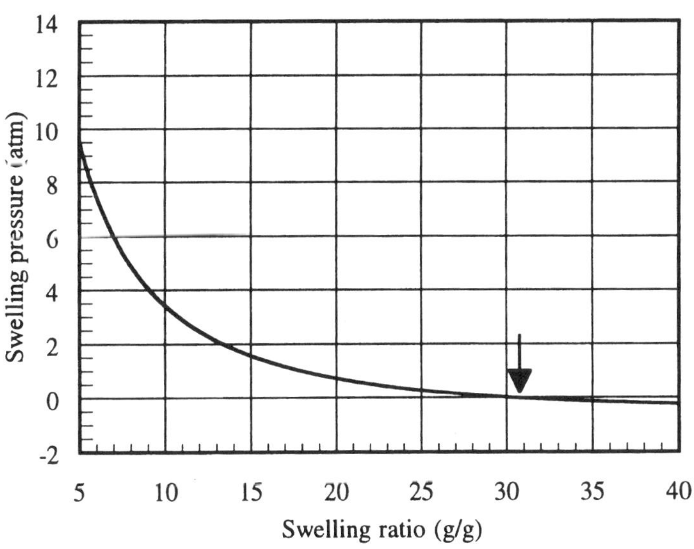

Figure 5.9. The calculated swelling pressure of a polycarboxylate gel as a function of its swelling extent in 0.9% NaCl solution. Equilibrium swelling is identified where the curve passes through a swelling pressure of zero (arrow).

ionic concentration of the external liquid fixed at 0.154 mol/L, and the extent of neutralization of the polymer fixed at 0.65 (conditions that are typical for superabsorbent polymer swelling in urine). The equilibrium swelling is given at a swelling pressure of zero. If an external pressure is applied, by mechanical means or by osmotic means, the polymer will swell to a lower value corresponding to the abscissa value at the applied pressure. The calculations at other values of M_c indicate little difference in the swelling pressure behavior of materials with different crosslink densities, and the curves are grouped closely to the curve shown in Figure 5.9. One method of applying pressure to a gel is by surrounding the gel with a polymer solution of known osmotic pressure. The swelling equilibrium in a number of different polymer solutions is measured. This has been done with polyelectrolyte gels in dialysis bags surrounded by solutions of poly(ethylene oxide).[28] Similar measurements have been conducted in our laboratories with superabsorbent polymers immersed in solutions of partially neutralized poly(acrylic acid) of high molecular weight. The large molecular size and the high charge density of the soluble polymer chains prevents them from penetrating the polyelectrolyte gel. A number of granular superabsorbent polymers of different crosslink densities were immersed in polymer solutions of different concentrations, and the swelling capacity in 60 min was taken as close to the equilibrium value. The results are shown in Figure 5.10. Because the osmotic pressure of the partially neutralized poly(acrylic acid) solutions is linearly proportional to their concentration, the ordinate axis of Figure 5.10 corresponds to the coordinate axis of Figure 5.9 (the linear pressure axis). Then the decrease in swelling of the gels with increasing concentration of soluble polymer confirms the

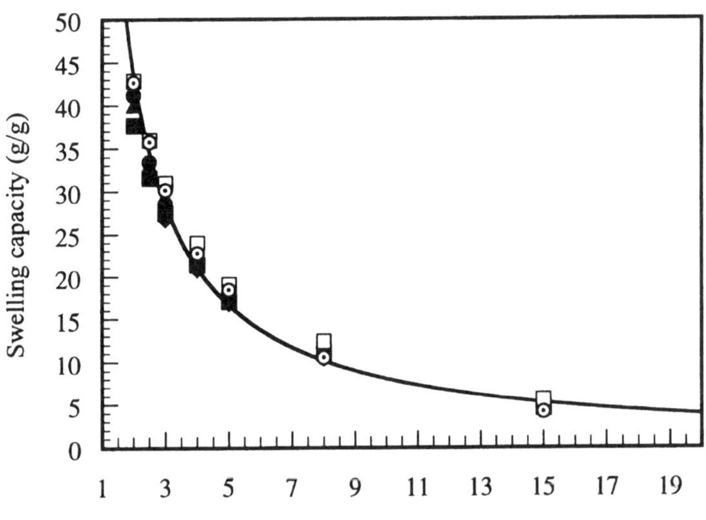

Figure 5.10. Swelling capacity of superabsorbents in solutions of partially neutralized poly(acrylic acid). The samples vary in their crosslink density.

calculated results from Eq. 5.17. Both the calculations and the experimental results show that moderately high pressures are required to restrict the swelling of superabsorbent polymers. The light pressure (up to 3.5 kPa) exerted by an infant on the superabsorbent polymer in a diaper should only decrease the swelling of a typical commercial superabsorbent polymer (centrifuge capacity = 30 g/g) by about 7% to 27.8 g/g. As is evident below, sometimes the swelling under a compressive load can be much lower.

5.1.2.3. Absorbency Under Load. An experimental technique that measures the effect of mechanical compression on the swelling process has been used in the study of commercial superabsorbent polymers. This is known as the *absorbency under load*.[29] The details of the analytical method are given in Chapter 4, section 4.6.4.2. An early test method using the same principles is described by Reid.[30] The swelling extent of the polymer sample is measured as a function of a load, which is applied to the polymer through a weighted piston. The polymer is in contact with free liquid phase through a small, rigid, mesh screen. The sample swells from the dry state to a gel state while the load is present. As a result of the granular nature of the polymer sample, there are pores between the particles, and a small amount of free liquid is contained in the pores. This amount of liquid is frequently included in the reported value of the liquid absorbed, but may be eliminated by blotting the gel phase through the fine mesh screen at the bottom of the measurement cell. (In a diaper, the swollen gel particles are usually in contact with each other. The inclusion of the pore water in the measurement has validity in this case of practical application.) In the next section, it will be shown that this measurement of gel swelling extent is extremely sensitive to the crosslink density and modulus of the polymer, and very small loads depress the swelling far more than expected from either theory or the osmotic deswelling experiments.

Representative results for the absorption under load phenomenon are shown in Figures 5.11–5.15. Figure 5.11 shows that the absorbency under load increases with increasing primary chain molecular weight of the network. Figure 5.12 shows that the absorbency under load at higher loads increases with increasing crosslink density even while the equilibrium swelling capacity at zero load decreases. Figure 5.13 shows how the absorbency under load changes with increasing monomer concentration during synthesis of the polymer. The absorbency under load rises rapidly with increasing monomer concentration and then levels off and even drops slightly. This should be contrasted to the effect of monomer concentration on the swelling at no added load, which shows a smooth decrease as monomer concentration increases. The increasing modulus that results from increased monomer concentration (section 5.1.1.3) thus has a significant effect on absorbency under load. Figure 5.14 shows that the absorbency under load increases slightly as the extent of neutralization of the polymer is increased, but that once the polymer has been neutralized to 50 mol%, little additional improvement in absorbency under load is seen. Figure 5.15 shows that the swelling capacity at small load increases with decreasing salt concentration in the swelling liquid and that the absorbency under load at higher load is less affected by the change in salt concentration. The decrease in absorbency

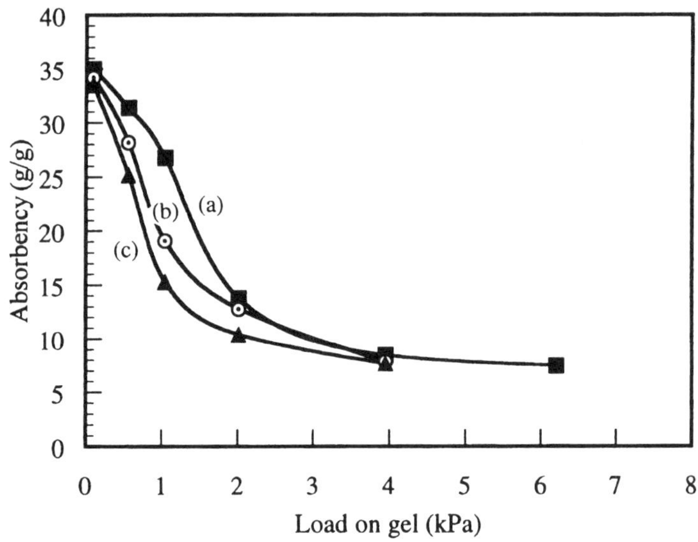

Figure 5.11. The absorbency under load of superabsorbent polymers as a function of the applied load. The primary chain length of the network was shortened by polymerizing in the presence of 0 ppm **(a)**, 800 ppm **(b)**, and 1,600 **(c)** ppm of sodium hypophosphite as chain transfer agent. The equilibrium swelling capacity of each sample was kept constant by changing the crosslink density.

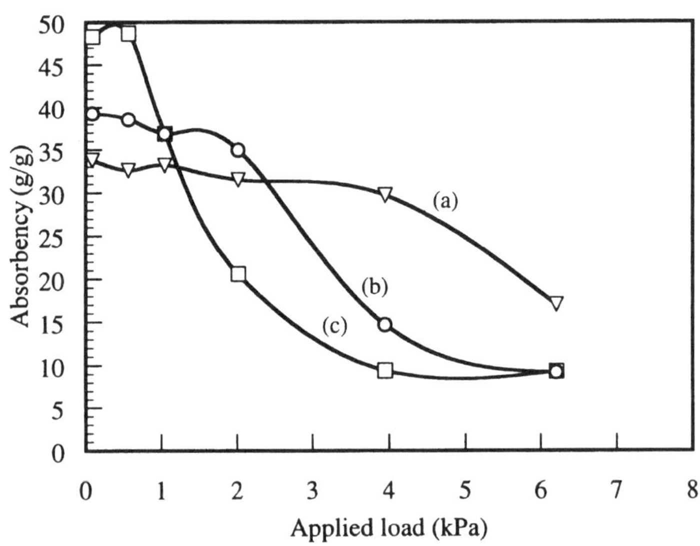

Figure 5.12. Gel swelling versus applied compression for three different crosslink densities, 16,000 ppm **(a)**, 8,000 ppm **(b)**, and 4,000 ppm **(c)**, based on weight of acrylic acid. All samples are 33 mol% neutralized poly(acrylic acid), swollen in 0.225% NaCl solution.

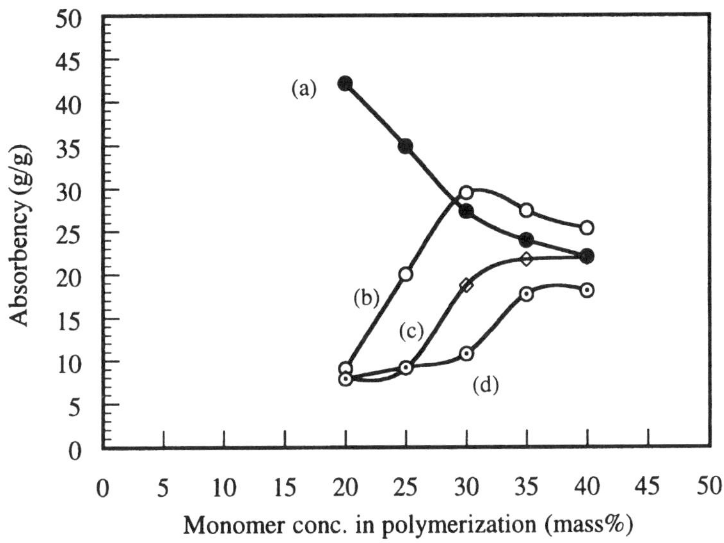

Figure 5.13. The absorbency under load as a function of the monomer concentration used during preparation of the sample. The ratio of crosslinker to acrylic acid is constant between samples. Loads are 0.09 **(a)**, 2.01 **(b)**, 3.95 **(c)**, and 6.21 kPa **(d)**.

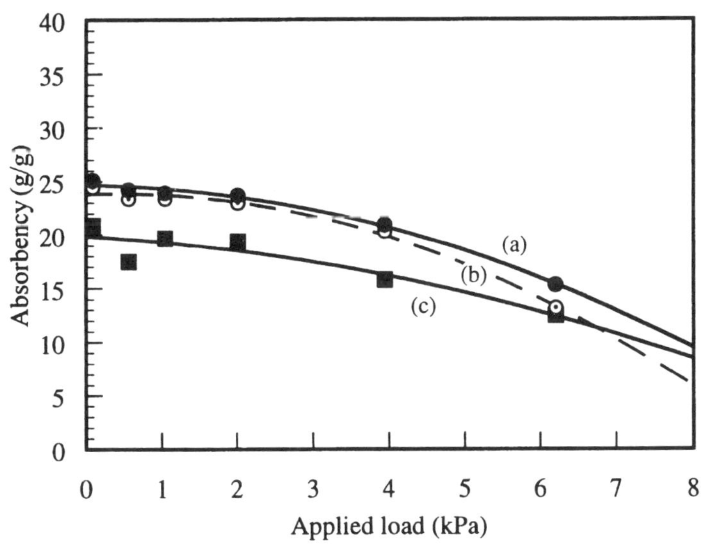

Figure 5.14. Gel swelling versus load and as a function of the extent of neutralization for a sample with relatively high crosslink density, swollen in 0.9% NaCl solution: **(a)** 66%, **(b)** 50%, and **(c)** 33% neutralized.

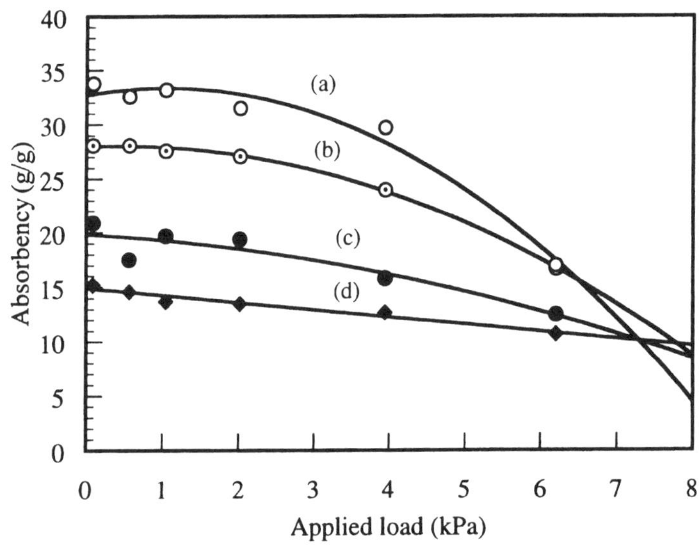

Figure 5.15. Gel swelling versus applied load for the highest crosslink density sample in various concentrations of NaCl solution: **(a)** 0.225%, **(b)** 0.45%, **(c)** 0.9%, and **(d)** 1.8%. The sample is 33 mol% neutralized.

with increasing salt concentration, which results from the larger osmotic pressure of the salt solution, is similar to that seen in Figure 5.7. At higher salt concentration, the concentration of polymer in the gel phase is larger, the modulus of the gel phase is larger, and the gel phase is then more difficult to compress (Fig. 5.9). All of these relationships have in common a dependence of the absorbency under load on the elastic modulus of the gel at the conditions of the absorbency under load test. In general, the absorbency under load increases as the modulus of the swollen gel increases. It must be appreciated, however, that the particles deform gradually during the swelling process. The measurement is inherently a dynamic one, which depends on the time rate of change of the modulus of an individual particle and on the time-dependent interaction of the particles. A mathematical description of this phenomenon is developed in section 5.3, after a discussion of the kinetics of free swelling. But first, one general, synthetic technique used to increase absorbency under load in superabsorbent polymers is described.

5.1.3. The Structure of Surface Crosslinked Superabsorbents

A synthetic technique used in the preparation of superabsorbent polymer particles is surface crosslinking.[31–35] In this method, a low molecular mass reagent, capable of reacting with the functional groups of the base polymer, is spread onto the surface of the polymer particles and caused to react by applying appropriate reaction conditions. In one early example, a solution of glycerol, water, and isopropanol was spread onto particles of a partially neutralized, crosslinked poly(acrylic acid).[36] The coated particles were then heated at temperatures in the range of 150°–190°C, which caused the esterification of the glycerol and carboxylic acid groups of the polymer and subsequent crosslinking of the polymer chains. The ratio of water to isopropanol was adjusted to allow the penetration of the glycerol into the surface of the particle. By controlling this ratio and the amount of glycerol, the depth of penetration and the amount of extra crosslinks can be controlled over a wide range. The structure of the particle is thus altered to consist of a surface region or shell that is more highly crosslinked than the center or core region. This particle structure has several implications for the kinetics of swelling and the time-dependent elastic properties, as is discussed in more detail below.

An alternative synthetic technique has been developed to make a similar core and shell structure. This approach involves polymer chain reactions in addition to the reactions of the functional groups of the polymer, but needs no separate addition of a surface-reactive crosslinker. In one example, the polymer chains in the center of the particle are preferentially reacted with an oxidizing agent to cleave the chains, presumably by a free-radical mechanism.[37] As a result, the polymer in the center has shorter primary polymer chains connected by the crosslinks. From Eqs. 5.6, 5.8, and 5.10, it is apparent that reducing the primary chain length M_n has a similar effect to reducing the crosslink density μ_e/V_0. The content of pendent chains is increased, which reduces the modulus and increases the extent of swelling. Shortening the length of the primary chains also reduces the extent of entangling of the polymer chains and increases the extent of free swelling. Using the swelling approximation of Eq. 5.10, the effect of this chain cleavage may be calculated. In Figure 5.16, each curve shows how changing the primary chain molecular mass at constant crosslink

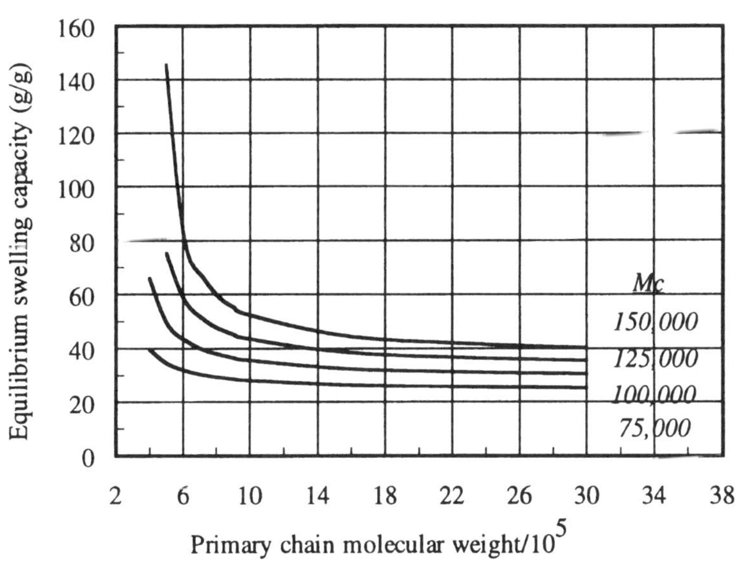

Figure 5.16. The effect of the primary chain molecular weight on swelling capacity for different values of the molecular weight between crosslinks. Values are calculated from the phantom network approximation using Eq. 5.10.

density affects the equilibrium swelling capacity. Each of the curves has a different values of M_c. The values in Eq. 5.10 were taken as $\chi = 0.48$, $V_1 = 0.018$ L/mol, $\rho_0 = 1.67$ g/mL, and $\upsilon_{2}0 = 0.3$, with varying M_c and M_n. In the range of primary molecular masses of 1–1.5 million g/mol (typical for poly[acrylic acid]), the swelling capacity can increase by a factor of 1.2–1.5 if the primary chain molecular mass is cut in half.

The swelling capacity and the elastic modulus of a core and shell particle vary radially through the particle. The swelling behavior of core and shell structures has been studied in the case where the core is rigid and the shell is soft.[38,39] In these cases the swelling of the shell is restricted due to the stresses that develop at the interface of the soft shell and the rigid core. The core and shell structure of superabsorbent polymers is the inverse system, with a softer, more swellable core and a somewhat more rigid shell. In this case, restriction of the swelling of the core by the more highly crosslinked shell is expected. In the first surface crosslinking technique described above, a high swelling polymer particle is converted to a lower swelling particle by reacting the surface region of the particle with more crosslinking agent. Examples from the literature are collected in Table 5.3.

The surface crosslinked superabsorbent polymers share the feature that the swelling capacity of the particles decreases with decreasing particle size, whereas homogeneously crosslinked superabsorbent polymers show no particle size dependence of the swelling capacity. Under the assumption that the added crosslinker is absorbed into different-sized particles to the same depth, the thickness of the surface-crosslinked region (shell) can be determined by measuring the particle-size dependence of the swelling capacity and finding the particle size at which the capacity again becomes independent of particle size, as in Figure 5.17. The line with the steepest slope represents the swelling of the particles that have a distinct shell because the added reagent has not penetrated to the center of the particle. The dashed line of low slope represents the swelling of particles that are crosslinked more or less homogeneously by the added reagent because the reagent penetrates to the center. At the particle size indicated by the intersection of the lines, the added reagent has penetrated just to the center of the particle, and the thickness of the shell may be taken as the radius of this particle. Some particle-size dependence of swelling remains for the smallest particles because the smaller particles absorb more of the applied surface crosslinker per unit volume of particle as a consequence of their higher ratio of surface area to volume.

Table 5.3. Effect of surface crosslinking on the swelling capacity of polymers having different initial crosslink densities

Sample Characteristics	Swelling Capacity (g/g)	Reference No.
Stock polymer (a)	59 (0.9% NaCl)	40
(a) after surface crosslink	52	
Stock polymer (b)	620 (water)	41
(b) after surface crosslink	190	
Stock polymer (c)	62 (0.9% NaCl)	42
(c) after surface crosslink	43	

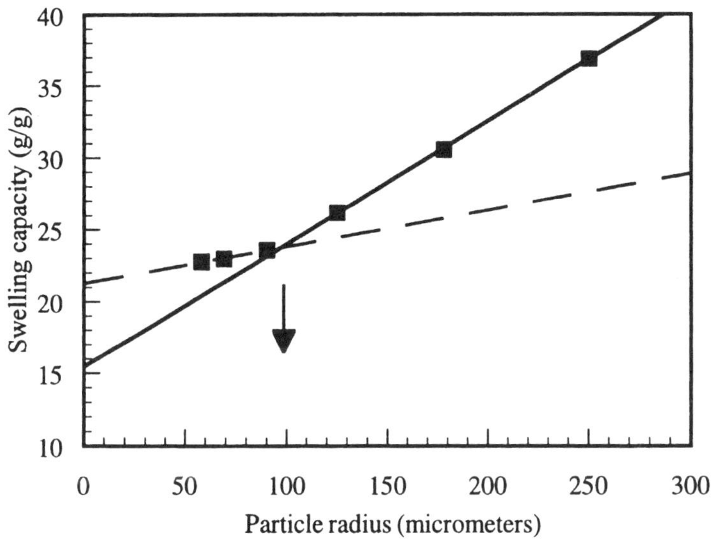

Figure 5.17. The particle-size dependence of the swelling capacity for core and shell superabsorbent particles. A breakpoint (arrow) may be used to estimate the thickness of the shell.

The core and shell structure may have two effects on the swelling behavior, especially when the polymer swells under a compressive load. First, the shell may impose membrane-like surface properties to the particle that affects its swelling behavior. A more highly crosslinked membrane-like surface may alter the kinetics of salt and water transport into the particle due to a change in the Donnan potential at the surface. Second, and more significantly, the higher crosslink density at the surface modifies the mechanical rigidity of the surface layer of the particle. During the swelling process, as the surface of the particle moves outward, the outermost shell is more rigid in the early stages of the swelling process than the same particle without the crosslinked shell. Due to the lower crosslink density in the core, the average swelling capacity is still larger than if the entire particle were homogeneously crosslinked. The crosslink density of the particle as a whole may be unchanged, but the distribution of the crosslinks has been shifted toward the surface. If a collection of gel particles deforms too much under a mechanical load, the transport of liquid to the particles is severely restricted and the swelling practically stops because the rate-controlling mechanism of swelling has shifted from liquid convection to diffusion. This phenomenon is further discussed in section 5.3.

5.2. NETWORK STRUCTURE AND SWELLING KINETICS

The requirements for swelling kinetics of a superabsorbent polymer vary depending on the particular design used for the diaper or other personal care device. In diapers the overall absorption rate of the liquid will be a complex function of the relative amounts of cellulose fluff and superabsorbent polymer used as well as how they are mixed together in the diaper, or even if they are mixed at all but reside instead in separate layers within the structure of the diaper. As absorbent products for personal care evolve to even thinner designs with less cellulose fluff, the fluff will play a decreasing role for the initial absorption. The behavior of the superabsorbent polymer when it is in the matrix of fluff and when it is under the influences of compressive stress in the device is of critical importance to the appropriate swelling kinetics. Additionally, when an application, such as adult incontinent devices, requires a fast rate of absorption, the fundamental aspects of the swelling process must be understood to endow the superabsorbent polymer with faster absorption rate. A number of methods that are useful for measuring the swelling kinetics are described in Chapter 4, section 4.6.2. In the next several sections the attributes of the polymer and swelling liquid that affect the kinetics of free swelling are discussed. Then the influence of the compressive stresses during use are included in order to gain a clearer understanding of how superabsorbent polymers behave.

5.2.1. Maximum Swelling and the Rate Constant of Swelling

The swelling process is a diffusive process, with water diffusing into the polymer as the polymer chains relax and diffuse outwardly into the liquid. The polymer chains must diffuse and move for the volume of the polymer to increase. Because of the diffusive mechanism, a principal factor in swelling kinetics is the size of the polymer particle, its morphology (e.g., monolithic, porous, smooth or rough surface) and its shape (e.g., spherical, irregular granular, flakes). In Table 5.4 is a comparison of the swelling kinetics of two different manufactured superabsorbents, as measured by a characteristic swelling time from the vortex time analysis (described in Chapter 4, section 4.6.2).

Sample A is principally a monolithic, granular product, and the swelling time changes smoothly as a function of the particle size. Sample B is an agglomerated sample and is quite different in its swelling rate character as shown by the nonregular dependence of swelling time on particle size.

Because the swelling rate is critical to the performance of the superabsorbent

Table 5.4. Dependence of the sorption time, t_v, as measured in the vortex time analysis, on particle radius, r, for two different samples

Mesh Size	r (cm)	Sample A (t_v, sec)[a]	Sample B (t_v, sec)[b]
20–25	0.0387	392.2	8.2
25–30	0.0325	280.7	316.7
30–35	0.0274	171.1	215.7
35–40	0.0230	130.2	5.5
40–50	0.0177	86.3	4.2
60–70	0.0115	46.6	38.3

[a]Sample A is from a homogeneously crosslinked, granular product.

[b]Sample B is from a granular, partially agglomerated, rapid swelling product.

polymer in many applications, and especially when the polymer is under compression, the kinetics of swelling are reviewed first for particles that are not under load, and then the effects of an added load are considered. The swelling kinetics of superabsorbent polymer can be studied by application of a first-order kinetic expression based upon Fick's second law of diffusion,[43] as shown in Chapter 4, section 4.6.2. In first-order swelling processes, the rate of swelling is first-order in the amount of swelling capacity remaining at any time, as given by Eq. 5.18:

$$\frac{dQ}{dt} = k(Q_{max} - Q) \tag{5.18}$$

where Q_{max} and Q are the swelling capacities at equilibrium and at any time t, and k is the swelling rate constant, which depends on particle size and on the diffusion coefficient. The values of Q are given as the ratio of the mass of liquid absorbed to the dry mass of the polymer. First-order swelling behavior is commonly observed for the swelling of networks.[44–46] Taking the definite integral between the limits of swelling capacity equal to zero and Q_{max} yields

$$Q(t) = Q_{max}(1 - e^{-kt}) \tag{5.19}$$

In this model of the dynamic swelling process, the structural features of the polymer and system that affect the maximum swelling capacity influence the rate through the Q_{max} term. Other structural features of the polymer and all the physical characteristics of the system, such as the particle size, affect the swelling through the swelling rate constant k. The inverse of the swelling rate constant is also called the characteristic time τ of the swelling process, and is defined as[47]

$$\tau = \frac{r^2}{\pi^2 D} \tag{5.20}$$

for spherical particles, where r is the radius of the unswollen particle and D is a diffusion coefficient. The value of D depends on the structure of the polymer network. For ionic gels, the value of D also depends on the extent of ionization of the network, the ionic strength of the swelling liquid,[48] and any buffering capacity of the swelling liquid.[49] The diffusion coefficient will also increase with the temperature of the system. Temperature fluctuations are an important aspect of the performance of superabsorbent polymers in personal care sanitary goods, because the polymer can be flooded with warm body liquids, which then cool gradually during and after absorption by the polymer. The effects of these parameters on the swelling kinetics are now shown, based on experimental results.

5.2.2. Particle Size and Distribution

Based on Eq. 5.20 for the characteristic time τ, a plot of the square of the particle radius versus the sorption time should be linear with a slope that is dependent on the diffusion coefficient D. The functional relation between the square of the radius and sorption time is shown in Figure 5.18 for three samples of granular poly(acrylic acid)-based superabsorbent polymers separated into narrow particle-size fractions by means of sieving using U.S. Standard sieves.[50] The linear slope of t_v versus r^2 plots is consistent with a diffusion-controlled absorption rate. However, the three chemically similar samples have different swelling rates as evidenced by the different slopes of the lines. The diffusion or relaxation coefficients D may be calculated from the slopes of the lines in Figure 5.18. These are given in Table 5.5 along with the swelling capacity of the samples in 0.9% sodium chloride solution. The diffusion coefficients increase as the maximum swelling capacity decreases. Smaller swelling capacity results from a larger crosslink density, which also yields shorter elastic chains between the crosslinks. The shorter chains relax more quickly and yield a larger value of D, as shown in Table 5.5.

The particle-size distribution also affects the swelling rate of granular superabsorbents. The number-average particle radius r_{av}, calculated by means of Eq. 5.21, may be used to normalize swelling rate results for different particle-size distributions.

$$\frac{1}{r_{av}} = \sum_i \frac{x_i}{r_i} \tag{5.21}$$

In Eq. 5.21, x_i is the weight fraction of polymer in fraction i with radius r_i.[51] A series of blends of particles was synthesized by mixing narrow particle-size cuts, and the swelling time t_v according to the vortex method was determined as a function of the

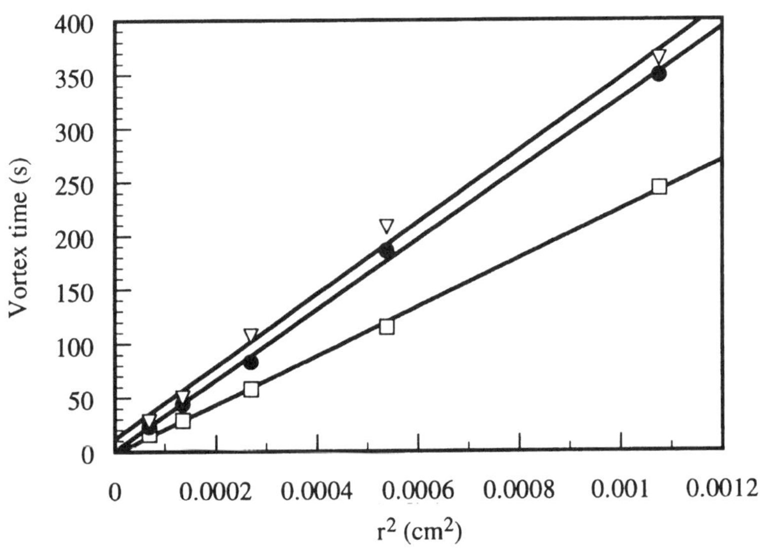

Figure 5.18. The vortex time of samples with different crosslink densities, taken as a function of the particle radius *r*.

Table 5.5. Dependence of the diffusion coefficient *D* on the crosslink density as reflected in the equilibrium swelling capacity *S*

S (g/g)	D (cm^2/s)
22.0	4.47×10^{-7}
34.5	3.10×10^{-7}
41.9	3.02×10^{-7}

average particle radius. A plot of t_v for these mixed fractions versus the square of their average radius is shown in Figure 5.19, superposed with the data for the narrow fractions from above. Excellent normalization is achieved by using the value of r_{av} calculated in this way.

5.2.3. Polymer Density and Surface Area

The various commercially available superabsorbent polyacrylates vary in their bulk densities, reflecting differences in particle-size distribution, particle packing efficiency (related to particle morphology), and the apparent density of the individual polymer particles. In addition, the particles may have different specific surface areas due to the presence of cracks and fissures and also differences in surface roughness. The model of the swelling kinetics is based on spherical particles, and some adjustment is necessary for the differences in real particles compared with

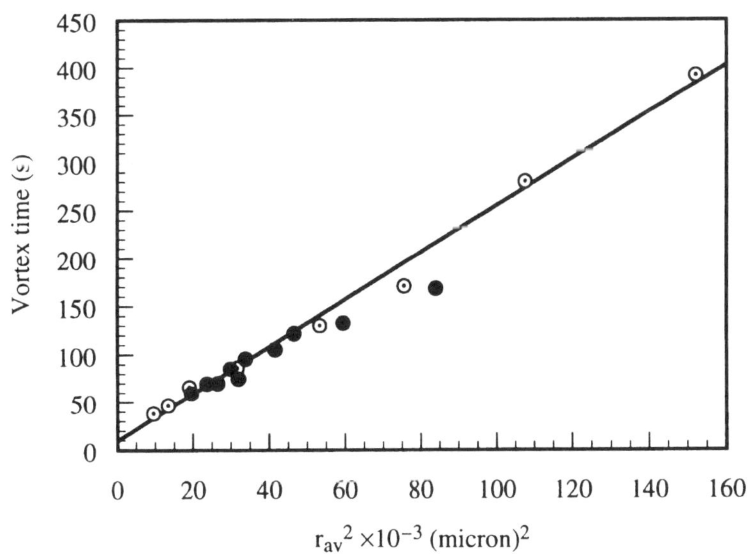

Figure 5.19. The swelling time measured by the vortex method as a function of the number average particle radius for narrow particle size cuts (open) and blends (closed) of superabsorbent polymer.

spheres. To illustrate these effects, the specific surface area of 80–100 mesh samples of various polymers was determined along with the apparent polymer density. The specific surface area was determined by means of the B.E.T. method with argon gas as the adsorbent, and the apparent polymer density was obtained by immersing a weighed sample of powder into a volume of methanol and noting the volume increase due to polymer. These data are given in Table 5.6. If the apparent density is measured as a function of particle size, the density reaches an asymptotic value of 1.667 g/cm^3, which may be taken as the true density of the sample. The differences of apparent density for samples in Table 5.6 from 1.667 g/cm^3 are a measure of the fraction of internal pores or bubbles in the sample.

The dependence of absorption rate on these additional properties was determined as follows. For any spherical particle mixture, the surface area per unit mass times the density is equal to $3/r$, determined from purely geometric considerations. For particles of known macroscopic radius (via mesh size) but with different polymer densities and porosities, the absorption rate is changed because the microscopic distance across which a molecule of water must diffuse has changed. This is accounted for by calculating an effective radius r_{eff} for the polymer particle. A polymer particle with a lower density (achieved perhaps by incorporating bubbles into the particle) will behave as if it has a smaller effective radius. A particle with a higher surface area must have open pores and will then also behave as if it has a smaller effective radius. Equation 5.22 is used to calculate the effective particle radius:

$$r_{eff} = r_0\left(\frac{A_0}{A}\right)\left(\frac{\rho}{\rho_0}\right) \tag{5.22}$$

In this equation, r_0 is the radius associated with the mesh size, A_0 is the surface area (per gram) for equivalent spheres of radius r_0, A is the actual surface area (per gram) for the mesh size with r_0 as determined by the B.E.T. method, ρ is the polymer density as determined by the immersion method, and ρ_0 is the polymer density in the absence of any physical defects (such as bubbles). This value of r_{eff} is then used, in

Table 5.6. Swelling rate–related properties of several superabsorbent polymers (40–60 mesh cut)

Polymer Sample[a]	Surface Area (m^2/g)	Sorption Time (s)	Swelling Capacity (g/g)	Bulk Density (g/cm^3)	Apparent Polymer Density (g/cm^3)
Arasorb 720	3.49	3.7	30.3	0.6098	1.500
Aridall 1078	0.16	3.1	44.4	0.3596	1.250
Aquakeep	0.25	2.0	65.4	0.4430	1.667
Sanwet IM-1000	0.10	17.1	58.3	0.5624	1.500
Drytech 510	0.08	56.0	39.9	0.8989	1.667

[a]The following trademarks are used: Arasorb (Arakawa Chemical), Aridall (American Colloids), Aquakeep (Sumitomo Seika), Sanwet (Sanyo Chemical), Drytech (Dow Chemical).

place of the particle radius as determined from the mesh size, in the diffusion equations. This substitution makes all the samples behave similarly as a function of their effective radii.

5.2.4. Temperature

Superabsorbent polymers are used under conditions in which the system temperature may change over time. For example, in a diaper, the superabsorbent polymer will first be bathed in a salt and urea solution that is at the internal temperature of the human body, but the resulting gel will cool slowly in contact with the external environment. The extent and rate of cooling will depend on the climate and other environmental factors such as the use of building air-conditioning systems. The diffusion coefficients of polymers in solution are temperature dependent, and this should be reflected in the absorption rate of superabsorbent polymers. The diffusion coefficient D may be related to the temperature T and molecular weight M of the polymer by the following equation:

$$D = \frac{kT}{CM^a} \tag{5.23}$$

where k is the Boltzman constant, C is a number characteristic of the polymer structure, and a depends on the solvent quality.[52] Based on Eq. 5.20, which relates the diffusion coefficient D to the swelling time, and on Eq. 5.23 for D as a function of temperature, the swelling time should be inversely proportional to the absolute temperature. The experimental verification of this is shown in Figure 5.20. Values of t_v were measured as a function of the temperature of the swelling liquid. The line in the figure was obtained by means of the linear least-squares method. The slope represents the temperature sensitivity of the swelling rate (as measured by the vortex method) for the sample. The value of the slope in this case is –4.81 s/K, which represents a change in rate of about 6% per degree at temperatures near body temperature.

5.2.5. Ionic Strength

Changing the ionic strength of the swelling liquid causes the maximum swelling extent of the polyelectrolyte gel to change, as noted earlier. If a superabsorbent polymer is swollen in aqueous solutions of different ionic strength, the rate appears to be faster in the solution with the lower salt concentration. This is a result of the larger maximum swelling extent that is achievable in the solution with less salt. However, the molecular weight between the crosslinks does not change when the sample is swollen in different salt solutions. The diffusion coefficient D and the swelling time behavior as a function of particle size should be affected only slightly because the value of a only varies slightly with changing salt concentration. This effect is shown in Figure 5.21. The plots of t_v versus r^2 are nearly co-linear, indicating almost no change in D over this fourfold change in NaCl concentration. We expect that the ex-

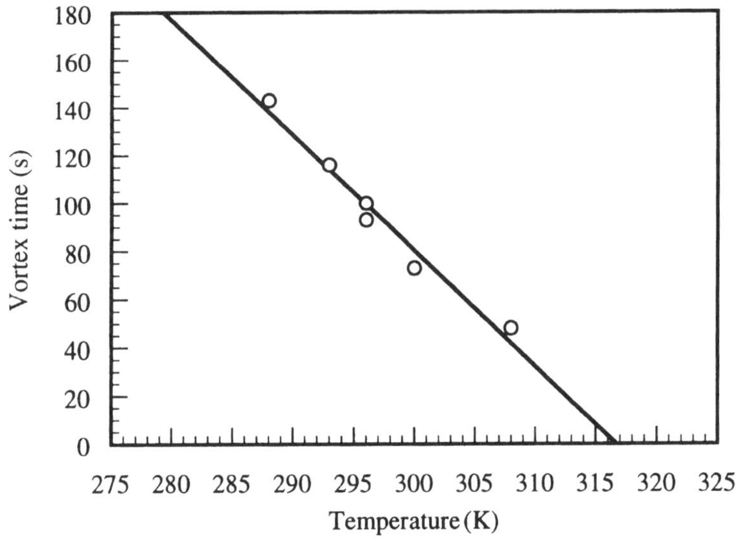

Figure 5.20. The effect of temperature on the swelling rate as measured by the vortex time test. For this sample the equation of best fit is $t_v = 429{,}153/T - 1{,}351$.

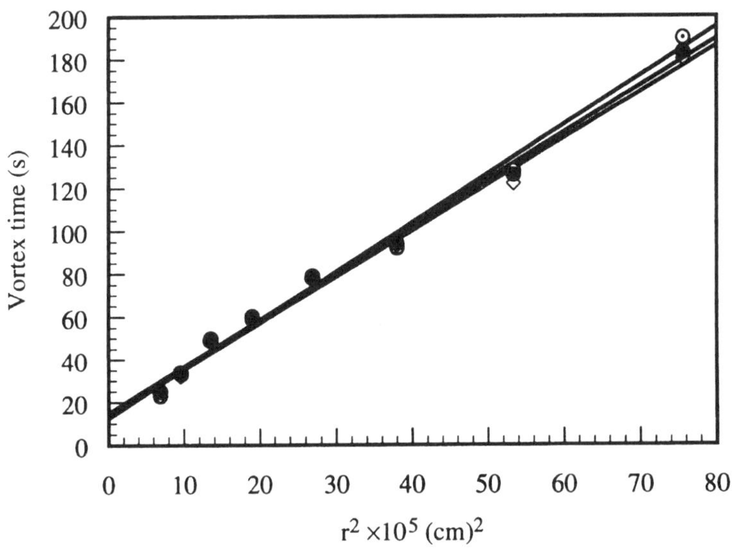

Figure 5.21. Vortex time of a sample in three different NaCl solutions, as a function of particle size: 0.225% NaCl (○), 0.45% (●), and 0.9% (◆).

tent of neutralization of the polymer also will affect the swelling rate in a similar fashion. As the extent of neutralization is increased, the principal effect on the swelling rate in a given salt solution will be the result of the change in maximum swelling capacity, and not a result of an appreciable change in the diffusion coefficient of the network polymer.

5.3. EQUILIBRIUM SWELLING AND KINETIC EFFECTS OF MICROPARTICLES UNDER LOAD

The absorption rate of liquid by superabsorbent polymers is also important under compressive load conditions, for example, when an infant lies or sits on the superabsorbent polymer in the diaper. The swelling kinetics under load is more complicated than in the case of free swelling described above, because the effects of the elastic deformation of the gel must be taken into account. In this section, a model of the swelling of superabsorbent polymers under load is given in terms of the structural attributes of the polymer.

5.3.1. Swelling Equilibrium Under Pressure

In section 5.2 the free-swelling rate was related to the equilibrium swelling capacity, particle size and particle-size distribution, specific surface area, and density of the superabsorbent polymer. In this section, these ideas are developed further to obtain a model of the rate of swelling under load in terms of the equilibrium swelling capacity (centrifuge capacity), the shear modulus of the gel, the applied pressure, the particle size and distribution, and the porosity of the collection of gel particles. As a result, the absorbency under load, the residual porosity of the mass of gels, as well as the hydraulic permeability of the resulting gel mass may be calculated.

Such a model must account for at least the following physical variables: (1) the applied pressure and (2) the porosity of the gel mass. It must also account for the following sample variables: (1) swelling capacity of the polymer, (2) elastic modulus of the sample as it swells, and (3) the particle size and distribution. The time dependence of the absorbency under load and its relationship to the other common measure of absorption speed, vortex time, is also of considerable interest.

The driving force for swelling is the inequality of the chemical potential of water (and ions) between the liquid and gel phases, and the extent of swelling is limited by the buildup of elastic stress in the expanding network. The elastic stress develops because the polymer chains are forced to occupy more and more volume as the polymer absorbs liquid. The resulting tension in the network increases the hydrostatic pressure in the gel phase. The pressure in turn increases the chemical potential of water in the gel, balancing it with the chemical potential of the water in the external liquid phase, ultimately limiting the swelling of the gel. When the crosslink density is higher, the critical tension is reached at lower content of water, and the maximum swelling is lower. A simple measure of maximum swelling is the centrifuged capacity.[53] Here, the 60-min centrifuge capacity is employed as the measure of maximum swelling of the gel in the absence of load. The change in swelling from this value, under a compressive load, is developed below.

The rate of swelling under a compressive load is simple to measure. An exponential curve may be fit to the swelling extent versus time data with excellent correlation coefficients. Exponential rate behavior may be simply described as a first-order process according to Eqs. 5.18 and 5.19. Fitting the experimental data to Eq. 5.19 yields values of Q_{max} and the rate constant k. Values of Q_{max} determined in this way are typically smaller than the centrifuge capacity. The rate constant, k, varies with the applied load, the modulus of the gel, and the particle size and distribution.

The values of Q_{max} and k may be calculated in terms of the fundamental properties of the gels according to the following concepts. Swollen networks are known to deswell under compression by a mechanism of reverse osmosis[54–56] as described in section 5.1.2. The added compressive pressure acts on the gel in a manner similar to additional crosslinks to increase the hydrostatic pressure in the gel phase and to reduce the maximum swelling. The gel exudes water spontaneously until the chemical potential of water is once again equal between the phases. The change in swelling of the gel that accompanies the change in chemical potential of water in the gel may be calculated from a theory of network behavior. This may be accomplished by using a scaling theory of the behavior of gels.

The scaling theory relates the swelling pressure Π (increases swelling) and the elasticity of the network G (restricts swelling) to the externally applied pressure P (also restricts swelling). The resulting Eq. 5.24 has the same form as Eq. 5.17 if the effects of mixing of chains and the ionic effects are lumped together:

$$P = \Pi - G \tag{5.24}$$

The swelling pressure is a function of polymer concentration:

$$\Pi = Ac^{9/4} \tag{5.25}$$

where A is the scaling factor (constant of proportionality), and the exponent $9/4$ (derived in the theory[57]) accounts for the interactions or correlations between polymer chains in a good solvent. The concentration c is expressed as mass fraction of polymer in the gel (gram polymer per gram gel) and is approximately the reciprocal of the swelling capacity (gram fluid absorbed per gram dry gel). The elastic modulus G is a function of polymer concentration and crosslink density (as in Eq. 5.7):

$$G = G_0 c^{1/3} \tag{5.26}$$

where the factor G_0 represents the reference modulus in the "as prepared" state and is proportional to the crosslink density. Substituting Eqs. 5.25 and 5.26 into Eq. 5.24 yields

$$P = Ac^{9/4} - G_0 c^{1/3} \text{ or } Ac^{9/4} = P + G_0 c^{1/3} \tag{5.27}$$

Rearranging Eq. 5.27 yields

$$c^{9/4} = 1/Q^{9/4} = \frac{P + G_0 c^{1/3}}{A} \tag{5.28}$$

where Q is the swelling capacity defined as the ratio of gel mass to the dry mass of the polymer. Taking the ratio of polymer concentration of a gel under pressure to its concentration in the absence of pressure, to normalize the result, yields Eq. 5.29:

$$\frac{c^{9/4}}{c_e^{9/4}} = \frac{P + G_0 c^{1/3}}{G_0 c^{1/3}} \tag{5.29}$$

Expressing the reference modulus, G_0, in terms of the fully swollen modulus in the absence of pressure (Eq. 5.26) yields equation 5.30:

$$G_0 = \frac{G_e}{c_e^{1/3}} \tag{5.30}$$

Combining Eqs. 5.29 and 5.30 yields

$$\left(\frac{c}{c_e}\right)^{9/4} = \frac{P + G_e(c/c_e)^{1/3}}{G_e} \tag{5.31}$$

Letting $m = c/c_e$ in Eq. 5.31, where m is a multiplier ($m \geq 1$) of the equilibrium concentration of polymer (zero applied pressure), and rearranging yields

$$m^{9/4} - m^{1/3} = P/G_e \tag{5.32}$$

To simplify this expression to a single-termed power law in m, values of P/G_e were calculated from Eq. 5.32 for values of m between 1 and 10. Then a simple power law in m was fit to the calculated values of P/G_e. The result of the fit is Eq. 5.33

$$1 + P/G_e = 0.9532m^{2.26} \text{ or } m = 1.02(1 + P/G_e)^{0.44} \tag{5.33}$$

Therefore, given an applied load P and the modulus of the fully swollen polymer G_e, the factor m, which is the ratio of polymer concentrations in the loaded versus unloaded state, may be calculated. If the concentrations are stated instead as swelling capacities Q, which are approximately the reciprocal of the polymer concentrations, Eq. 5.34 results:

$$Q = \frac{Q_e}{1.02(1 + P/G_e)^{0.44}} \tag{5.34}$$

Equation 5.34 indicates that the pressure behavior of swelling scales in the same way for all samples of superabsorbent polymer, as shown earlier in Figures 5.9 and 5.10. Ignoring the minor factor of 1.02 (introducing only a 2% deviation in the result), Eq. 5.34 can be rewritten as a ratio of the swelling capacity, Q, at some pressure to that at zero applied pressure, Q_e. The result is shown graphically in Figure 5.22. The graph demonstrates that the relative decrease in swelling capacity of any superabsorbent polymer is given by the ratio of the shear modulus to the sum of the modulus and the pressure. This aspect of the model is demonstrated in Figure 5.23 for hypothetical superabsorbent polymers with centrifuge capacities of 30 or 40 g/g, each with three different moduli. This figure shows the expected change in the absorbency under load with respect to pressure when only the modulus differs between samples and no time-dependent effects are present.

5.3.2. Time Dependence of the Elastic Modulus

The elastic modulus of the swollen gel is exactly proportional to the crosslink density of ideal polymer networks, as established in theory. For nonideal networks, the inverse relationship between the modulus and extent of swelling is approximate because the theories do not yet adequately treat network imperfections. Therefore, the elastic modulus of real networks must be measured in addition to their swelling capacity, for adequate understanding of the system. Also, the modulus of a particle of superabsorbent polymer changes continuously during the process of swelling. How the modulus changes during the swelling process and how the modulus affects the *rate* of swelling are described below.

The change of modulus during swelling of a given sample of superabsorbent polymer is easily formulated by combining Eqs. 5.26 and 5.30. This result, Eq.

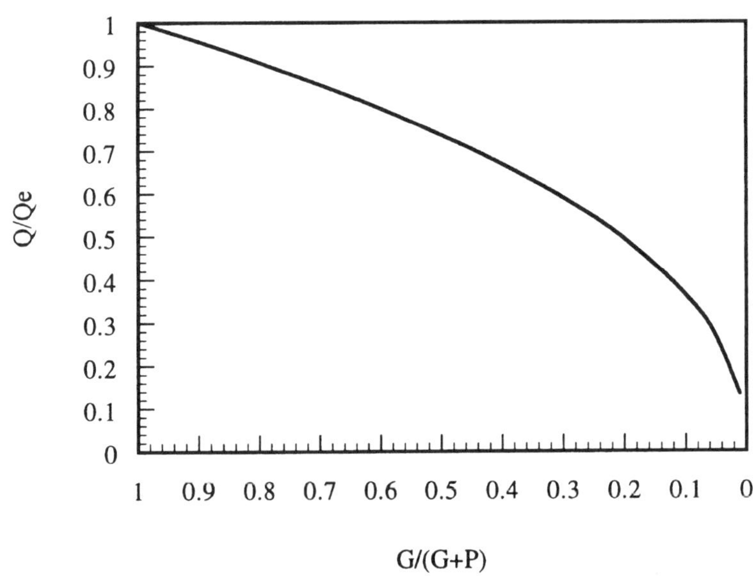

Figure 5.22. Normalized swelling capacity of superabsorbent polymer as a function of the modulus of the gel and the applied pressure. The porosity of the gel mass is not included.

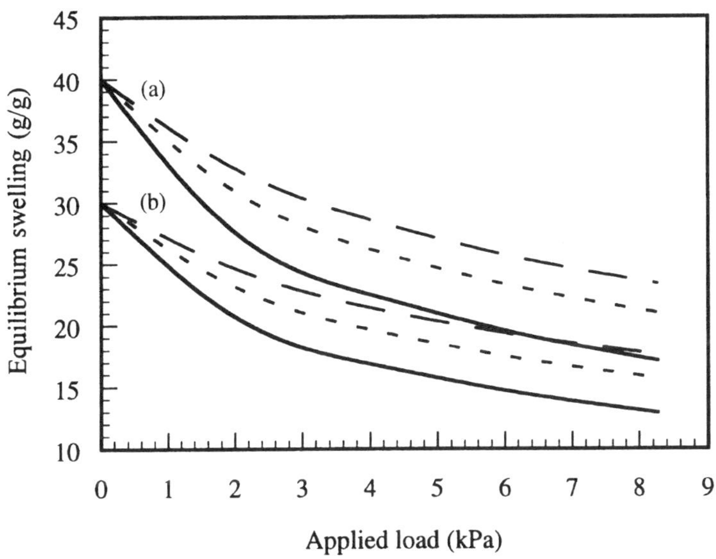

Figure 5.23. The pressure response of swelling for hypothetical superabsorbent polymers, **(a)** centrifuge capacity = 40 g/g and **(b)** centrifuge capacity = 30 g/g, with three different moduli: solid line = 1.5 kPa, short dashes = 2.5 kPa, and long dashes = 3.5 kPa.

5.35, is valid during the intermediate stages of the swelling process if the particle is uniformly swollen at all times.

$$G = G_0 c^{1/3} = \frac{G_e}{c_e^{1/3}} c^{1/3} = \frac{G_e Q_e^{1/3}}{Q^{1/3}} \tag{5.35}$$

Assuming that the gel particle swells uniformly throughout its volume during the swelling process, Eq. 5.35 may be rewritten in terms of a time-dependent modulus $G(t)$ and swelling ratio $Q(t)$. Equation 5.19 is used to express $Q(t)$, and Eq. 5.34 is used for the pressure-dependent value of Q_{max}. Making these substitutions, Eq. 5.36 is obtained for the modulus as a function of swelling time:

$$G(t) = \frac{G_e Q_e^{1/3}}{\left(\dfrac{Q_e}{(1 + P/G_e)^{0.44}}(1 - e^{-kt})\right)^{1/3}} \tag{5.36}$$

The value of Q_e needed in Eq. 5.36 is the ratio of gel mass divided by the mass of dry polymer. However, the swelling ratio commonly used in industrial practice (such as the centrifuged capacity method described in section 4.6.1.2) is a ratio of liquid mass absorbed to the dry polymer mass, S. The relationship between Q and S is:

$$Q = 1 + S \tag{5.37}$$

where S is the centrifuge swelling capacity in g/g. Equation 5.37 may be used to replace the values of the various Q in Eq. 5.36 with the corresponding values of S, if desired.

5.3.3. Modulus Dependence of the Diffusion Coefficient

The modulus also affects the swelling rate of a deformable gel under load.[58] Without external influences, the gel swells isotropically (maintains its shape). Nonporous gels, such as most commercial superabsorbents, swell by a diffusion mechanism. The water penetrates the gel in a direction toward lower water concentration and typically perpendicular to the surface of the sample. For a spherical particle, water penetrates radially. The polymer chains expand in a direction generally opposite to the direction of the water penetration, and the relative motion of the solvent and the chains is in opposite directions only. But an elongated particle of gel behaves somewhat differently.

Consider a long rod of gel rather than the spherical bead. Water penetrates each surface of the rod (i.e. the ends, and the cylindrical surface) to the same depth in a given time increment. If the only transport mechanism that occurs is diffusion of the small molecule, water, the relative change in dimensions of the rod is not equal in the radial and axial directions. For example, if the length of the rod is four times its diameter, a penetration depth equal to 20% of the diameter is equal to only 5% of the length. However, a long rod of gel does not behave quite in this way. Actually, the rod of gel increases each of its dimensions the same relative amount (isotropic swelling) because the motion of all parts of the gel are coupled through the crosslinks. Therefore, the relative motion of solvent molecules has a component parallel to the surface through which the solvent initially penetrates. In effect, an individual solvent molecule travels a longer diffusional path in a rod than a sphere, and the swelling rate therefore is slower. This phenomenon affects the rate of swelling under load.

Particles of gel swell under load and deform under compression. If the shape of the particle initially is approximately spherical, the shape becomes disk-like under compression and the swelling rate must decrease because of the longer diffusion path in the flattened particle. The magnitude of deformation depends on the modulus and therefore on the initial crosslink density and the extent of swelling. A more rigid particle of gel resists deformation and maintains a faster swelling rate under load than will the same size particle of a softer, lower modulus, more deformable gel. The slowing of diffusion in gels as a result has been modeled by Li and Tanaka[58] as a decrease in the diffusion coefficient with increasing change in axial ratio of the shape. Following their concepts, the diffusion coefficient of the swelling gel may be scaled by the term $(1 + P/G_e)$, which is a measure of the relative deformation of a gel of modulus G_e under a compression P:

$$D = \frac{D_0}{1 + P/G_e} \tag{5.38}$$

In Eq. 5.38, D is the apparent diffusion coefficient of the gel system, D_0 is the diffusion coefficient in the absence of nonisotropic deformation, P is the applied pressure, and G_e is the modulus of the gel in its fully swollen state. The diffusion coefficient D calculated by this method is then an average value taken over the entire swelling process.

Several theories of the swelling rate of gels are available, from which the diffusion coefficient may be related to the rate constant or to the characteristic time.[59] Following Li and Tanaka,[60] the rate constant of swelling of a sphere depends of the diffusion coefficient, D, of the system and the particle radius, r:

$$k = \frac{1}{\tau} = \pi^2 D/r^2 \tag{5.39}$$

Based on the shape-dependent diffusion coefficient concept that was discussed above, the rate constant must depend on the applied pressure and on the modulus of the gel particle. Replacing D in Eq. 5.39 with its value from Eq. 5.38 yields

$$k = \frac{\pi^2 D_0}{r^2(1 + P/G_e)} \tag{5.40}$$

where the diffusion coefficient of the system in the absence of pressure and modulus effects, D_0, must be directly related to the swelling rate under zero load. Therefore, using Eq. 5.40, the rate constant of swelling under load may be calculated from a knowledge of the diffusion coefficient in free swelling, the fully swollen modulus, and the applied pressure. The rate constant then may be used in Eq. 5.36 to estimate the swelling-time-dependent modulus $G(t)$. The temperature dependence of the swelling under load may be obtained by using the temperature dependence of D_0, which can be determined through the temperature dependence of the vortex time, as demonstrated in Figure 5.20.

5.3.4. Gel Porosity and Permeability

5.3.4.1. Gel Porosity. In the case of a collection of particles, the value of Q_{max} is taken to be the sum of the volume of the swollen particles (assuming a gel density of 1) and the volume of the pores, which is accounted for through the porosity, ϕ, of the gel mass according to Eq. 5.41:

$$Q = Q_{gel} + Q_{voids} = \frac{Q_{gel}}{1 - \phi} \tag{5.41}$$

The change in porosity of the gel mass as swelling proceeds is accommodated by considering the swelling mass of gel as a viscoelastic medium undergoing a creep process.

This type of process can be envisioned by using the Voight–Kelvin mechanical model of viscoelastic creep. The Voight–Kelvin model couples an elastic spring and a viscous dashpot in parallel to describe the stress–strain relationship. In our system, the elasticity of the gel at any time is the equivalent of the spring, and the change in elasticity with swelling is the equivalent of the viscous dashpot. The elasticity of the gel effectively dissipates over time as the gel swells and softens. This basic model of viscoelastic behavior has been used to describe the sponginess of bread,[61] the compression of open-celled foam,[62] and the settlement of saturated soils.[63] In these systems, the strain varies exponentially with time and the "viscosity" of the system and depends on the ratio of stress to the "elasticity."

For a collection of swollen gel particles, the porosity decreases as pressure is added, as in a sponge. Gels of high modulus resist deformation better than low modulus gels. Therefore, the porosity of a collection of swollen gel particles with high modulus decreases slower with added pressure than does a collection of low modulus gels. Of course, as any gel swells, the modulus decreases with time as the volume of gel increases. The change in porosity with time should also be characterized by a rate constant, which is taken as the same rate constant used earlier for the swelling change with time (Eqs. 5.19 and 5.20) because the porosity change depends on the extent of swelling and the resulting modulus. Using the swelling rate constant within the Voight–Kelvin model of the compression of open-celled foam, the porosity change in a collection of gel particles may be described as follows.

If an open-celled foam is compressed vertically while being confined in the lateral direction (constant cross-sectional area), the fluid in the cells is displaced and the resulting strain of the foam is equal to its change in vertical dimension. The change in vertical dimension is proportional to the volume change of the foam, which is equivalent to the change of porosity. Therefore, the change of porosity is equivalent to the strain in this geometry, as given in Eq. 5.42, where G is the value of the modulus at any time and ϕ_0 is the initial porosity.[61]

$$\phi_0 - \phi = strain = \frac{\phi_0(P/G)}{(1 + P/G)} \tag{5.42}$$

Rearranging Eq. 5.42 yields an expression for the porosity as a function of the pressure and the time-dependent modulus $G(t)$ (which is given by Eq. 5.36).

$$\phi = \frac{\phi_0}{1 + P/G(t)} \tag{5.43}$$

Equation 5.43 has the property that the porosity decreases to a minimum level with time. The minimum porosity depends on the modulus of the particles and the applied pressure. Thus, samples of low modulus may collapse to zero porosity (one aspect of gel blocking) at sufficiently high applied pressure. The actual volume of any voids is determined from the porosity and the volume of the swollen gel that exists at any time:

$$V_{voids} = V_{gel} \frac{\phi}{(1 - \phi)} \tag{5.44}$$

The swelling rate of the mass of particles also depends on the volume of the pores, because the applicability of the particle radius as a scale factor for the rate constant (Eq. 5.40) presumes that the particles have free access to liquid at all times. If the volume of pores or voids decreases to zero, the effective diffusion length in the collection of particles is the thickness of the bed, rather than the particle radius, and the swelling rate appears to slow down dramatically.

5.3.4.2. Gel Permeability. The swelling rate under load then depends on the porosity and the permeability of the particle mass. The permeability of the mass of particles is a measure of how rapidly liquid flows through the mass of swollen particles. Low permeability indicates that liquid cannot flow readily through the gel (gel blocking) and that any forced flow of liquid (such as a second insult of urine during use of a diaper) must take an alternate path (e.g., diaper leakage). The permeability of the mass of swollen particles may be determined by experiment[64,65] and the application of Darcy's law (Eq. 1.1). It is known to be a function of the porosity, the particle size of the grains that make up the medium, and the connectedness of the open pores. A number of models exist that allow for the prediction of permeability as a function of the porosity, geometry, and packing of the particles that make up the porous medium.[66] These semi-empirical models describe packing of rigid particles, and their direct application to packs of very deformable particles is questionable. Nevertheless, we can use one of the existing models to calculate permeability and compare those values to our experimental observations. The Rumpf-Gupta relationship (Eq. 5.45), which is a power law model that relates permeability of a collection of particles κ_{RG} to the particle diameter d_p and the porosity ϕ, may be used for random packs of beads. All existing models have a similar form, relating the grain size and a porosity function to the permeability.

$$\kappa_{RG} = \frac{d_p^2 \phi^{5.5}}{5.6} \tag{5.45}$$

Assuming that Eq. 5.45 holds as well for random packs of soft, deformable particles, the trend of permeability of a pack of swollen particles as a function of applied pressure is as shown in Figure 5.24. The permeability at maximum swelling is plotted as a function of applied pressure. The hypothetical mass of gel had a centrifuge capacity of 35 g/g and a modulus of 1.5, 2.5, or 3.5 kPa. The curves show that permeability should be maintained better for a sample of higher modulus, but that all samples will eventually decrease in permeability as pressure increases. However, for samples with high modulus relative to the applied load, the permeability may in-

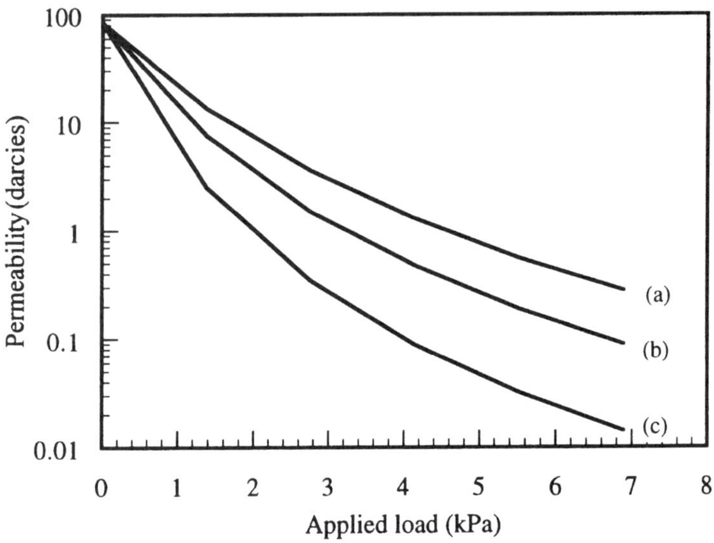

Figure 5.24. Permeability of the gel mass versus the applied pressure. The curves show the effect of increasing the modulus on the permeability: **(a)** 3.5, **(b)** 2.5, and **(c)** 1.5 kPa.

crease in the initial stages of swelling because the square of the particle size increases faster than the modulus decreases. This behavior is modeled in Figure 5.25.

5.3.5. Dynamic Swelling Under Load

Combining the result for porosity (and therefore permeability) of the gel mass (Eq. 5.43) with the expression for the gel swelling at any pressure (Eq. 5.34), and substituting them into the expression for the total swelling of gel and voids (Eq. 5.41), an expression for the total swelling under load at any time, $Q(P,t)$, is obtained:

$$Q(P,t) = \frac{Q_e}{(1 + P/G_e)^{0.44}} \frac{(1 - e^{-kt})}{\left(1 - \dfrac{\phi_0}{1 + P/G(t)}\right)} \tag{5.46}$$

In Eq. 5.46, the first term represents the maximum swelling of the gel (from Eq. 5.34), and the numerator of the second term represents the time function of swelling, which includes the shape-dependent and particle-size-dependent rate constant (from Eq. 5.40). The denominator of the second term represents the fraction of the gel mass that is gel particles (from Eq. 5.41) and includes the changes in porosity arising from the decrease in gel modulus that accompanies swelling.

The value of $Q(P,t)$ of several superabsorbent polymers having different crosslink densities was measured, and the results were compared with swelling curves generated from the current model. In each case, the parameters needed in the model were derived from measurements of the fundamental properties of the sample. Figure

5.26 shows the comparison of experiment to model for three polymer samples having different shear moduli. The comparison is generally good. However, samples of the low modulus sample A at small particle sizes and at high loads (compared with the swollen modulus) are fit very poorly by the present model. These samples were observed to gel-block in the measurement cell, characterized by dry polymer remaining at the top of the column of gel, next to the plastic piston. This is a different sort of gel-blocking than the loss of permeability discussed here. The loss of permeability is modeled as a homogeneous property of the mass of gel, whereas the dry polymer type of gel-blocking is a nonhomogeneous phenomenon.

5.3.5.1. Porosity. A simple test of the model given in Eq. 5.44 may be obtained by measuring the porosity of the gel mass by squeezing the gel in the measurement cell with the pressure of a finger to remove any free liquid and remeasuring the mass of the sample cell. The free liquid may be absorbed on a lab tissue paper to help draw the excess liquid out of the mass. This measurement was performed for the three different superabsorbent polymers used above, at various applied loads, and the experimental values are compared with the predictions of the Eq. 5.44 in Table 5.7.

As predicted, the volume of pores decreases with increasing load, and higher modulus samples, such as sample C, have larger pore volumes. However, the predicted values of void volume are consistently larger than the measured values, despite the relatively large errors associated with the measurement method. A more

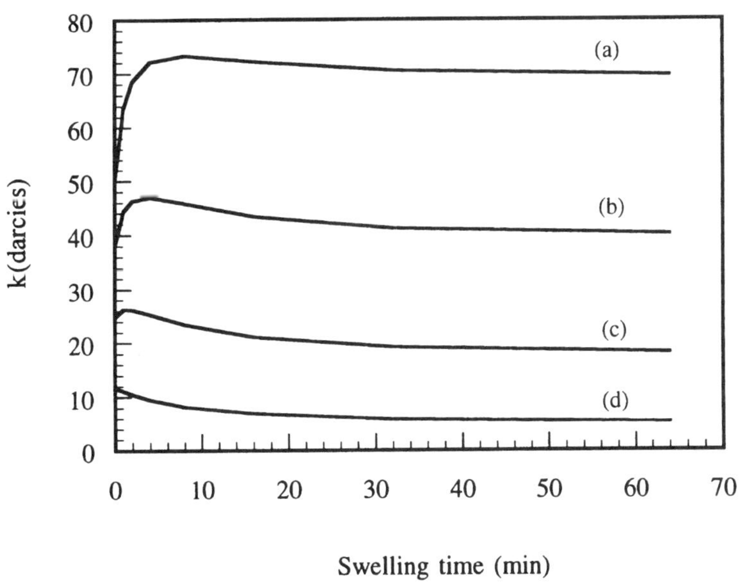

Figure 5.25. Permeability change in a mass of particles during swelling, calculated with the Rumpf-Gupta model (Eq. 5.49), as a function of the modulus of fully swollen gel: **(a)** 2.5, **(b)** 2.0, **(c)** 1.5, and **(d)** 1.0 kPa. All calculations are with a load of 2.07 kPa.

Table 5.7. Comparison of measured and calculated void volumes of various polymer samples and particle sizes (given in mL/g polymer)

		Void Vol. at 2.07 kPa		Void Vol. at 4.14 kPa		Void Vol. at 6.21 kPa	
Sample	Mesh Size	Measured	Calculated	Measured	Calculated	Measured	Calculated
A	25–30	1.4	1.7	nd	nd	nd	nd
A	45–50	1.7	1.9	nd	nd	nd	nd
B	25–30	2.0	3.8	0.5	2.1	nd	nd
B	45–50	1.5	3.9	0.8	2.2	nd	nd
B	20–100	2.1	3.8	0.6	2.1	nd	nd
C	25–30	2.4	5.7	0.8	3.2	0.2	2.0
C	45–50	3.9	6.0	1.6	3.6	0.8	2.5
C	20–100	2.4	6.0	1.3	3.5	0.6	2.4

precise experiment is necessary to adequately test the model prediction. The lowest value of porosity of a pack of deformable particles will occur when the polymer is fully swollen and shear modulus G is at its smallest value. Particles of lower crosslink density yield particle beds of lower porosity.

5.3.5.2. Permeability. The permeability may be predicted based on a knowledge of the particle size and porosity of the bed of gel and measured by timing the flow of a given amount of liquid through a given volume of swollen gel, as described in Chapter 4, section 4.6.4.3. Comparisons of the predicted and measured permeabilities of the three samples of superabsorbent polymer are shown in Table 5.8. The data and predictions in Table 5.8 are not close, unsurprisingly, but the trends are similar. The model, based on a random packing of rigid spheres, predicts much larger permeabilities than we observed for soft, deformable gels. At this time, not enough data exist to propose a new empirical permeability model of flow through deformable gels, but a power-law fit of the experimental data from sample C at two different particle sizes and three applied loads suggests that an equation of the form of the Rumpf-Gupta model, Eq. 5.45, will fit the data if a larger exponent of the porosity is used.

5.3.5.3. Particle Size. In the present model of swelling under pressure, the particle size and distribution are incorporated through the use of the average effective

Figure 5.26. Swelling under pressure for polymer samples A, B, and C versus time as a function of the applied load. Lines are model fits with the following parameters: **(a)** sample A, centrifuge capacity = 30 g/g; shear modulus, G = 1.2 kPa, r_{av} = 325 μm; ϕ_0 = 0.22, D= 2.2 × 10^{-7} cm^2/s. **(b)** sample B, centrifuge capacity = 29 g/g, G = 2.2 kPa, r_{av} = 212 μm, ϕ_0 = 0.28, a D = 1.3 × 10^{-7} cm^2/s. **(c)** Sample C, centrifuge capacity = 30 g/g; G = 2.99 kPa, r_{av} = 212 μm, (0 = 0.33, D = 1.0 × 10^{-7} cm^2/s.

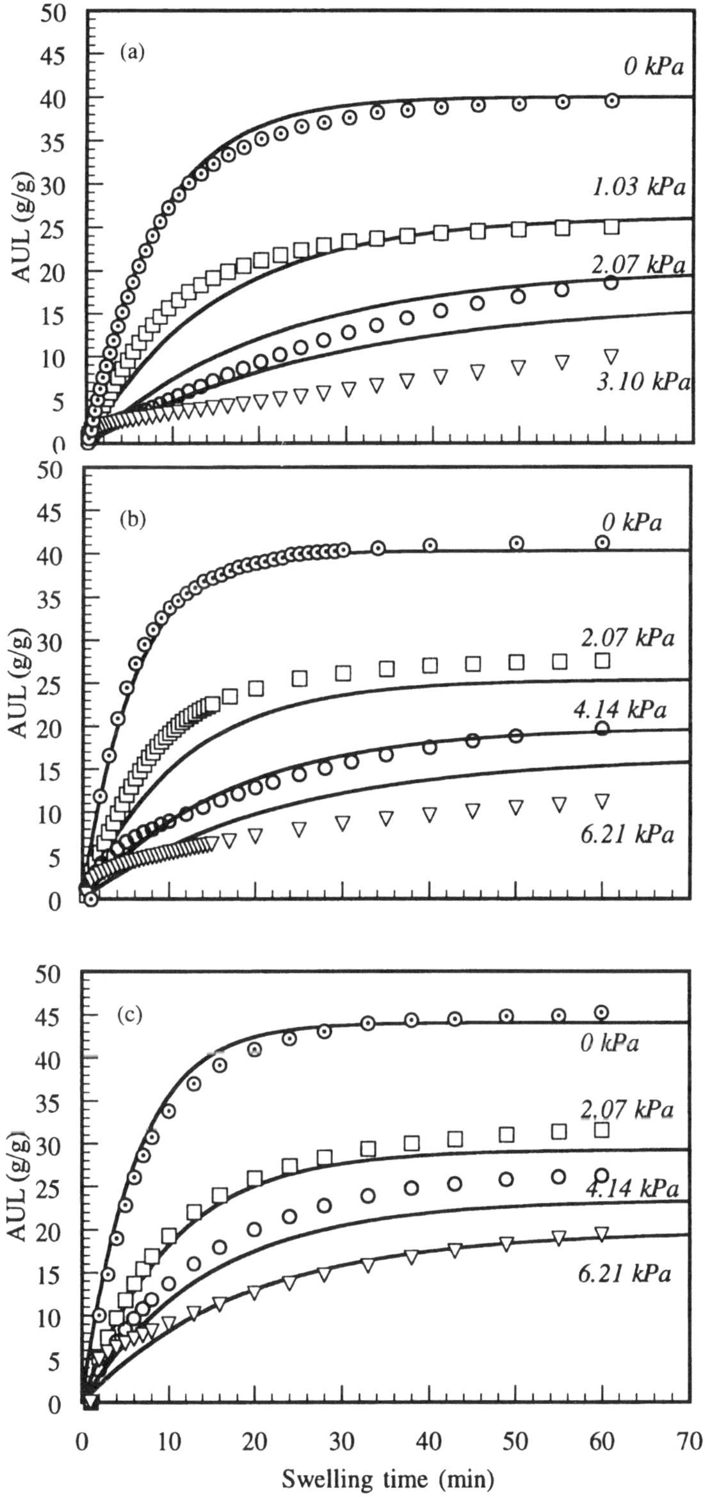

(a)
0 kPa
1.03 kPa
2.07 kPa
3.10 kPa
(b)
0 kPa
2.07 kPa
4.14 kPa
6.21 kPa
(c)
0 kPa
2.07 kPa
4.14 kPa
6.21 kPa
AUL (g/g)
Swelling time (min)

Table 5.8. Comparison of measured and calculated gel permeabilities of various polymer samples and particle sizes

		Permeability at 2.07 kPa		Permeability 4.14 kPa		Permeability at 6.21 kPa	
Sample	Mesh Size	Measured	Calculated	Measured	Calculated	Measured	Calculated
A	25–30	0.293	0.9	nd	nd	nd	nd
A	45–50	0.128	0.2	nd	nd	nd	nd
B	25–30	4.12	18.6	0.086	2.7	nd	nd
B	45–50	0.550	4.6	0.018	0.6	nd	nd
C	25–30	2.38	94.3	0.148	16.9	0.011	5.0
C	45–50	2.27	23.7	0.185	4.1	0.020	1.1

particle radius, similar to the case of free swelling rate. The average, effective particle radius may be calculated using the above equations with measured values of the nominal particle size, the density of the powder the specific surface area of the powder, and a knowledge of the particle size distribution (sieve analysis). This process may be useful to sort out which fundamental measurement is affecting the effective particle size. A simple method to obtain the average effective particle size of a sample when the magnitude of the component parts is unimportant is to measure the vortex time and centrifuge capacity of the sample and calculate the average effective particle radius from these measurements and the vortex time constant.

In general, the swelling rate increases as the particle size decreases, as given by Eq. 5.39. However, several other factors affect $Q(P,t)$ as particle size changes. Even for homogeneous gel samples, such as sample A, whose swelling capacity and gel modulus are practically invariant with particle size, the rate under pressure is not precisely linked to a change in the mesh size of the dry particles. This is due to a change in the density of the dry particles caused by air bubbles and cracks in the particle. These properties affect the swelling rate through Eq. 5.22. In addition to these effects, particles of sample B and sample C are thought to have a core–shell structure. As a result of the process that yields this structure, the different particle sizes have different swelling capacities and moduli that also affect the rate of swelling. We attempted to separate particle density and particle structure in the following way.

First, the particle size dependence of $Q(P,t)$ for sample A was measured. To accurately model the rate as the particle size (screen mesh) was changed, it was necessary to adjust the value of the diffusion coefficient, D_0, in Eqs. 5.38 and 5.39. The diffusion coefficient must be made larger with larger particle size to compensate for bubbles and cracks that are present in the larger particles. The adjusted values of D_0 at several particle sizes were then used to model the behavior of sample C at the same particle sizes. In other words, the density differences between particle sizes of sample A were assumed to be the same in sample C. Of course, the changes in capacity and modulus of sample C with particle size were also included in the calculations. The comparison of measured rate to calculated rate is shown in Figure 5.27.

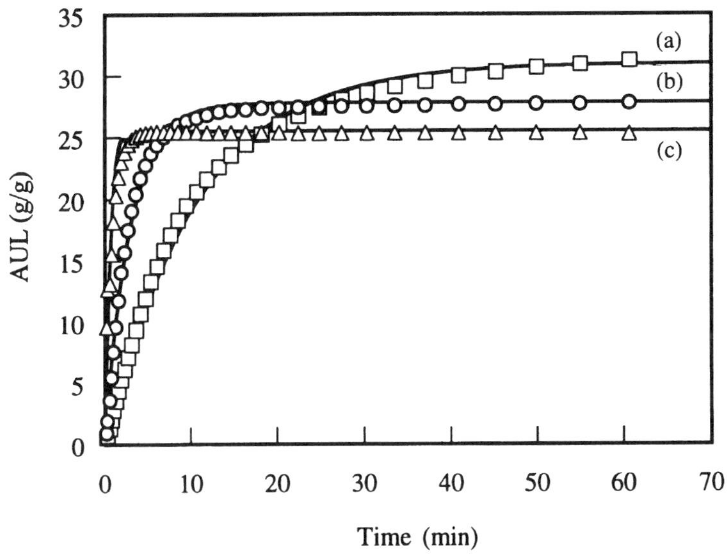

Figure 5.27. Comparison of measured absorbency under load rate to calculated values for three different particle sizes of sample C: **(a)** 25–30 mesh, **(b)** 45–50 mesh, and **(c)** 80–100 mesh.

The smaller particles have faster swelling rates as a result of the particle size and density differences but smaller maximum swelling as a result of their smaller centrifuge capacities. Ideally the density differences, and surface area differences (needed in Eq. 5.21) would be measured on the sample as a function of particle size to judge the adequacy of this approach.

5.3.5.4. Soluble Polymer. Another potential factor that can affect the rate of swelling is the amount of soluble polymer present in the polymer sample. Because of an increased driving force for swelling, the swelling rate of nonionic gels is increased by the presence of soluble polymer in the gel.[67] The molecular mass of the soluble polymer, relative to the molecular mass between crosslinks in the network, determines whether these molecules will diffuse from the gel and how rapidly they will be extracted[68] in the presence of excess external liquid phase. However, once extracted from the gel, the soluble polymer can depress the swelling of the sample by reducing the chemical potential of water in the external liquid phase[69,70] and may slow the flow of liquid through the mass of particles by increasing the viscosity of the external liquid phase. However, due to the complex nature of soluble polymer in gels, in this instance any positive or negative contributions of soluble polymer to the swelling rate have been ignored.

5.3.5.5. Areal Polymer Distribution. Typically, the swelling under pressure is determined with a constant mass of polymer added to the cell. In most cases, the amount of polymer is 0.16 g, which corresponds to a loading per unit area of cell of

0.0316 g/cm^2. This is about the same areal loading as the average loading in a modern thin diaper. If diapers become thinner yet, it may be necessary to test absorbent polymers at higher areal loading, unless the values of swelling under an applied pressure are independent of the extent of the system. However, such values can depend on the areal loading, becoming smaller as the amount of polymer per unit area increases. The decrease in swelling with pressure also can depend on the particle size as shown in Figure 5.28. Gels with higher modulus have a smaller dependence

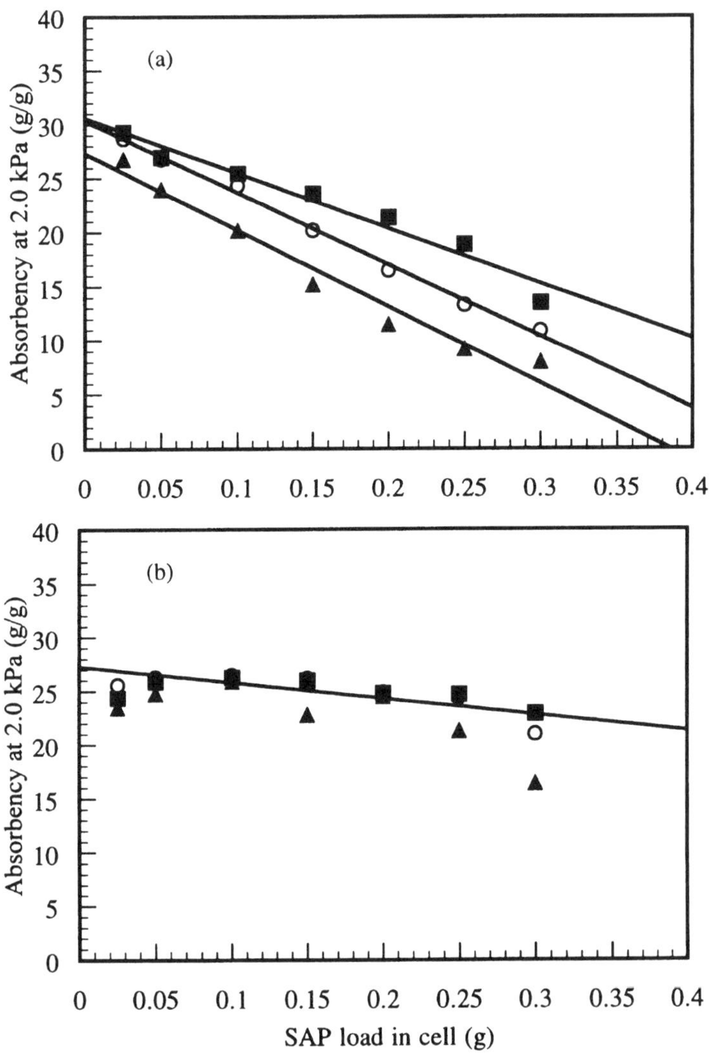

Figure 5.28. The effect of superabsorbent polymer (SAP) loading in the cell on the absorbency under load value obtained for a low modulus polymer **(a)** and for a high modulus polymer **(b)**. ■, 20–25 mesh; ○, 40–45 mesh; △, 70–80 mesh.

on particle size than gels with low modulus, as shown by a comparison of parts a and b in Figure 5.28. These results show that the extent of the system (number of particles per unit area) and the rigidity of the gels play important roles in the measured values of swelling under a compressive load and that the measurement does not yield information that is identical to the swelling pressure measurement discussed in section 5.2.

5.3.5.6. Rate Constant of Swelling Under Load. The rate constant of the swelling process is inversely related to the vortex time as shown by Eq. 5.20. The rate constant is related to the diffusion coefficient of the system by Eq. 5.39 and depends on the modulus of the polymer according to Eq. 5.40 when the polymer is deformed. The rate constant of the process measured in the measurement cell with no added load should therefore be proportional to the rate process in free swelling, and the vortex time must be related inversely to the rate constant of swelling in the unloaded measurement. In Table 5.9, the vortex time for a variety of samples and particle sizes is listed along with the rate constants calculated from the swelling kinetics in the measurement cell. The very good correspondence of the two measures of swelling rate is shown in Figure 5.29.

From the proportionality between the vortex time and the reciprocal of the swelling rate constant, a scaling factor may be formulated for the vortex time (no load) that will give a measure of the swelling rate constant under load. Equation 5.47 may be obtained from Figure 5.29 and Eq. 5.40.

$$\frac{1}{k_0} = 2.352 t_v = \frac{r^2}{\pi^2 D_0} \quad \text{at } P = 0 \tag{5.47}$$

Using this result in Eq. 5.40 at pressures other than $P = 0$ yields an expression that demonstrates the inverse proportionality of the swelling rate constant at any load to the vortex time at zero load.

Table 5.9. Comparison of vortex time and swelling capacity at that time in the AUL (0 kPa) analysis for different particle size cuts of different polymers

Sample	Mesh Size	Vortex Time (s)	Rate Constant (min^{-1})
A	25–30	170	0.121
A	45–50	52	0.386
A	20–100	122	0.180
B	25–30	349	0.080
B	45–50	103	0.266
B	20–100	127	0.260
C	25–30	229	0.103
C	45–50	74	0.370
C	20–100	162	0.152

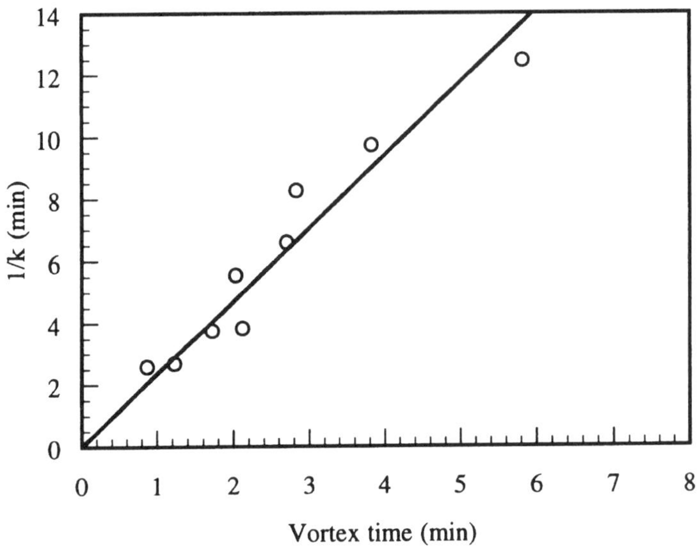

Figure 5.29. Correlation of the rate constant derived from absorbency under load under zero load to the vortex time of samples with varying swelling capacity, different particle sizes, and particle-size distributions. The line is $y = 2.352x$.

$$\frac{1}{k} = 2.352t_v(1 + P/G_e) \tag{5.48}$$

Thus, from the simple measurement of a characteristic swelling time (the vortex time) and the shear modulus of the superabsorbent polymer swollen to equilibrium, the rate constant of swelling under pressure can be determined.

In the previous section, the effects of compression on the swelling of a collection of particles of superabsorbent polymer are discussed in the light of a mathematical model of the swelling process coupled with a deformation process. The equilibrium swelling of the individual gel particle is of course decreased by compression; however, the swelling of a group of particles under compression is often much less than expected on the basis of thermodynamics alone. This phenomenon is a consequence of the dynamically changing elastic modulus and its effect on the porosity and permeability of the gel mass. Softer gels are more prone to deform into the pores during swelling under a load, and the swelling rate becomes limited by the thickness of the gel layer and the diffusion coefficient of the swelling liquid in the gel substance. The current model still assumes that the swelling of an individual particle is homogeneous throughout at each instant of swelling. This is clearly not true for some samples, as the outer layers of the gel relax before the centrally located sections. This feature is visually observable as a relaxation front moving into the swelling particle. The synthetic technique of surface crosslinking is an effective way to deal with the negative attributes of the soft relaxation front, by making the outer layers of the gel more rigid.

This chapter is an overview of the properties of superabsorbent polymers that are important to their performance in personal care sanitary products. The most important properties are the swelling capacity and the elastic modulus and how these change with time, volume, and compression during the swelling process. The structure of the polymer network is responsible for these properties, which may be altered by changing the structure of the network by application of the appropriate synthetic techniques.

REFERENCES

1. J. Hasa, M. Ilavsky, K. Dusek, *J. Polym. Sci. Polym. Phys. Ed.,* **13,** 253 (1975).
2. J. Ricka, T. Tanaka, *Macromolecules,* **17,** 2916 (1984).
3. W. Oppermann, *Angew. Makromol. Chem.,* **123/124,** 229 (1984).
4. H.H. Hooper et al., *Macromolecules,* **23,** 1096–1104 (1990).
5. M. Zrinyi, F. Horkay, *Polym. Bull.,* **3,** 665–670 (1980).
6. F. Horkay, M. Zrinyi, *Polym. Bull.,* **4,** 361–368 (1981).
7. F. Horkay, M. Zrinyi, *Macromolecules,* **15,** 1306–1310 (1982).
8. P.J. Flory, *Principles of Polymer Science,* Cornell University Press, Ithaca (1953).
9. P.J. Flory, B. Erman, *Macromolecules,* **15,** 800–806 (1982).
10. Y.L. Yin, *Swelling Behavior of Polyelectrolyte Gels,* Ph. D. Dissertation, Princeton University, 1993.
11. H. Tobita, *Macromolecules,* **26,** 5427–5435 (1993).
12. J.E. Mark, B. Erman, *Rubberlike Elasticity, A Molecular Primer,* John Wiley and Sons, New York (1988).
13. I. Bahar, B. Erman, *Macromolecules,* **20,** 1696 (1987).
14. A,G, Mikos, C.G. Takoudis, N.A. Peppas, *Polymer,* **28,** 998 (1987).
15. Y. Yin, R.K. Prud'homme, F. Stanley, In *Polyelectrolyte Gels: Properties, Preparation, and Applications,* R.S Harland, R.K. Prud'homme, eds., American Chemical Society, Washington, D.C. (1992).
16. W. Oppermann, In *Polyelectrolyte Gels: Properties, Preparation, and Applications*, R.S. Harland, R.K. Prud'homme eds., American Chemical Society, Washington, D.C. (1992).
17. A. Eisenberg, M. King, *Ion-Containing Polymers, Physical Properties and Structure,* Academic Press, New York (1977).
18. M. Rubinstein et al. *Macromolecules,* **29,** 398–406 (1996).
19. R. Skouri et al., *Macromolecules,* **28,** 197–210 (1995).
20. S. Pyschnyi, R. Rydzewski, *Continuum Mech. Thermodyn.,* **3,** 27–37 (1991).
21. J.S. Bastide, C. Picot, *J. Polym. Sci. Polym. Phys. Ed.,* **17,** 1441 (1979).
22. G. Maurer, J.M. Prausnitz, *Fluid Phase Equilibria,* **115,** 113–133 (1996).
23. R. Wehn, D. Woermann, *Polymer,* **28,** 1729 (1987).
24. T. Komori, H. Takehashi, N. Okamoto, *Coll. Polym. Sci.,* **266,** 1181 (1988).
25. S. Sakohara, F. Muramoto, M. Asaeda, *J. Chem. Eng. Jpn.,* **23,** 119 (1990).
26. E. Vasheghani-Farahani et al., *Ind. Eng. Chem. Res.,* **29,** 554–560 (1990).

27. L.S.A. Dubrovskii et al., *Polym. Bull.,* **24,** 107–113 (1990).
28. M. Silberberg-Bouhnik et al., *J. Polym. Sci. Part B Polym. Phys.,* **33,** 2269–2279 (1995).
29. S.R. Kellenberger, *Eur. Patent Appl. 339,461* (1989).
30. A.R. Reid, *U.S. Patent 4,051,086* (1977).
31. S. Obayashi et al., *Jpn. Kokai Tokkyo Koho 57–44,627* (1982).
32. T. Ito et al., *Jpn. Kokai Tokkyo Koho 58–42,602* (1983).
33. S. Obayashi et al., *U.S. Patent 4,507,438* (1985).
34. L.R. Parks, *U.S. Patent 4,295,987* (1981).
35. T. Tsubakimoto, T. Shimomura, Y. Irie, *U.S. Patent 4,666,983* (1987).
36. T. Tsubakimoto et al., *U.S. Patent 4,734,478* (1988).
37. J. Burgert et al., *WIPO Int. Appl. 94–20,547* (1994).
38. L.R.G. Treloar, *Polymer,* **17,** 142–146 (1976).
39. J. Klier, N.A. Peppas, *Polymer,* **28,** 1851–1859 (1987).
40. K. Saotome, U.S. *Patent 4,783,510* (1988).
41. M. Makita, S. Tanioku, *U.S. Patent 4,587,308* (1986).
42. K. Kimura, T. Hatsuda, K. Nagasuna, *Eur. Patent Appl. 450,924* (1991).
43. J. Crank, *The Mathematics of Diffusion,* 2nd Ed. Clarendon Press, Oxford, England (1975).
44. A.R. Berens, H.B. Hopfenberg, *Polymer,* **19,** 489–496 (1978).
45. A. Peters, S.J. Candau, *Macromolecules,* **19,** 1952–1955 (1986).
46. G.D. Cody, R.E. Botto, *Macromolecules,* **27,** 2607–2614 (1994).
47. F. Schosseler et al., *J. Dispersion Sci. Technol.,* **8,** 321 (1987).
48. I. Ogawa, H. Yamano, K. Miyagawa, *J. Appl. Polym. Sci.,* **47,** 217–222 (1993).
49. L.Y. Chou et al., *J. Appl. Polym. Sci.,* **45,** 1411–1423 (1992).
50. F.L. Buchholz, In *Absorbent Polymer Technology,* L. Brannon-Peppas, R. Harland, eds., Elsevier Science Publishers, New York (1990).
51. R.H. Perry, C.H. Chilton, eds., *Chemical Engineer's Handbook,* 5th Ed., McGraw-Hill, New York (1973).
52. H. Morawetz, *Macromolecules in Solution,* 2nd Ed., John Wiley and Sons, New York, pp. 290–299 (1965).
53. T. Tsubakimoto, T. Shimomura, Y. Irie, *Br. Patent Appl. 2,146,343* (1985).
54. A.M. Hecht, E. Geissler, *J. Chem. Physics,* **73,** 4077–4080 (1980).
55. I. Bahar, B. Erman, *Macromolecules,* **20,** 1696–1701 (1987).
56. W. Borchard, *Prog. Coll. Polym. Sci.,* **86,** 84–91 (1991).
57. P.G. deGenne, *Scaling Concepts in Polymer Physics,* Cornell University Press, Ithaca (1979).
58. Y. Li, T. Tanaka, In *Polymer Gels,* D. DeRossi et al., eds., Plenum Press, New York (1991).
59. Y. Li, C. Li, Z. Hu, In *Superabsorbent Polymers: Science and Technology,* F.L. Buchholz, N.A. Peppas, eds., American Chemical Society, Washington, D.C. (1994).
60. Y. Li, T. Tanaka, *J. Chem. Physics,* **92,** 1365–1371 (1990).
61. S. Swyngedau, A. Nussinovitch, M. Peleg, *Polym. Eng. Sci.,* **31,** 140–144 (1991).

62. E.A. Meinecke, R.C. Clark, *Mechanical Properties of Polymeric Foams,* Technomic Publishing Co., Westport, CT (1973).
63. J.E. Bowles, *Physical and Geotechnical Properties of Soils,* McGraw-Hill, New York (1984).
64. M. Plischke, M. Schmidt, C.P. Bewick-Sonntag, *Eur. Patent Appl. 640,330* (1993).
65. T. Amiya et al., *Br. Patent Appl. 2,267,094* (1993).
66. F.A.L. Dullien, *Porous Media, Fluid Transport and Pore Structure,* Academic Press, New York (1979).
67. A.B. Scranton, J. Klier, N.A. Peppas, *Polymer,* **31,** 1288–1293 (1990).
68. M. Zrinyi, F. Horkay, *J. Polym. Sci. Polym. Phys. Ed.,* **20,** 815–823 (1982).
69. V.V. Vasilevskaya, A.R. Khokhlov, *Macromolecules,* **25,** 384–390 (1992).
70. J. Kiefer et al., *J. Coll. Polym. Sci.,* **271,** 253–261 (1993).

6

OTHER SUPERABSORBENT POLYMER FORMS AND TYPES

David S. Allan

6.1. INTRODUCTION

As new demands are placed on superabsorbent polymers to provide new functions in existing applications and to meet the needs of new applications, improvements in the products will be required. This chapter explores some of the advanced performance properties that have been identified as potentially valuable for superabsorbents' emerging roles. The primary focus is on advanced superabsorbent polymers for absorbent personal care devices. Advances supporting other potential or burgeoning applications are mentioned as appropriate. Other applications for superabsorbent polymers are discussed in Chapter 7.

The components in an absorbent core must perform three principal functions in absorbing fluid: fluid acquisition, distribution, and retention. (See Chapter 1 for discussion of the absorbency model.) In early absorbent products containing superabsorbent polymers, the polymer functioned merely to absorb and contain fluid. However, the role of superabsorbent polymers has expanded with improved engineering of the absorbent core and improved ability of superabsorbent polymer producers to tailor specific superabsorbent polymer properties. For example, the ability of superabsorbent polymers to retain fluid under an externally applied load is now at least as important as the overall fluid capacity for advanced absorbent cores. In some applications, such as adult incontinence devices, a rapid rate of fluid uptake is of particu-

Modern Superabsorbent Polymer Technology. Edited by Fredric L. Buchholz and Andrew T. Graham.
ISBN 0-471-19411-5.

lar importance, so polymers with appropriately matched absorption rate have been developed (see Chapter 7, section 7.1).

There are several reasons for the development of advanced superabsorbent polymers. Diaper manufacturers would like to reduce their manufacturing costs. A superabsorbent that can take on other roles in the absorbent core might displace some other components in the core, perhaps reducing raw material costs and simplifying construction. For example, a superabsorbent fiber that can provide rapid fluid absorption and adequate wicking and transport of fluid could displace cellulose fluff in the absorbent core, leading potentially to a simpler or less expensive diaper. On the other hand, manufacturers of premium diapers seek to differentiate their products in order to command a higher price. An advanced superabsorbent that confers marketable performance advantages—for example, biodegradability—to an absorbent product may be of special value to diaper manufacturers. These advances will probably be in new performance dimensions not currently appreciated rather than simply improvements in the current performance properties.

This chapter considers some of the new performance dimensions that might be provided by advanced superabsorbents. Candidate materials, including nonconventional chemistries and physical forms, are described. Comparative property data are presented when available, but, because of the seminal nature of some of the technologies presented, detailed information and property data are often unavailable.

6.2. ALTERNATIVE SUPERABSORBENT POLYMER FORMS

The vast majority of superabsorbent materials are used in granular form. These granules are very effective, and technologies for accurately dosing and placing them in absorbent products are well known. However, granular solids exhibit a number of deficiencies associated with their physical form. Unless the granules are immobilized in the absorbent core, they can migrate in the structure. Since the placement of the superabsorbent polymer in the absorbent core contributes significantly to its effectiveness, migration of superabsorbent polymer may reduce the structure's overall effectiveness. Superabsorbent polymers are hard, brittle materials that can damage other components in the absorbent product (e.g., they can cause pinhole formation in the waterproof backsheet of a diaper) or can be damaged themselves during the manufacturing process. Alternative physical forms may be packaged more conveniently, for example, as roll goods, making them easier to handle. They also may provide additional functionality in service. Alternative forms to granules that have been investigated include fibers, films, foams, coatings, sprays, and laminates. Of these, fibers, films and laminates, and foams have received the most attention.

6.2.1. Fibers

Over the years many companies have shown interest in superabsorbent fibers. Fibers promise a tantalizing array of performance and handling advantages over conventional powdered superabsorbent polymers. Since their chemical nature is similar to conventional superabsorbent polymers (lightly crosslinked polyelectrolytes), the advantages offered by fibers derive from their physical form and dimensions. Fibers typically have a diameter of about 30 μm,[1] while granules are usually between 200 and 800 μm. The larger surface-to-volume ratio of fibers over granules presents much greater surface area to the fluid, leading to enhanced fluid absorption rates and diminished loss of absorbency due to surface fouling by some fluids, e.g., blood or protein solutions. Superabsorbent polymer fibers may participate in wicking of fluid and therefore displace some portion of the wicking component (usually cellulose fluff) in an absorbent structure. Superabsorbent polymer fibers are less likely to migrate in an absorbent core because they become intertwined with cellulose fibers in the air-laid fabric. The superabsorbent polymer fibers can also contribute to the integrity of the fabric.[2] The soft fibers may lend a more comfortable feel to an absorbent product or be used in a thinner product than a granule-based product.

Along with the promise of fibers come several significant obstacles to their success. The performance of fibers lags granular superabsorbents in several areas. The ability of fibrous superabsorbent polymers to retain fluid under an externally applied load is generally less than that of granules. Because the bulk density of fibers typically is lower than that of granules, storage and handling facilities for fibers must be larger. Additional floor space for fiber opening (separation), metering, and blending will also be required.[3] Their high surface area makes fibrous superabsorbent polymers quite moisture sensitive, so the humidity in processing areas should be controlled below 60% relative humidity. Finally, the present cost of superabsorbent fibers exceeds by several times the cost of granular superabsorbents. Until the cost/performance ratio of fibers surpasses conventional superabsorbent polymers, or until a high-value application develops for which fibers are uniquely suited, the market for superabsorbent polymer fibers is expected to remain relatively small.

The use of superabsorbent fibers in bandages, training pants, sanitary napkins, diapers, and incontinence garments is well known. Both The Kimberly–Clark Corporation[4] and The Procter & Gamble Company[5] (P&G) have been evaluating the performance of superabsorbent polymer fibers in these products. In addition to personal care products, superabsorbent polymer fibers have been used in cable wrap for undersea communications cables to prevent water penetration if the cable covering is damaged. Air-laid fabrics containing superabsorbent polymer fiber may be used in filtration devices to remove water from nonaqueous fluids, in meat packaging trays, absorbent pads for beds, and swellable gaskets, as well as personal care products. These products typically are made from superabsorbent fibers that have been previously spun, crosslinked, cut, and packaged.

6.2.1.1. Preparation. Currently there are several companies with commercial efforts in the superabsorbent fiber market (Table 6.1). Technical Absorbents Limited, a joint venture of Allied Colloids and Courtaulds Fibers Limited, produces Oasis superabsorbent fiber; Camelot Superabsorbents Limited produces Fiberdri superabsorbent fibers; and Toyobo Company Limited makes Lanseal superabsorbent fibers. The first two producers supply mostly the personal care industry, while the Lanseal product is used primarily for cable wrap applications. The products are based on

Table 6.1. Commercial superabsorbent fibers

Trade name	Chemistry	Manufacturer
Oasis	Acrylic acid/methyl acrylate co-polymer	Technical Absorbents
Fibersorb	Isobutylene/maleic anhydride co-polymer	Camelot Superabsorbents
Lanseal	Surface-hydrolyzed poly(acrylonitrile)	Toyobo Corporation

completely different technologies from one another. All of these companies have semicommercial plants making the superabsorbent polymer fibers. The sales price of fibers is roughly three to four times that of powdered superabsorbent polymer because of the small production facilities, the additional capital required for spinning fibers, and the low production rate of these products.

Oasis fiber is based on the chemistry and process developed by Allied Colloids and the fiber-making expertise of Courtaulds. The fiber is dry-spun from an essentially linear co- or terpolymer of acrylic acid, methyl acrylate, and hydroxyl-containing monomers, such as hydroxyethyl acrylate, and hydroxypropyl methacrylate.[6–10] The methyl acrylate is incorporated as a "flexibilizing" agent or internal plasticizer, and the hydroxyl-containing monomer is present to form crosslinks during an elevated temperature, postspinning, heat treatment step. (Self-crosslinking of hydroxyl groups with carboxyl groups is discussed in Chapter 2, section 2.7.3.) For example, a co-polymer consisting of 78 mol% acrylic acid (75% neutralized as sodium salt), 20 mol% methyl acrylate, and 2 mol% hexapropylene glycol monomethacrylate was prepared as a 38% aqueous solution.[11] Poly(vinyl alcohol) was added in the proportion of 2% relative to the total polymer. The dispersion was spun into filaments at 90°–100°C through a spinneret into a cell, where the water was evaporated in an air bath at about 150°C. The filaments were wound to yield a fiber of approximately 15 dtex (1 dtex = weight in grams of 10,000 m of fiber), or about 150 μm, and subsequently crosslinked by heating in air for 6–10 min at 210°C. The free swell capacity reportedly ranged from 40 to 50 g/g and the absorbency under 1.7 kPa load from 24 to 27 g/g (0.9 wt% NaCl). Courtaulds also has several recent patents that continue the theme of combining these monomers, by means of water phase polymerizations, into linear polyelectrolytes, spinning the viscous solutions into fibers, and then thermally postcrosslinking them to form superabsorbent fibers.[11–13] This technology can also be used to prepare superabsorbent films.

Camelot Superabsorbents' Fibersorb and Fiberdri superabsorbent fibers are based on chemistry and process developed by Arco and DuPont. Maleic anhydride and isobutylene are co-polymerized with a crosslinker selected from the group of alkanolamines (e.g., tris(hydroxymethyl)aminomethane) or polyhydroxylic compounds (e.g., di- or oligosaccharides or poly(alkylene glycol)). The fibers are formed by a dry spinning process and crosslinked by curing at about 180°C.[14–16] For example, ISOBAM 10 isobutylene/maleic anhydride co-polymer (46.6 mol% maleic anhydride, Kuraray Isoprene Chemical Company, Ltd.) was dissolved in water to form a 39 wt% solution. About 54% of the pendant carboxylate groups were neutralized with 50% sodium hydroxide solution. Propylene glycol was added in an amount of 3–10 parts per hundred parts of resin (phr), and water was removed to yield a 45% solids composition suitable for fiber spinning. Fibers having diameters of 5–25 μm were produced by dry spinning and were subsequently cured at 210°C for 30 min. Presumably the hydroxyl groups form ester crosslinks with the free carboxylic acid groups at elevated temperatures, fixing the fiber structure.

Another approach to superabsorbent fibers is based on surface modification of a nonsuperabsorbent acrylic fiber. Toyobo Corporation's Lanseal fiber is a composite fiber made by hydrolyzing the surface of poly(acrylonitrile) fibers with concentrated sodium hydroxide solution. The resulting fiber has both absorbency and a strong textile fiber core. The absorbency of 0.9% sodium chloride solution is 40–50 mL/g. Tensile strength is 1.9–2.4 g/denier, and the dry elongation of the dry fiber is 15%–35%.[17] This product is produced on a semicommercial scale, primarily for cable wrap applications (see Chapter 7, section 7.5).[18–20]

A variety of other approaches to superabsorbent polymer fibers have been described in the patent literature, although none of these is practiced commercially. Several are based on the approach of forming fibers from an essentially linear, hydrophilic polymer followed by a postspinning crosslinking step. Hirakawa et al.[21] described the formation of fibers based on carboxyl-modified alcohol-containing polymer such as poly(vinyl alcohol) or poly(ethylene-co-vinyl alcohol). The polymer fibers were melt spun using conventional equipment. The carboxyl modification was accomplished by acetalizing the polymer hydroxyl groups with a carboxyaldehyde at 40°–90°C in the presence of a strong acid such as H_2SO_4, as shown in Scheme 6.1.

Superabsorbent fibers can also be prepared by conventional free-radical co-polymerization of maleic anhydride, styrene or other aromatic comonomer, and optionally a third co-monomer such as N-vinylpyrrolidone or isobutyl acrylate.[22,23] The resulting polymer can be dissolved in a spinning solvent containing a crosslinker capable of reacting with the anhydride group in the amount of 0.5%–2% based on the co-polymer weight. Suitable crosslinking agents are polyols, aminoalcohols, and polyamino compounds, such as ethylene glycol, diethylene glycol, tetraethylene glycol, and *o*-aminophenol. The solution is pumped through a spinneret and wet-spun into a coagulant bath or dry-spun into a hot air bath to yield filaments. These

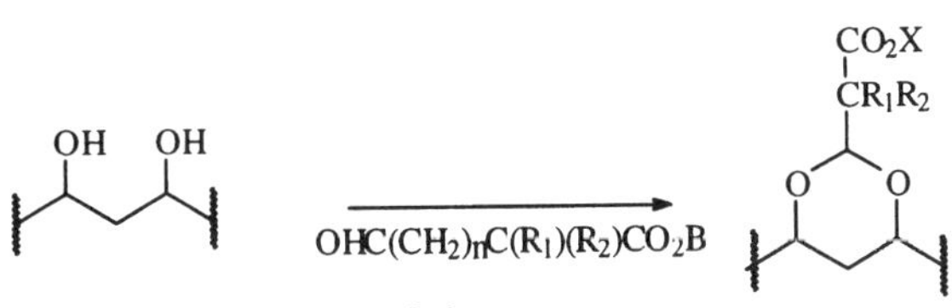

n = 0, 1
R_1, R_2 = H or alkyl
(at least one is alkyl)
B = H or alkyl
X = H or monovalent cation

Scheme 6.1

filaments may then be crosslinked by a thermal treatment. The crosslinked fibers are postneutralized with gaseous or liquid ammonia or sodium hydroxide solution, or both, to yield fibers with an absorbency of physiological saline solution of about 20–40 g/g. These compositions may also be used to prepare superabsorbent films.

Hybrid superabsorbent mats have also been made by combining streams of melt-blown thermoplastic fibers and aqueous fiber-streams of superabsorbent polymers.[24] The intermingled fibers of the two components form a mat while the heat from the thermoplastic fiber evaporates the water from the superabsorbent and thermally activates a crosslinking reaction. The superabsorbent fiber is spun from a 60% solids solution of poly(acrylic acid). A 3% solution of benzoyl peroxide is added in-line to the spinning dope just before the dope passes through the spinneret. The superabsorbent polymer fibers exit the spinneret vertically downward and are impacted with a high velocity stream of hot melt-blown thermoplastic fibers, with which they entangle. The heat from the melt-blown fibers dries the acrylic acid polymer and initiates decomposition of the benzoyl peroxide. The peroxide radicals presumably cause crosslinking of the acrylic acid by a grafting mechanism. The fabrics prepared from these hybrid fibers had superabsorbent polymer concentrations ranging from 17% to 25%. The ability to absorb deionized water increased in proportion to the superabsorbent polymer content. The advantage claimed for this method was a reduction in "dusting out" of the superabsorbent component in a web that could be easily handled.

6.2.1.2. Properties. The commercial superabsorbent polymer fiber producers claim a variety of property enhancements over conventional granular superabsorbents, especially an enhanced rate of absorption. The usual property analyses that are applied to granules, e.g., absorbency under load and centrifuged capacity, are not well suited to evaluating fibrous superabsorbents. The fibers are not readily loaded into standard testing cells. Particle size effects that are minimized by testing narrow size cuts of granules have ambiguous impact on fiber testing. And fiber–fiber interactions may be significantly different from particle–particle interactions in a bed of superabsorbent in a test cell. For these reasons, comparisons of the measured performance properties of fibers and granular superabsorbent polymers should be viewed with caution. For example, the claimed absorbency under load for two commercial superabsorbent fibers are compared in Table 6.2 with values measured in our laboratories. A large discrepancy between the claimed values and measured values exists. The free-swell capacity overstates the actual superabsorbent polymer capacity due to inclusion of interstitial water in the measurement; this overstatement is greater for fibers, with their relatively large interstitial volumes and strong capillary forces. The differences in absorbency under load may be ascribed in part to the different loads applied in the published data versus our measurements.

A more realistic assessment of superabsorbent fibers may be made by testing composite absorbent cores. We have compared Oasis 101 superabsorbent fiber with conventional granular superabsorbents in composite structures with cellulose fluff (Henton et al., The Dow Chemical Co., unpublished results). Tests included the fluid intake–flowback evaluation (FIFE) test,[25] the vertical wicking test,[26] and a pad

Table 6.2. Comparison of claimed with measured absorbency properties

Fiber	Capacity (g/g)		AUL (g/g)	
	Claimed	Measured	Claimed	Measured
Oasis 101 (Technical Absorbents)	45[a,b]	16.4[c]	21[d]	17.7[d]
Camelot fiber (unknown type)	80–85[b,e]	29.8[c]	55[f]	8.9[d]

[a]Property data taken from Technical Absorbents product literature.
[b]"Free-swell absorbency."
[c]"Centrifuged capacity."
[d]2.0 kPa.
[e]Data from Dan.[2]
[f]3.4 kPa retention.

absorbency under load (AUL) test. The pad AUL test was similar to the conventional AUL test of polymer granules except that a 4 × 4 in square of composite was tested under a 2.0 kPa load. Loadings of superabsorbent polymer ranged from 25 to 50 wt%.

The vertical wicking of pads containing superabsorbent polymer fiber was similar to that of pads containing granular superabsorbent polymer. The enhanced absorption rate of composite pads containing superabsorbent polymer fibers over conventional superabsorbent polymer is evident in results of the FIFE test, shown in Figures 6.1–6.3. The three figures show fluid uptake times for three 80-mL insults

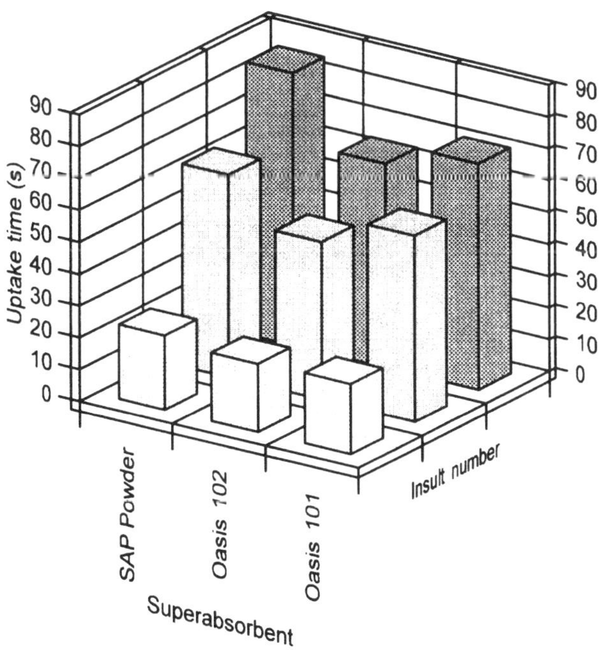

Figure 6.1. Comparison of FIFE performance of superabsorbent fiber with granular superabsorbent, 25% superabsorbent polymer (SAP) loading.

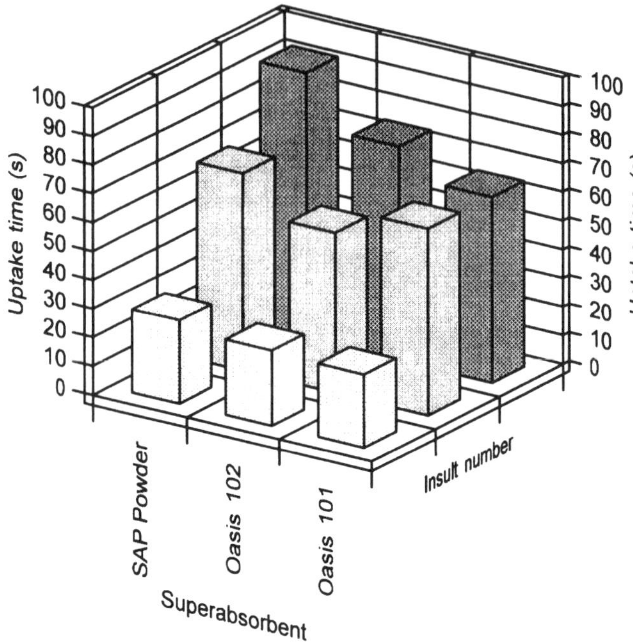

Figure 6.2. Comparison of FIFE performance of superabsorbent fiber with granular superabsorbent, 37% superabsorbent polymer (SAP) loading.

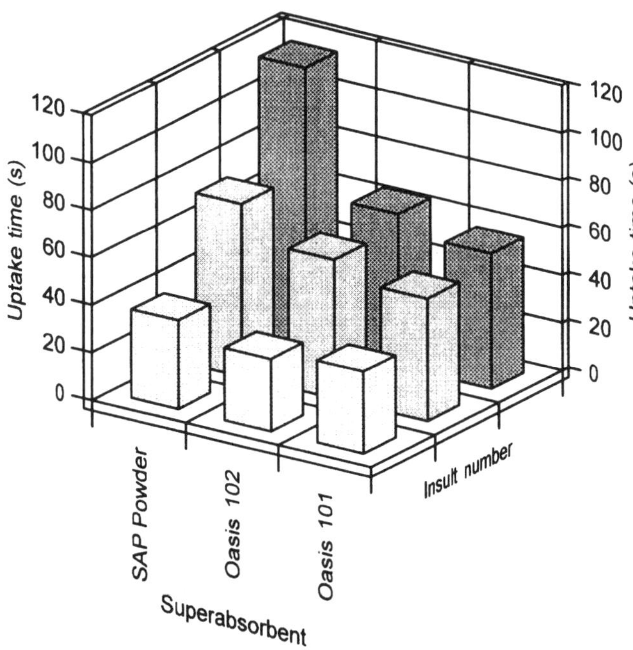

Figure 6.3. Comparison of FIFE performance of superabsorbent fiber with granular superabsorbent, 50% superabsorbent polymer (SAP) loading.

of 0.9% NaCl solution for composite pads containing superabsorbent polymer at loadings of 25%, 37%, and 50%. As the polymer loading is increased, the difference between granular and fiber superabsorbent polymer is more pronounced.

The AUL testing of pads reveals some other differences in the performance of superabsorbent polymer fibers versus powders, as shown in Figure 6.4. The fiber-containing pad rapidly absorbs fluid up to its maximum value, while the powder-containing pad continues to absorb for a much longer time. However, the total absorption of the powder is significantly greater than that of the fiber. This difference may be due to the lower capacity and AUL of the fiber (Table 6.2).

6.2.2. Films and Laminates

Superabsorbent films are another physical form that may offer some processing and performance advantages over existing commercial superabsorbent polymers. Superabsorbent films will not migrate in an absorbent core, and they may be less likely to create dust, depending on their brittleness. However, superabsorbent films have not been demonstrated to have a cost or performance advantage over existing superabsorbent polymers. While research into superabsorbent polymer films continues, no commercial contenders appear to be imminent.

The Dow Chemical Company was the first to commercialize a superabsorbent polymer film with Dow Water Absorbent Laminate (DWAL). This product was prepared by casting a film of an uncrosslinked anionic latex, thermally crosslinking the film, and transferring it to a tissue carrier.[27–32] In this process a copolymer latex of ethyl acrylate and methacrylic acid was mixed with an aqueous solution of sodium hydroxide to neutralize the carboxylic acid and partially saponify the acrylate ester. The solution of hydrolyzed polymer was mixed with 1%–5% of a crosslinker capa-

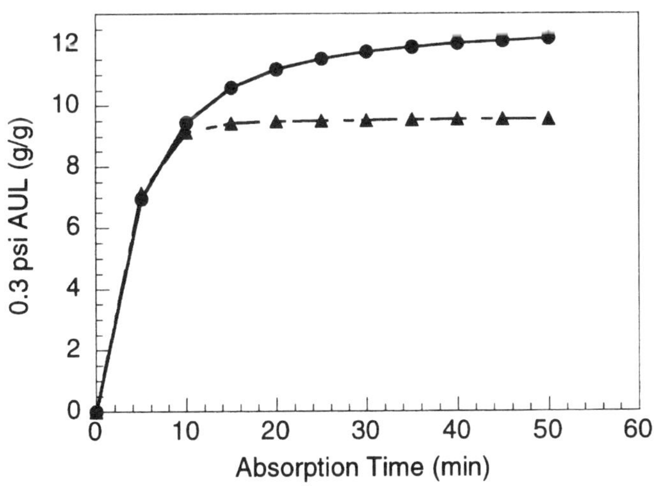

Figure 6.4. Composite pad AUL comparison of Dow superabsorbent powder (●) and Oasis 101 superabsorbent fiber (▲).

ble of reacting with carboxyl groups at relatively low temperatures, such as a polyepoxide. The viscous syrup was spread on a substrate that was coated with a release agent, then heated to dry and crosslink the film. The film was wetted lightly with a water spray, and tissue was applied to the wet film to support it and keep it from adhering to itself when wound up. The same composition could also be used to spin fibers or to "print" superabsorbent polymer onto a wicking or nonwicking substrate, thereby increasing the absorbency of the substrate.[29]

In another approach, a finely powdered conventional superabsorbent is formed into a flexible sheet or film of thickness between 0.3 and 5 mm by the mutual adhesion of the particles in the presence of water.[33] Superabsorbent polymer granules of size between 150 and 1,000 μm are blended with 0.1%–10% of fine, water-insoluble particles (e.g., silica), surfactants, or fibers. The blend is spread into a thin layer, then exposed to water mist or steam in an amount of 15%–150% based on polymer. The polymer particles adhere to form a flexible sheet. Alternatively, a polyhydric alcohol such as glycerol or ethylene glycol can be added in an amount equal to or greater than the water to impart improved flexibility and storage stability to the sheet. When the sheet is fully swollen, as in service in a diaper, it loses its integrity and returns to the particulate state.

Similarly, Gelok Corporation sells a superabsorbent laminate prepared by adhering conventional superabsorbent particles between two layers of nonwoven material such as tissue paper or polyester fabric. The laminate thickness ranges from 0.015 to 0.030 in. The 5-min absorbency of 0.9% NaCl solution ranges from about 20 to 35 g/g by an unspecified test method, with approximately 75% of this absorbency achieved in the first 30 sec.[34] The absorbency of the laminate is determined principally by the superabsorbent polymer loading.

A superabsorbent polymer suitable for coating or impregnating a porous substrate, such as a fibrous web or a foam, can be prepared by co-polymerizing an emulsion of a carboxylic monomer (typically acrylic acid), a "reactive, self-crosslinking monomer," and a hydrophobic monomer.[35] This composition can undergo in situ crosslinking by application of heat. The reactive, self-crosslinking monomer has the general structure shown in Scheme 6.2, where the reactive group X can react by self-condensation or with the carboxylic acid by condensation or ring opening, e.g., glycidyl functions. The linear co-polymer is polymerized in an emulsion. The emulsion then can be broken by neutralizing the carboxylic acid groups with a base, for example, sodium hydroxide. The resultant clear, viscous solution can be cast into a film, dried, and crushed to yield a superabsorbent powder, or it can be applied to a substrate such as a cellulosic or other fibrous web, or polyurethane or regenerated cellulose foam. Subsequent drying and heating causes the reactive monomer to crosslink the polymer, thereby forming the superabsorbent.

The advantage of this composition is the long shelf-life of the co-polymer. The

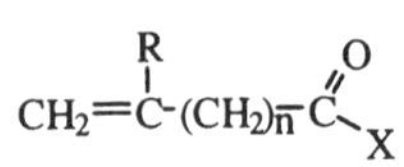

Scheme 6.2

hydrophobic co-monomer is present to prevent premature gelation of the mixture by maintaining the co-polymer in the dispersed phase of the aqueous emulsion. The crosslinkable groups are thought to be buried in the interior of the colloidal particles and less accessible for the crosslinking reaction, thereby conferring a long shelf life to the polymer. The crosslinking reaction can be induced when desired. In one example a polyester web modified with superabsorbent had an absorbent capacity for 1% NaCl solution of about 18 g/g.

In a novel approach, a co-polymer suitable for fiber and film formation is prepared by the photo-polymerization of partially neutralized acrylic acid with a co-monomer selected from alkyl acrylate or methacrylate, C_6–C_{18} α-olefin, or a substituted styrene.[36–39] A mixture of 65–95 wt% acrylic acid (neutralized 60%–80%), 5–35 wt% co-monomer such as 2-hydroxyethyl methacrylate (HEMA) or dialkylamino(meth)acrylate, 0.01–5 wt% of a photo-initiator, and 0–5 wt% of a crosslinker are dissolved in enough water so that the resulting mixture is 25–45 wt% water. The monomer mixture is spread, sprayed, or otherwise delivered to a suitable substrate in the desired thickness of 0.001–0.025 in, then subjected to ultraviolet light for a short time. The exposure time depends on the film thickness, the photo-initiator, and the exact composition of the monomer mixture. Alternatively, the film can be polymerized by means of an electron beam. This approach has the advantage of employing aqueous chemistry rather than organic solvents.

However, surfactants or dispersants must be used to obtain a stable, homogeneous dispersion of the monomers. The incorporation of a hydrophobic monomer (e.g., styrene) results in greater hydrolytic stability. The absorbent capacity of simulated urine claimed for the polymer is approximately 35–40 g/g.

6.2.3. Foams

Superabsorbent foams have attracted considerable attention as candidates for replacing multiple components of an absorbent core. They hold the potential of offering some of the same advantages sought for film or fiber forms: performing the absorbent functions of multiple components of a conventional diaper, not migrating in the product, and not creating dust. Several different approaches to superabsorbent polymer foams have been described in the patent literature. One approach seeks to embed conventional superabsorbent granules in a nonsuperabsorbent foam. A second option is to utilize capillary suction forces, instead of or in addition to osmotic pressure, to create an absorbent material.

6.2.3.1. Conventional Superabsorbent Polymers in Hydrophilic Foams. The incorporation of a conventional granular superabsorbent polymer into a foam structure is a relatively obvious extension of superabsorbent polymer technology. For any composite structure comprising a foam and a superabsorbent polymer to have enhanced performance, a number of technical challenges must be met. The foam must convey the fluid to the superabsorbent polymer particles rapidly and effectively. As the superabsorbent polymer swells, it must not block the passage of fluid through the foam, nor should it become detached from the foam structure. The foam should have an appropriate texture or "feel" if it is to be used in a personal care application. Of course, the overall performance properties of the composite structure (e.g., absorbent capacity, AUL) must not be compromised.

In an early example of a superabsorbent foam, a conventional superabsorbent powder is mixed in a 50:50 ratio with a polyurethane support just before the polyurethane is foamed to produce a superabsorbent foamed sheet.[40] The foam is claimed to be suitable for diapers, bandages, and a number of other applications.

Alternatively, a foamed latex can be used as the supporting foam.[41] In this case the latex is foamed first, and then the superabsorbent polymer is sprayed as a powder into the foamed structure at a loading of 5%–45%. Since this product is intended as a wound or surgical dressing, chitin or chitosan may also be added. The chitin biopolymer depolymerizes in the wound and its monomeric components enter into a metabolic reaction in the wound to speed healing.

Superabsorbent polymer powder can be immobilized in a flexible foam structure by mixing the superabsorbent polymer with a polyhydric alcohol and a blowing agent and heating the mixture to fix the foamed structure.[42] About 25–125 parts of a conventional superabsorbent powder and about 2–30 parts of a blowing agent are mixed with 100 parts of polyhydric alcohol such as ethylene glycol or glycerol. The liquid-coated, superabsorbent polymer composition can be spread on a substrate to form a sheet, or it can be cold extruded into strips. The composition is cured by allowing it to stand at temperatures up to 80°C for 30 min to 24 h or by heating at temperatures up to 230°C for times up to 15 min. If the blowing agent is omitted, the composition is suitable for making absorbent films.

In 1990, Hoechst-Celanese introduced Sanfoam superabsorbent foam, which is based on a superabsorbent polymer powder.[43] This foam is described as a "reticulated polyurethane foam" in sheet form containing 20% superabsorbent polymer powder. It is made by mixing polyisocyanate, polyol, blowing agent, and other conventional additives (e.g., catalyst, surfactant) and the desired amount of superabsorbent polymer. In the preferred embodiment, the superabsorbent polymer particles are under 30 μm in size and are added at a level of 25–50 parts per hundred parts of polyurethane. The foam is then subjected to a thermal reticulation process, causing the windows or membranes to be removed from individual cells or bubbles and leaving an open cell or skeletal structure. A commercial sample of the foam had the following properties (Thomas et al., unpublished results): thickness, 0.3 cm; bulk density, 0.035 g/cm^3; 0.9% NaCl absorbency, 55 g/100 cm^2 (52 g/g); and 0.5 lb/in^2 retention capacity, 40 g/100 cm^2 (38 g/g).

In an interesting hybrid of the first two approaches, Griesbach[44] has dispersed a superabsorbent powder in a starch foam, resulting in a structure that is inherently hydrophilic and also contains superabsorbent in the foam cell walls. The foams are made by combining starch, superabsorbent polymer, a plasticizer such as poly(vinyl alcohol), and water in an extruder at temperatures between 100° and 200°C. The starch matrix foams when the pressurized melt is exposed to the atmosphere. Alternatively, the starch mixture can be extruded into pellets for subsequent shaping and processing. Typically, the foams contained 30%–50% starch, 20%–40% superabsorbent polymer, and 15%–25% plasticizing agent. The most desirable moisture content is between 12% and 17% water, based on dry starch weight. Commercial starch contains 7%–12% residual moisture, so little or no additional water is required to produce the starch foam. Significantly, when a biodegradable superabsorbent such as sodium carboxymethylcellulose or starch-based superabsorbent is used, the entire foam structure is highly degradable (see Chapter 6, section 6.3). The capacities of example compositions ranged from 6 to 17 g/g.

6.2.3.2. Capillary Foams. While conventional superabsorbent polymers utilize osmotic forces to retain aqueous fluids (Chapter 1), most other common absorbents depend on capillary forces, or the tendency of an aqueous fluid to spread and wet a high energy surface. Paper towels, sponges, and cellulose fluff are examples of capillary absorbers.

The capillary absorbers have several features in common. Because they are not, in general, polyelectrolytes, capillary absorbers are relatively insensitive to the ionic strength of the aqueous fluid. The capillary pressure that drives the absorption of fluid is not as great as the osmotic pressure operating in a superabsorbent polymer. Therefore, the absorption of fluid is more reversible for capillary absorbers, and fluid may be displaced more easily when a mechanical load is applied. Depending on the pore size and surface characteristics of a capillary material, it may be relatively sensitive to other components in the aqueous matrix, such as cellular debris or proteins and other biopolymers.

Recent patent literature shows a significant amount of activity on microcellular foams as superabsorbent structures. Lever Bros. (Lever) and P&G have patented absorbent materials made using high internal phase-ratio emulsions (HIPEs). The reported performance of these materials is comparable with conventional osmotic superabsorbents. An important aspect of the P&G foams is that they are inherently lipophilic, requiring a post-treatment in order to absorb aqueous solutions. The novelty and high performance of polymerized HIPE foams as absorbent materials make them attractive to the absorbent industry.

Much of the current HIPE technology is due to K.J. Lissant, who is listed as an inventor on 13 of the 17 patents in the field of HIPEs issued between 1970 and 1978. A broad range of applications of HIPEs have been patented since 1970, including cleaning solutions,[45–47] transport fluids,[48] crowd control materials,[49] fire extinguishing media,[50–52] low density polymers,[53–71] fuels,[72–74] topical delivery systems,[75,76] and oral hygiene solutions.[77] Of these, the low density polymeric materials are of greatest interest for absorbent applications.

Absorbent materials can be made from HIPEs containing a continuous monomeric phase by polymerizing, crosslinking, and drying the emulsion to form a low density, open-celled foam. The initial patents for HIPE absorbent materials were issued to Lever beginning in 1985.[58–63] Lever's first foam material consists of crosslinked vinyl monomers and was designed as an oil absorbent.[58] Aqueous absorbing materials were patented subsequently.

The general method employed by Lever to prepare hydrophilic foams is to functionalize the surfaces, throughout the foam, by sulfonation,[59] chemical introduction of nitrogen salts,[60] or introduction of one of many other polar or ionic groups.[61,62]

These HIPE materials are reported to have absorbent capacities up to 80 g water/g polymer.[62] The use of hydrophilic monomers is taught only to the extent that they can be made into a stable, polymerizable HIPE.[62] It is believed that the majority of the materials cited in the above patents are hard, inflexible materials. Lever was also issued a patent covering a more flexible, but lipophilic, HIPE foam recipe.[63]

In 1992, P&G began receiving patents for articles containing HIPE absorbent polymers[64–66] and for processes to make HIPE polymers.[67–69] In 1993, P&G was issued their first patent for a hydrophilic HIPE material.[70] The distinguishing feature of P&G's material seems to be that it remains collapsed in the dry state and expands as it absorbs liquid. Additionally, P&G's material is not chemically functionalized to make it hydrophilic. Instead, an emulsifier is coated on the walls of the foam in a postpolymerization rinse. The performances of these foams are reported to be comparable to conventional superabsorbent polymers, as indicated in Table 6.3. Another patent was issued to P&G in 1994 for an aqueous absorbent HIPE foam.[71] This second foam is reported to be particularly resistant to compression when full of liquid.

6.3. BIODEGRADABLE SUPERABSORBENTS

Because about 90% of all superabsorbent materials are used in disposable articles, most of which are disposed of in landfills (in the United States) or by incineration (in northern Europe), there is a perceived environmental problem with superabsorbent polymers. In the late 1980s bans or taxes on disposable diapers were being considered in at least 20 states.[78] However, life-cycle analyses of both disposable and cloth diapers manufactured in the late 1980s have shown that there is no clearly superior choice in terms of environmental impact.[79–81] Since then, disposable diapers have been modified to use less raw material, which should result in a reduced solid waste burden, reduced packaging costs, and reduced transportation costs. Despite the technical analysis, consumers clearly perceive disposable absorbent products, specifically diapers, as having a negative impact on the environment.[82] Therefore, superabsorbent polymer producers have been interested in developing biodegradable superabsorbent polymers to serve as a component in a fully biodegradable diaper or other absorbent product. Articles incorporating biodegradable superabsorbents might be disposed of in municipal composting facilities or flushed down the toilet to degrade in domestic septic tanks or at municipal wastewater treatment plants. Several diapers claiming biodegradability have been market-

Table 6.3. Properties of HIPE foam

		AUL				
Pore Volume	Average Cell Size	0.0 psi (0 kPa)	0.2 psi (1.37 kPa)	0.7 psi (4.8 kPa)	1.0 psi (6.9 kPa)	Vertical Wicking Rate (5 cm)
36.8 ml/g	40 μm	35.9 ml/g	34.0 ml/g	23.4 ml/g	13.0 ml/g	105 s

Data from Des Marais et al.[70]

ed, but none has enjoyed commercial success.[83] It should be emphasized that the superabsorbent polymer is only one component of a disposable absorbent article. Significant challenges in the development of biodegradable alternatives to other components in the absorbent article, such as water-impermeable back sheet and adhesive tape, remain to be solved.

Some of the issues pertinent to biodegradable superabsorbent polymers are also relevant to low molecular weight water-soluble polymers. Therefore, new developments of biodegradable detergent builders and soil anti-redeposition agents may have implications for biodegradable superabsorbent polymers. Some of the examples cited below relate to biodegradable polyelectrolytes in general, while others are specific examples of biodegradable superabsorbent materials.

6.3.1. Degradability Versus Chemical Structure

For the purposes of this chapter, the term *biodegradable* describes a polymer that can be reduced to carbon dioxide, methane, water, and biomass under biochemical action. Biodegradability may be contrasted with the more generic term *degradable*: A degradable polymer undergoes decomposition or degradation under unspecified environmental influences, including biochemical and photochemical influences. The final products of a "degradable" polymer are unspecified, while a biodegradable polymer degrades ultimately to carbon dioxide, methane, and water ("mineralizes"). There are several useful reviews on biodegradable water-soluble polymers.[84–86]

Based on experience and testing, several generalizations have been made relating biodegradability to chemical structure[84]:

1. Naturally produced polymers biodegrade.
2. Chemically modified natural polymers may biodegrade, depending on the extent of modification.
3. Synthetic addition polymers with carbon–carbon backbones do not biodegrade at molecular weights greater than about 500 g/mol.
4. Synthetic addition polymers with heteroatoms in their backbones may biodegrade.
5. Synthetic step-growth or condensation polymers are generally biodegradable to a greater or lesser extent, depending on the chemical nature of the chain coupling, molecular weight, morphology, and hydrophilicity.
6. Water solubility does not guarantee biodegradability.

With these generalities in mind, two approaches to achieving truly biodegradable superabsorbent materials are apparent: modification of a superabsorbent to enhance its biodegradability or modification of a biodegradable polymer (e.g., a natural polymer) to enhance its superabsorbency. The first approach would entail incorporating some nonbiochemical mechanism to degrade the polymer molecular weight to below 1,000–2,000 g/mol. The degraded polymer could then pass the cell membrane where a host of intracellular enzymes could biodegrade it. There are several significant challenges to the success of this approach, though some progress has been made. The latter approach would involve incorporating charged functional groups into a hydrophilic biopolymer to enhance its absorbency, then crosslinking the polymer to achieve desired absorbent properties. These modifications may impact the biodegradability of the polymer. A third approach of mixing biodegradable fillers with nonbiodegradable superabsorbents has also been attempted; however, this approach has not led to a fully biodegradable product (Chapter 6, section 6.3.5).

6.3.2. Poly(Acrylic Acid)-Based Superabsorbent Polymers

As is true for nearly all carbon–carbon backbone polymers, high molecular weight poly(acrylic acid) does not biodegrade.[87] Microbes have not evolved the extracellular enzymes necessary to cleave these polymers into a manageable size. However, poly(acrylic acid) of molecular weight below about 1,500 g/mol can traverse the microbial cell membrane, where it can be degraded completely.[87] Therefore, some mechanism must be provided whereby the molecular weight of a functional synthetic superabsorbent polymer can be degraded to below 1,500 g/mol. Several approaches to this kind of solution have been investigated for low molecular weight poly(acrylates), such as detergent builders. If technology to render these low molecular weight polymers biodegradable can be developed, perhaps it can be transferred to high molecular weight superabsorbent polymers.

Any scheme to degrade a poly(acrylate) absorbent must incorporate some mechanism to degrade the backbone. The most obvious way to achieve backbone degradation is to introduce a "weak link" into the backbone that can degrade by some nonbiochemical mechanism, for example, hydrolysis. One approach that has been explored in the literature is to co-polymerize acrylic acid with a weak link comonomer.

This approach builds on chemistry pioneered by Bailey.[88] The polymers of this invention are low molecular weight acrylic acid co-polymers intended for detergent builder applications. Acrylic acid is co-polymerized with a ketene acetal (Scheme 6.3). The ketene acetal undergoes a rearrangement upon polymerization, resulting in the incorporation of an ester linkage in the polymer backbone. The ester linkage can be hydrolyzed, cleaving the polymer backbone and decreasing the polymer molecular weight. If the oligomeric products of hydrolysis are of low enough molecular weight, then they can be biodegraded. Bailey et al.[89] were able to prepare a biodegradable polystyrene using this approach.

As noted, the oligomer molecular weight of poly(acrylic acid) must be below about 1,500 g/mol for biodegradation to occur. For a 65% neutralized sodium

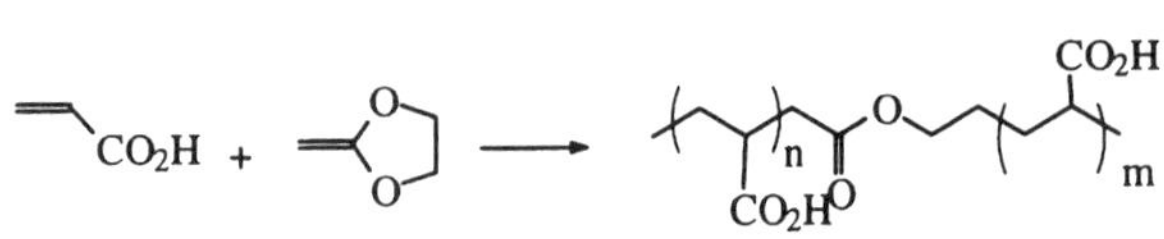

Scheme 6.3

poly(acrylate) polymer, this molecular weight corresponds to a degree of polymerization of about 18. Thus, the "weak link" co-monomer must be incorporated into the co-polymer at a concentration of at least about 5 mol%. Control of the ester linkage spacing in the co-polymer will be very difficult; therefore, the co-monomer will have to be incorporated at much higher concentrations to ensure that the hydrolysis will yield "bite-sized" biodegradable pieces. The ketene acetal illustrated in Scheme 6.3 is unstable in acidic media, so the polymerization must be carried out with sodium acrylate. In spite of these challenges, the concept is sound if these obstacles can be overcome.

6.3.3. Modified Polysaccharides

The modification of cellulose and other polysaccharides to render them superabsorbent has been well studied. Several commercial products based on this concept have emerged, though they have not been broadly successful. The history of commercial superabsorbents based on sodium carboxymethylcellulose (NaCMC) was reviewed briefly by Banyai et al.[90] Through the 1970s and 1980s NaCMC-based products were offered by Buckeye Cellulose (CLD-2 brand), Aqualon (Aquasorb brand), Akzo Zout Chemie (Akucell brand), Montedison (Celsocel brand), and Billerud (Cekosorb brand). The Buckeye Cellulose product was available in sheet, powder, and diced forms. The Aqualon product was available as a fiber or sheet, and the others were sold as granules. All of these developmental NaCMC-based superabsorbents were discontinued in the mid-1980s for a variety of reasons, including high cost and lack of large-scale availability. Purified, uncrosslinked NaCMC sells for nearly $2/lb, more than double the cost of conventional poly(acrylic acid) superabsorbents. Additional processing to convert the uncrosslinked material into a superabsorbent would disadvantage NaCMC even further. In spite of the commercial failure of these products, there is still a significant level of patent activity and some commercial activity concerning superabsorbent polymers derived from polysaccharides. Currently, an un-crosslinked carboxymethylcellulose absorbent product is available from Hercules (B. Banyai, Aqualon, personal communication), and a thermally crosslinked product is sold by Akzo for personal care and other applications (J. Schmitt, Dreeland, Inc., personal communication). A number of different polysaccharides and related natural polymers, including cellulose, starch, natural gums, chitin, and alginates, have been modified to add ionic groups and crosslinked to prepare superabsorbent polymers. Of these, carboxymethylcellulose is the most commonly used polysaccharide derivative as a result of its relatively low cost and ready availability.

Carboxymethylcellulose is the largest volume cellulose ether product and has historically been the lowest cost cellulose ether. Global production capacity is approximately 255,000 t/yr. NaCMC is used in detergents, the mining and oil industry, the cosmetics and personal hygiene industries, the paper industry, and the food industry.[91] The preparation of NaCMC has remained essentially unchanged since 1980. It is prepared, as shown in Scheme 6.4, by dissolving cellulose in 50% sodium hydroxide solution, slurrying the mixture in isopropanol, and reacting with

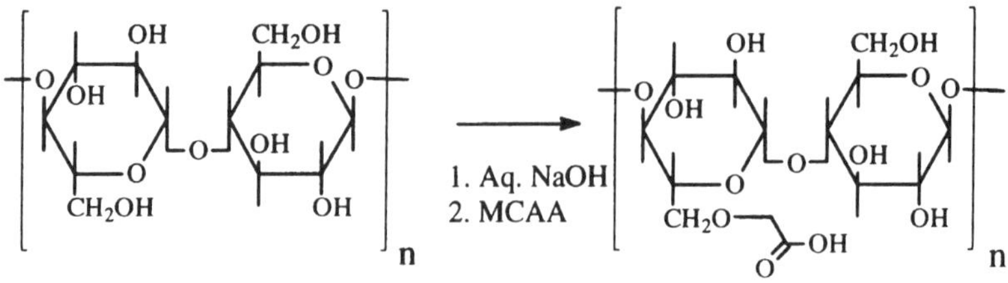

Scheme 6.4

monochloroacetic acid.[92] The product is neutralized with hydrochloric acid and dried. Depending on the application, the crude product can be sold, or it can be further purified by washing with alcohol to remove sodium chloride, sodium glycolate, and other reaction side-products. There are three hydroxyl groups per anhydroglucose unit available for substitution. The degree of substitution, or DS, refers to the average number of carboxymethyl ether groups per anhydroglucose unit. The DS is usually less than 1, typically about 0.7–0.9.

Studies by Wirick[93] of the biodegradability of NaCMC found that the rate of biodegradation and rate of microbial acclimation depended on the degree of substitution of the NaCMC. Degradation and bacterial acclimation proceeded faster for NaCMC with a DS of 0.4 than with a DS of 0.7. This result was confirmed by Sieger et al.[94]

To prepare an effective superabsorbent from NaCMC, the polyelectrolyte must be crosslinked by some means. A number of approaches have been disclosed in the patent literature. Ionic crosslinks can be introduced by mixing the polyelectrolyte with a multivalent metal ion. A simple heat treatment of the polymer causes formation of ester linkages between the carboxyl groups and the hydroxyl groups. Covalent crosslinks also can be introduced by adding reagents that react with hydroxyl or carboxyl groups on the polysaccharide chains.

The idea of crosslinking a polyelectrolyte with multivalent metal ions has been around for a long time. The sensitivity of conventional superabsorbents to divalent "hardness" ions (Ca^{2+} and Mg^{2+}) is well known: The divalent ions crosslink the polymer, resulting in contraction of the gel and loss of absorptive capacity (see Chapter 2, section 2.7.2). Multivalent ions of valency greater than 2, such as aluminum, have long been used to crosslink or gel NaCMC. In 1977 and 1978 Ganslaw and Katz[95,96] patented the surface crosslinking of anionic polyelectrolytes with polyvalent metal ions by dispersing the polymer and metal salt in a nonsolvent for the polymer, followed by isolation and drying. The preferred metal ions are aluminum, zirconium, chromium, titanium, and zinc, with aluminum being especially preferred. These patents expired in 1994 and 1995.

More recent competing technologies from Micro Vesicular Systems[97–100] and Isolyzer[101,102] differ in the process of distributing the metal ion crosslinker in the polyelectrolyte. The Micro Vesicular Systems process uses a "hydrophobicity agent" such as a C_1–C_4 carboxylic acid, acid chloride, or sodium salt or a dicarboxylic acid, apparently to modify the wetting or absorptive characteristics of the particle. A biodegradability enhancer such as cellulase can also be added to the composition. The crosslinker and hydrophobicity agents can be blended with the NaCMC in a wet process, using either organic or aqueous dispersants,[97] or dry-blended.[100] In the dry-blended process, the crosslinking of the polymer occurs simultaneously with its hydration. The examples given had absorbencies of 0.15 mol/L sodium chloride solution under a 1.47 kPa load of about 20–45 g/g. In the Isolyzer process a dry admixture of polyelectrolyte and metal salt is prepared. This mixture is slurried with a nonsolvent such as a C_1–C_3 alcohol.[101] A polysiloxane antifoaming agent is added to prevent particle clumping, and a fluorochemical nonionic wetting agent is added to improve particle wetting. Optionally, disinfecting agents such as phenols can be added at a level of 0.1%–3.0%. The process results in crosslinking of the particle surfaces. The products of this process are claimed to have saline absorbencies of 30–40 g/g.

Qin[103,104] utilized ionic crosslinking to prepare NaCMC particles having a core–shell morphology (see Chapter 2, section 2.7). The method involves forming a solution of a polyvalent metal ion, particularly an aluminum salt, adding the solution to the surface of NaCMC particles, and drying the product. This surface crosslinking improves the AUL relative to bulk crosslinked control samples. However, the reported 2.0 kPa AUL values of 15 g/g or less are quite modest relative to synthetic superabsorbent polymers.

The crosslinking of NaCMC by heat treatment at moderately high temperatures has been known for over 45 years. In a 1949 patent, Seymour and Schroder[105] applied a dilute aqueous solution of NaCMC to a fibrous sheet and heated the coated fibers to 150°–230°C to obtain a water absorbent mat. If the sample was heated to only 120°C, insolubilization of the NaCMC did not take place. Elliott[106] better defined the effect of heat treatment on the solubility of NaCMC as shown in Figure 6.5.

Several recent publications from Kimberly-Clark have explored the thermal crosslinking of NaCMC.[107–109] Their method generally comprises forming a mixture of 2% NaCMC in water and an optional crosslinker. Suitable crosslinkers are primarily diamines or polyamines, with chitosan glutamate being preferred. The solids are recovered from the solution by evaporation, and then the recovered material is ground into granules or flakes and heat treated at temperatures of 125°–150°C. The products obtained in this manner had greatly improved AUL, ranging from about 14 to 30 g/g under a 2.0 kPa load. These AULs approach values achieved routinely by synthetic superabsorbent polymers. However, the absorbent capacities of the NaCMC derivatives were not reported. The novel aspect of these patents appears to be the achieving of a certain minimum AUL.

6.3.4. Poly(Aspartic Acid)

A number of recent developments in the preparation of poly(aspartic acid) have led to renewed interest in this polymer as a biodegradable polyelectrolyte. The commercial production and uses of poly(aspartic acid) have been reviewed recently.[110] Poly(aspartic acid) is the homopolymer produced by the condensation polymeriza-

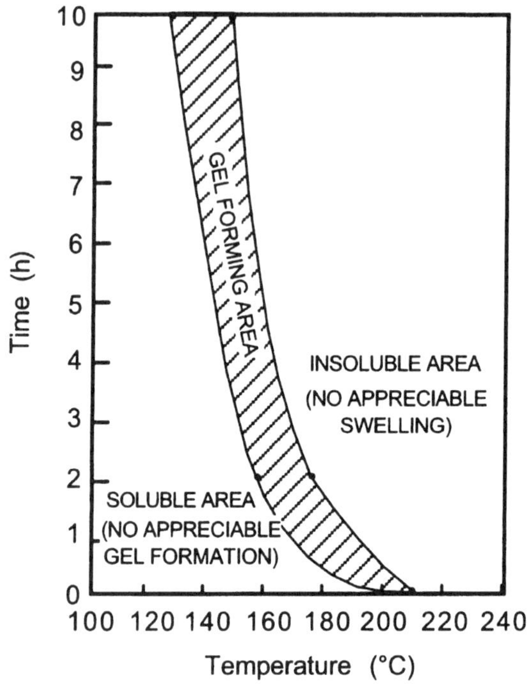

Figure 6.5. Time–temperature relationship for heat treatment of NaCMC.[106]

tion of the amino acid aspartic acid (Scheme 6.5). This polymer is prepared by heating aspartic acid (D and/or L isomers) to greater than 200°C to drive off water and form polysuccinimide. The imide ring is opened by hydrolysis with base, producing poly(aspartic acid) as a mixture of β-carboxyl (70%) and α-carboxyl (30%) material.[111] Molecular weights of 45,000–60,000 g/mol are readily achieved, and molecular weights as high as 200,000 g/mol have been reported.[112,113] However, these molecular weights may still be too low for application as superabsorbent materials. Importantly, regardless of whether the material is made using the D or the L isomer of aspartic acid, or whether it is connected to the chain by the β- or the α-carboxyl group, the material is reported to be completely biodegradable.[114]

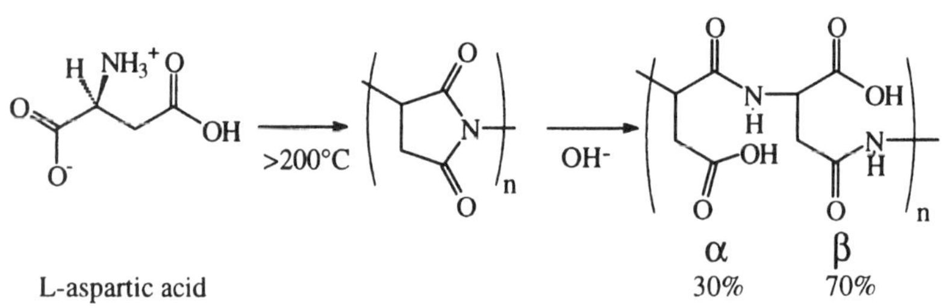

Scheme 6.5

Water-soluble polymer producers are interested in poly(aspartic acid) primarily for applications requiring low molecular weight products, such as soil anti-redeposition agents or scale inhibitors. However, as markets for low molecular weight material develop and economies of scale come into play, the use of poly(aspartic acid) as a superabsorbent polymer could develop. Patents for superabsorbent materials based on poly(aspartic acid) have been issued to Nippon Shokubai,[115] Kanayama et al.,[116] Donachy and Sikes,[117] and Ajioka et al.[118] Mitsui Toatsu recently announced the development of a biodegradable superabsorbent polymer based on poly(aspartic acid). Potential applications that were identified included seed-holding tape and water-retaining material for agricultural purposes. The projected selling price is two to three times that of poly(acrylic acid) superabsorbent polymer, but no plans for commercial production were announced.[119]

Poly(aspartic acid) is made by Rohm and Haas, Cygnus, Donlar, and BASF. These manufacturers all use L-aspartic acid as the starting material in their production of poly(aspartic acid). Most of the L-aspartic acid in the world is produced by a fermentation process in quantities too small to serve the superabsorbent polymer market. However, as the market for products based on L-aspartic acid increases, then it should become profitable to invest in new technology for aspartic acid production.

L-aspartic acid can also be prepared directly from maleic anhydride and ammonia (Scheme 6.6). Maleic anhydride first is dissolved in water to produce maleic acid. The maleic acid solution is heated at 100°–120°C for 2 h, isomerizing the maleic acid to fumaric acid. Ammonia is added, and the solution is cooled and passed through an enzyme column that has L-aspartase activity. The ammonium fumarate solution is converted into L-aspartic acid in 100% yield.[120]

Poly(aspartic acid) can be produced directly from ammonia and maleic anhydride (Scheme 6.7). Polysuccinimide is produced via the Michael addition of ammonia into the double bond of maleic anhydride to produce D,L-aspartic acid. The aspartic acid then homopolymerizes to yield polysuccinimide. Alternative methods for producing poly(aspartic acid) from maleic anhydride and ammonia have also been reported in recent years.[121–125]

Importantly, whether poly(aspartic acid) is produced from L-aspartic acid or directly from maleic anhydride and ammonia, the process that is used is very simple and should scale up cheaply. All that is involved is heating the material to above

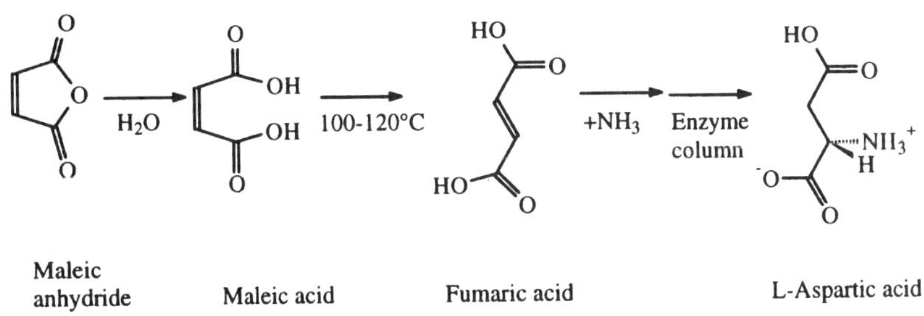

Scheme 6.6

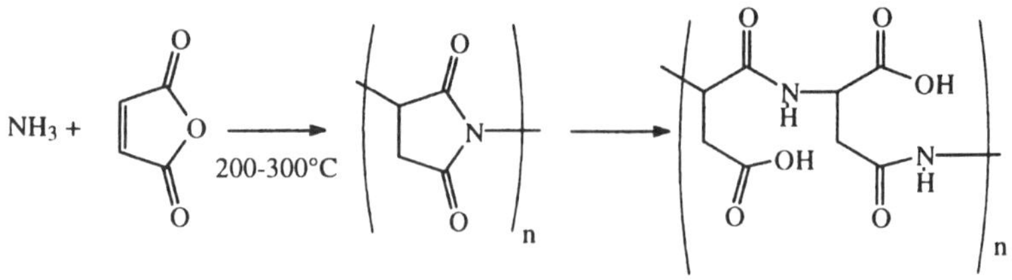

Scheme 6.7

200°C to drive off water, followed by hydrolysis of the polymer with a basic solution.

Poly(aspartic acid) may be crosslinked in several ways. Donachy and Sikes[117] added an amino acid bearing a functional group capable of reacting with the pendant carboxyl group of the poly(aspartic acid) to poly(aspartic acid) and heated the mixture for 4–36 h at 190°–250°C to render the polymer insoluble. Measured swelling capacities in 0.9% saline of the resulting polymer were 28-48 g/g, in the useful range for conventional crosslinked polyacrylate superabsorbents. Nippon Shokubai[115] teaches crosslinking poly(aspartic acid) with a basic polyamino acid, while Kanayama et al.[116] used a polyamine to crosslink polysuccinimide, followed by hydrolysis to crosslinked poly(aspartic acid). Wood and others have patented a number of approaches to crosslinked poly(aspartic acid) copolymers, mostly derived by ring-opening reactions of polysuccinimide using polymeric amines.[121–129] It remains to be seen whether these materials can function effectively as superabsorbents.

Poly(aspartic acid)–based superabsorbents may have special utility when used as a soil modifier, because aspartic acid (and other amino acids) can act as fertilizers for plants.[130] As the poly(aspartic acid) degrades, it should produce aspartic acid, which can then be used to fertilize the field.

6.3.5. Blends and Grafts of Poly(Acrylic Acid) With Biodegradable Substrates

Another approach that has been explored involves mixing the polyacrylate superabsorbent polymer with a biodegradable microbial food source, such as starch, with the intent that the starch will provide a locally high concentration of microbial activity, leading to degradation of the acrylate polymer. Generally speaking, this approach does not significantly improve the biodegradability of the host polymer, because this approach requires that the synthetic portion of the copolymer have low enough molecular weight to biodegrade. Such control of graft molecular weight is difficult, especially for a monomer that polymerizes as readily as acrylic acid.[85] The best that can usually be hoped for is that the host polymer loses its physical integrity and disappears into the environment. Sanyo Chemical described starch-grafted superabsorbent polymer in the 1970s.[131–133] The Sanyo patents disclose methods for the production of water-absorbing resins that comprise at least one type of starch or cellulose with at least one monomer containing a polymerizable double bond and a crosslinking agent. The products are said to have both excellent water absorption and biological degradation.

Chmelir[134] describes a process where natural polysaccharides (e.g., corn, wheat, or potato starch, cellulose and derivatives) or poly(vinyl alcohol) are added to a conventional synthetic superabsorbent polymer. The polysaccharides are added to the broken polyacrylate gel as free-flowing dry granules in the final stage of the polymerization, preferably after the conversion has reached 95%. The bonding between the polyelectrolyte and polysaccharide is improved by the addition of radical-forming reagents, such as sodium persulfate. The patent claims improved absorbent properties and improved biodegradability.

Several patents assigned to Starchem GmbH, a joint venture of Hüls AG and Cerestar in Germany, disclose starch-graft polymers prepared by inverse suspension polymerization.[135–139] (See Chapter 3 for more information on suspension polymerization of acrylic acid.) In the Starchem process a portion of the polysaccharide is suspended in a hydrophobic solvent. Then an aqueous solution of the unsaturated acid monomer (e.g., acrylic acid), the remainder of the polysaccharide, and initiator is added slowly. After polymerization is complete, a portion of the water is removed azeotropically and a postcrosslinker is added to yield an absorbent polymer with improved biodegradability.

6.4. CONCLUSIONS

Research in superabsorbent polymers remains active. Although granular polyacrylate superabsorbent polymers perform very well in existing absorbent product designs, absorbent product manufacturers are looking for ways to improve performance, reduce manufacturing costs, and differentiate their product in the marketplace. They are interested in superabsorbent polymers that offer unique performance characteristics, such as replacing one or more components in the absorbent core or offering biodegradability. To date, no alternative technology has emerged that could displace granular superabsorbent polymers in terms of cost or performance for personal care applications. However, recent developments of some alternate forms have been met with interest. It is possible that the significant technological challenges that face these materials can be overcome, and these materials may begin to displace conventional superabsorbent polymer in the future. Advanced technologies could also be developed for minor superabsorbent polymer applications. These technologies could then be expanded to the absorbent product applications.

Significant advances in biodegradable superabsorbent polymers have been made recently. In some cases the reported performance properties, as measured by laboratory tests, are comparable to today's poly(acrylic acid)–based superabsorbent polymers. However, the raw material and processing costs place these new polymers at a considerable disadvantage relative to conventional superabsorbent polymers. Furthermore, the advantages of a biodegradable superabsorbent polymer will be realized fully only in conjunction with a completely biodegradable structure, e.g., a diaper with biodegradable backsheet, tapes, adhesives, and elastics. Unless the performance/cost ratio of these new materials can be brought in line with existing materials or an outside influence such as government regulation requires them, biodegradable superabsorbent polymers will not displace conventional superabsorbent polymer to any significant extent in the near future.

REFERENCES

1. R. Heath, P. Akers, *Nonwovens World*, **5**, 66–69 (1996).
2. E. Dan, *Nonwovens World*, **5**, 63 (1996).
3. J. Hanson, *Nonwovens World*, Winter 1996, 109 (1996).
4. D.M. Jackson, B.J. Mathews, *U.S. Patent 5,350,370* (1994).
5. N.A. Ahr, D.M. Ooten, *WIPO Int. Appl. 92–07534* (1992).
6. A.S. Allen, D. Farrar, P. Flesher, *Eur. Patent Appl. 269,393* (1988).
7. A.S. Allen, D. Farrar, P. Flesher, *U.S. Patent 4,962,172* (1990).
8. A.S. Allen, D. Farrar, P. Flesher, *U.S. Patent 5,280,079* (1994).
9. A.S Allen, D. Farrar, P. Flesher, *U.S. Patent 4,861,539* (1989).
10. A.S. Allen, *U.S. Patent 5,147,956* (1992).
11. P.J. Akers, W. Brunskill, *WIPO Int. Appl. 93–24684* (1993).
12. P.J. Akers, W. Brunskill, *U.S. Patent 5,413,747* (1995).
13. P.J. Akers, W. Brunskill, *U.S. Patent 5,466,731* (1995).
14. B. Le-Khac, *U.S. Patent 5,026,784* (1991).
15. B. Le-Khac, *U.S. Patent 4,731,067* (1988).
16. B. Le-Khac, *Eur. Patent Appl. 272,074* (1987).
17. Lanseal-F product literature, Toyobo Co., Ltd. (1996).
18. K. Ban, S. Takegami, M. Wakitani, *U.S. Patent 4,442,173* (1984).
19. Japan Exlan, Ltd., *Jpn. Kokai Tokkyo Koho 55-132,754* (1980).
20. K.K. Toyoba, Ltd., *Jpn. Kokai Tokkyo Koho 04-161,513* (1992).
21. K. Hirakawa et al., *Eur. Patent Appl. 530,591,* (1993).
22. J.R. Gross, *U.S. Patent 3,954,721* (1976).
23. R.N. Bashaw, B.L. Atkins, *U.S. Patent 3,983,095* (1976).
24. E.C.A. Schwarz, *U.S. Patent 5,079,080* (1992).
25. S.R. Kellenberger, *U.S. Patent 5,147,343* (1992).
26. L.J. Bernardin, *Eur. Patent Appl. 399,564* (1990).
27. J.R. Gross, R.T. McFadden, *U.S. Patent 3,926,891* (1975).
28. J.R. Gross, J.L. Potter, F.B. Kassell, *U.S. Patent 4,026,932* (1977).
29. N.D. Burkholder, Jr., *U.S. Patent 4,076,673* (1978).
30. R.M. Krajewski, R.E. Erickson, *U.S. Patent 4,186,233* (1980).
31. R.E. Erickson, *U.S. Patent 4,293,609* (1981).
32. R.E. Erickson, *U.S. Patent 4,117,184* (1978).

33. T.H. Hatsuda, *Eur. Patent Appl. 624,618* (1994).
34. Gelok International Corporation product literature (1996).
35. S. Dabi, *Eur. Patent Appl. 206,358* (1986).
36. P.J. George, *U.S. Patent 4,167,464* (1979).
37. P.J. George, *WIPO Int. Appl. 81-01850* (1981).
38. P.J. George, *Eur. Patent Appl. 47,009* (1982).
39. P.J. George, *U.S. Patent 4,486,489* (1984).
40. B.G. Harper, R.N. Bashaw, B.L. Atkins, *U.S. Patent 3,669,103* (1972).
41. S.F. Nielsen, D.W. Kim, *U.S. Patent 5,011,864* (1991).
42. R. Korpman, *Br. Patent Appl. 2,083,487* (1982).
43. D.J. Kelly, D.W. Kim, *U.S. Patent 4,985,467* (1991).
44. H.L. Griesbach, *Can. Patent Appl. 2,138,062* (1995).
45. K. J. Lissant, *U.S. Patent 3,523,826* (1970).
46. K. J. Lissant, *U.S. Patent 3,700,594* (1972).
47. K. J. Lissant, *U.S. Patent 3,732,166* (1973).
48. K. J. Lissant, *U.S. Patent 3,617,095* (1971).
49. K.J. Lissant, *U.S. Patent 3,983,213* (1976).
50. L. R. Hopper, *U.S. Patent 3,984,334* (1976).
51. L.R. Hopper, *U.S. Patent 4,037,665* (1977).
52. T.J. Falkler, R.E. Mertz, U. S. Patent 4,168,239 (1979).
53. M.P. Aronson, M.F. Petko, *U.S. Patent 4,606,913* (1986).
54. D.C. Sherrington, P.W. Small, *U.S. Patent 4,965,289* (1990).
55. A.R. Emes, K. Hammond, D.C. Sherrington, *U.S. Patent 4,985,468* (1991).
56. D. Barby, Z. Haq, *Eur. Patent Appl. 60,138* (1982).
57. D.P. Gregory, M. Sharples, I.M. Tucker, *Eur. Patent Appl. 299,762* (1988).
58. D. Barby, X. Haq, *U.S. Patent 4,522,953* (1985).
59. Z. Haq, *U.S. Patent 4,536,521* (1985)
60. K. Jones et al., *U.S. Patent 4,611,014* (1986).
61. K. Jones et al., *U.S. Patent 4,668,709* (1987).
62. K. Jones et al., *U.S. Patent 4,612,334* (1986).
63. C.J.C. Edwards, D. P. Gregory, M. Sharples, *U.S. Patent 4,788,225* (1988).
64. T.D. LaVon, G.W. Taylor, G.A. Young, *U.S. Patent 5,147,345* (1992);
65. T.A. DesMarais et al., *U.S. Patent 5,268,224* (1993).
66. T.D. LaVon, G.W. Taylor, G.A. Young, *U.S. Patent 5,318,554* (1994).
67. T.A. DesMarais, S.T. Dick, T.M. Shiveley, *U.S. Patent 5,149,720* (1992).
68. T.A. DesMarais, S.T. Dick, T.M. Shiveley, *U.S. Patent 5,198,472* (1993).
69. T.A. DesMarais, S.T. Dick, T.M. Shiveley, *U.S. Patent 5,250,576* (1993).
70. T.A. DesMarais et al., *U.S. Patent 5,260,345* (1993).
71. T.A. DesMaraiset al., *U.S. Patent 5,331,015* (1994).
72. V.D. Beaucaire, L.A. Marlin, *U.S. Patent 4,030,894* (1977).
73. K.J. Lissant, *U.S. Patent 4,084,940* (1978).

74. L.A. Marlin, *U.S. Patent 4,153,421* (1979).
75. R.C. Cucca, *U.S. Patent 4,385,049* (1983).
76. K.J. Lissant, *U.S. Patent 4,396,417* (1983).
77. A.A. Siciliana, E. Tabibi, *U.S. Patent 5,130,122* (1992).
78. *Chicago Tribune,* June 17, 1990, p. C-8, Associated Press.
79. C. Lehrburger, *Diapers in the Waste Stream: A Review of Waste Management and Public Issues,* P.O. Box 580, Sheffield, MA (1988).
80. *Disposable Versus Reusable Diapers: Environmental, Health and Economic Comparisons*, Report to Procter & Gamble, Arthur D. Little, Inc., Cambridge, MA (1990).
81. B. Sauer et al., *Environ. Tox. Chem.,* **13,** 1003–1009 (1994).
82. J.H. Michael, P.M. Smith, *Wood and Fiber Science,* **26,** 370–381 (1994).
83. Film Technologies Ltd. introduced Harmonies, a flushable diaper, in 1994. *Brand Strategy* (February 25, 1994).
84. G. Swift, *Acc. Chem. Res.,* **26,** 105–110 (1993).
85. Y.H. Paik, E.S. Simon, G. Swift, *Adv. Chem. Ser.,* **248,** 79–98 (1996).
86. G. Swift, *Polym. Degrad. Stab.,* **45,** 215–231 (1994).
87. T. Hayashi et al., *Biosci. Biotech. Biochem.,* **58,** 444–446, (1994).
88. W.J. Bailey, *U.S. Patent 4,923,941* (1990).
89. W.J. Bailey, B.K. Kuruganti, J.S. Angle, In *Agricultural and Synthetic Polymers: Biodegradability and Utilization, ACS Symp. Ser. No. 433,* (1990).
90. B.E. Banyai, E.C. Linsay, D.C. Zecher, *Nonwovens World*, **2,** 87–90 (1993).
91. R. Calder et al., *Chemical Economics Handbook,* Cellulose Ethers Report 582.5000, SRI International, Menlo Park, CA, (1995).
92. W.S. Fong, *Process Economics Program,* Water Soluble Polymers, Report No. 130 Supplement A, SRI International, Menlo Park, CA, (1993).
93. M. Wirick, *J. WPCF,* **46,** 512–21 (1974).
94. C.H.N. Sieger et al., *Carbohydrate Polymers,* **27,** 137–143 (1995).
95. S.H. Ganslaw, H.G. Katz, *U.S. Patent 4,043,952* (1977).
96. S.H. Ganslaw, H.G. Katz, *U.S. Patent 4,090,013* (1978).
97. D.F.H. Wallach, A.C. Chang, *S. Afr. Patent Appl. 901,923* (1990).
98. D.F.H. Wallach, *U.S. Patent 4,944,734* (1990).
99. D.F.H. Wallach, *U.S. Patent 4,952,550* (1990).
100. D.F.H. Wallach, *U.S. Patent 4,959,341* (1990).
101. T.W. Honeycutt, *U.S. Patent 5,252,340* (1993).
102. T.W. Honeycutt, *U.S. Patent 5,417,977* (1995).
103. J. Qin, *U.S. Patent 5,470,964* (1995).
104. J. Qin, *U.S. Patent 5,498,705* (1996).
105. R.B. Seymour, G.H. Schroder, *U.S. Patent 2,486,805* (1949).
106. E.S. Elliott, *U.S. Patent 2,639,239* (1953).
107. X. Ning, S. Tong, *Eur. Patent Appl. 538,904* (1993).
108. J. Qin, *Eur. Patent Appl. 566,118* (1993).
109. J. Qin et al., *WIPO Int. Appl. 95-11925* (1995).

110. K.C. Low, A.P. Wheeler, L.P. Koskan, *Adv. Chem. Ser.*, **248,** 99–111 (1996).
111. A.M. Atencio et al., *U.S. Patent 5,373,086* (1994).
112. J. Knebel, K. Lehmann, *U.S. Patent 5,142,062* (1992).
113. D.E. Adler et al., *U.S. Patent 5,556,938* (1996).
114. M.B. Freeman et al., in *Hydrogels and Biodegradable Polymers for Bioapplications, ACS Symp. Ser. No. 627,* R.M. Ottenbrite, S.J. Huang, K. Park, eds., American Chemical Society, Washington, D.C., pp. 118–136 (1996).
115. Nippon Shokubai, *Jpn. Kokai Tokkyo Koho 80-59820* (1996).
116. K. Kanayama et al., *Eur. Patent Appl. 683,177* (1995).
117. J. Donachy, S.C. Sikes, *WIPO Int. Appl. 92-17525* (1992).
118. M. Ajioka et al., *Eur. Patent Appl. 658,586* (1995).
119. *Japan Chemical Week,* **37,** 9 (7 Nov. 1996).
120. G.J. Calton, *Biocatal. Prod. Amino Acids Deriv.,* **3,** (1992).
121. L.L. Wood, *U.S. Patent 5,502,117* (1996).
122. L.L. Wood, *U.S. Patent 5,286,810* (1994).
123. L.P. Koskan, A. Meah, A.R.Y. Meah, *U.S. Patent 5,296,578* (1994).
124. L.L. Wood, *U.S. Patent 5,288,783* (1993).
125. M.B. Freeman et al., *U.S. Patent 5,393,868* (1995).
126. L.L. Wood, *WIPO Int. Appl. 94-03527* (1994).
127. L.L. Wood, *U.S. Patent 5,496,914* (1996).
128. L.L. Wood, *U.S. Patent 5,510,426* (1996).
129. L.L. Wood, *U.S. Patent 5,510,427* (1996).
130. A.M. Kinnersley et al. *U.S. Patent 5,350,735* (1994).
131. F. Masuda, A. Nakamura, K. Nishida, *U.S. Patent 4,076,663* (1978).
132. Sanyo Chemical Co., *Jpn. Kokai Tokkyo Koho, 52-059,690* (1977).
133. Sanyo Chemical Co., *Jpn. Kokai Tokkyo Koho 01-103,644* (1989).
134. M. Chmelir, *Eur. Patent Appl. 476,574* (1994).
135. K. Heidel, *U.S. Patent 5,032,659* (1991).
136. K. Heidel, *U.S. Patent 5,032,659* (1991).
137. K. Heidel, S. Scholz-Weigl, *U.S. Patent 5,1183,872* (1993).
138. K. Heidel, P. Hofmann, F. Krause, *U.S. Patent 5,180,800* (1993).
139. K. Heidel, F. Zimmerman, *U.S. Patent 5,219,971* (1993).

7

APPLICATIONS OF SUPERABSORBENT POLYMERS

Fredric L. Buchholz

Superabsorbent polymers possess a number of attributes that make them attractive in many different applications. Superabsorbent polymers have supplanted much of the traditional absorbents in infant diapers and have made significant improvements in the performance of feminine hygiene products and adult incontinence products as a result of superior water-absorbing properties. The basic property of water absorption has suggested the use of superabsorbent polymers in many other applications, including paper towels, surgical sponges, meat trays, disposable mats for outside doorways and in bathrooms, and for household pet litter, bandages and wound dressings. The ability of the swollen gels to release the water to the surroundings as vapor has also been used in various ways, for example, as humidity-controlling products or as soil conditioners. Superabsorbent polymers may also be used to release water-soluble substances from within the network structure into the surroundings as a solution. For example, pharmaceuticals and fertilizers may be incorporated into superabsorbent polymer to yield controlled release products. Another characteristic of the swollen polymer is its rubbery nature, which has been used to control the consistency of products as diverse as cosmetics or concrete or to contribute a soft, yet dry, feel to a product like a hot or cold pack for sore muscles. The soft, rubbery nature may also be employed to impart sealing properties to products that are in contact with water or aqueous solutions, for example, underground wires and cables. In previous chapters, the structure of superabsorbent polymers, as derived

Modern Superabsorbent Polymer Technology. Edited by Fredric L. Buchholz and Andrew T. Graham.
ISBN 0-471-19411-5.

from the synthetic conditions, and the physics of superabsorbent polymers, are described. In the following sections, many of the applications and emerging applications of superabsorbent polymers are described in more detail, with some attention paid to how the structure of the superabsorbent polymer affects its performance in the specific application.

7.1. PERSONAL HYGIENE PRODUCTS

The largest use of superabsorbent polymers is in personal hygiene products. These consumer products include, in order of volume of superabsorbents used, disposable infant diapers, children's training pants, adult incontinence articles, and feminine sanitary napkins. Of these, infant diapers accounted for over 90% of the total superabsorbent polymer sold in 1995.[1] Because of this, the development of superabsorbent properties in general has been focused on optimizing absorbency of urine in a disposable diaper.

7.1.1. Disposable Infant Diapers

Disposable diapers have changed greatly during the past 30 years, but three basic design components are used in any diaper. A diaper consists of an absorbent core that is sandwiched between a porous top-sheet and an impermeable back-sheet, as in Figure 7.1a.[2] The top-sheet, which is next to the baby's skin, helps contain the absorbent core, especially when the core is wet and easily torn. The top-sheet must allow the urine to flow through it and not allow the liquid to remain near the baby's skin. Usually it is made of a porous, hydrophobic substance, for example, polyester or polypropylene nonwoven fabric. The back-sheet helps keep the baby's clothing dry and is a nonporous, hydrophobic substance, for example, polyethylene film. The absorbent core takes in the liquid, distributes it to distal regions of the core, and holds the liquid under pressure from the baby.

The absorbent cores of the earliest disposable diapers were made of multiple layers of creped tissue paper.[3] Liquid in the absorbent core was held only in the void space between the fibers of the paper. To hold a large volume of liquid, the core needed lots of void volume. To provide capillary pressure to wick and retain the liquid, the size of the voids or pores had to be small. The solution was to make a thick core, to have many small pores. Even then, the essential problems in creating an efficient diaper were known. The importance of total absorptive area, absorptive mass, density of the mass, and the location of this mass relative to the user's anatomy were recognized. The limitations of fit to the human body require that absorptive mass be less in the crotch area and more in the front and back. As a result, the distribution of liquid from the crotch to the rest of the diaper is essential. Reduced leakage from the diaper was recognized as a principal design criterion. Liquid distribution was enhanced by running the creping ridges along the length of the diaper, by folding the diaper in the lengthwise direction to yield simple channels for liquid flow, and by gradations in the density of the absorbent material along the length and

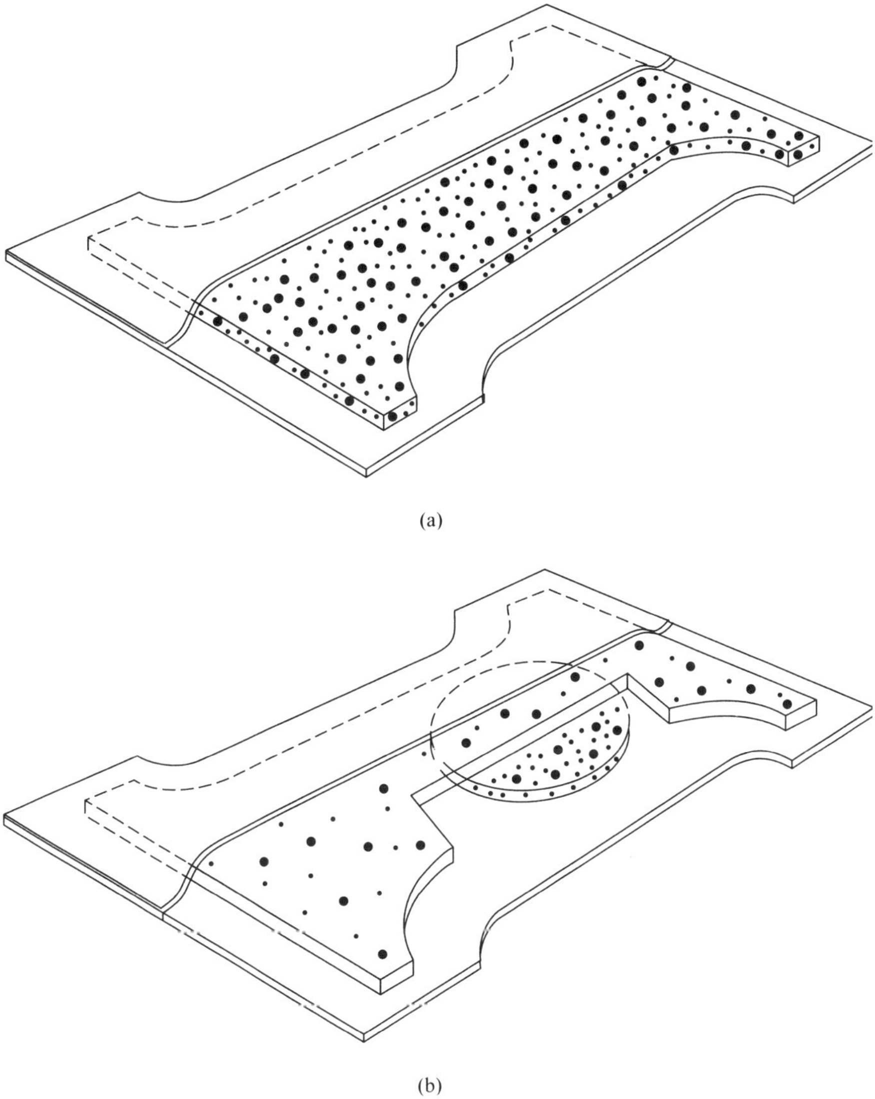

Figure 7.1. Schematic diagrams of diapers with **(a)** absorbent core of superabsorbent polymer and cellulose fluff sandwiched between the impermeable back sheet and the porous top sheet and **(b)** layered structure with higher density circular layer of superabsorbent polymer and fluff nearer the back sheet. (Adapted from *U.S. Patent 5,147,343* and *Br. Patent Appl. 2,215,609*.)

width of the diaper. The density gradations enhanced both the wicking of fluid (high density) and the liquid-holding, volumetric capacity (low density). This was accomplished by discrete layering of the fluff in regions having different densities or by graduating the density across the thickness of the pad. This trend toward layers is quite evident in modern diaper designs, which have numerous layers to accomplish the various tasks of liquid movement and storage. Layering has become more prevalent with the incorporation of superabsorbent polymer into absorbent cores.

The storage of liquid in a diaper was revolutionized by adding superabsorbent polymers to it.[4] Eventually this idea caught on with diaper manufacturers as a result of the superior liquid retention of superabsorbent polymer. Superabsorbent polymers absorb, and retain under a slight mechanical pressure (1–5 kPa in diapers), about 30 times their weight in urine. The swollen gel holds the liquid in a solid, rubbery state and prevents the liquid from leaking onto the baby's skin and clothing. In a diaper, the polymer is mixed with 1–10 times its mass of cellulose pulp fluff (processed wood fiber) to make up the absorbent core. In the first diapers containing superabsorbent polymer, the polymer amounted to about 12% of the mass of the absorbent core. More recently, diaper designs incorporate up to 60% of superabsorbent polymer[5] and more.[6,7] Cores containing superabsorbents can be made thinner because a smaller volume of dry superabsorbent polymer can absorb the same volume of aqueous liquid as a larger volume of fluff. The fraction of superabsorbent polymer in the absorbent core has increased as the properties of the polymer and the properties of the entire core become better understood.

The superabsorbent polymer must function in concert with the other components in the diaper to produce the desired performance. The absorbent core became a composite structure, with the fluff and superabsorbent polymer randomly mixed. Early in the development of superabsorbent-containing diapers (the early to mid-1980s), only the superior absorption and retention of water by superabsorbent polymers were recognized as important, and the absorption capacity for urine was maximized. This required superabsorbents having a low crosslink density. However, this also resulted in small gel modulus when swollen. As a result of their high deformability, early superabsorbents tended to "gel-block," or create swollen masses of gel that blocked the channels that allowed urine to penetrate into the diaper structure.[8] Because urine cannot flow as easily within a gel-blocked absorbent core, diapers containing superabsorbents with large capacity and small gel modulus are more likely to hold urine next to the wearer's skin and are more likely to leak. As the understanding of the new composite structures, containing both fluff and superabsorbent polymer increased, increased gel strength or modulus emerged as a means to prevent gel blocking. Increasing the crosslink density of the gel particles resulted in products with higher modulus and higher absorption under pressure, but lower free swelling capacity. Although the equilibrium swelling capacity was somewhat lower, the improved gel modulus led to an overall improvement in diaper performance.[9]

The swelling and deformation properties of the superabsorbent polymer caused the development of numerous designs either to work around the limiting properties of the composite or to work synergistically with improved superabsorbent polymer. The basic feature of low density of the core could be, at least partly, discarded because the absorbency function was largely taken on by the superabsorbent polymer instead of the void space between fibers. But the requirement of liquid flow and distribution remained as important as ever, and is complicated by the presence of superabsorbent polymer. Consider a dry, random mixture of superabsorbent polymer and cellulose fluff in a given ratio. As the superabsorbent polymer absorbs liquid, the volume and mass of each polymer particle increases, and the mass fraction of gel in the core increases. The change of the gel fraction with increasing swelling of the polymer in the core is shown for three initial ratios of polymer to fluff in Figure 7.2. The volume of the pores between the particles of superabsorbent polymer and the fibers of fluff is affected by the elastic properties of the core and the superabsorbent polymer, and different cores may be designed to optimize performance.

If the superabsorbent polymer is only slightly crosslinked, it is soft and deformable, and may expand into the pores between the fluff and polymer particle as it swells. In this case, flow is inhibited as the composite becomes less porous as swelling proceeds and the pore size decreases. Diaper absorbent cores may be designed and constructed with the limitations of the polymer in mind.[10] However, the large absorption capacity of the polymer still is desirable, and modified designs may be developed to alleviate the inhibition of flow during swelling. A general technique is to redistribute the superabsorbent polymer from a random distribution into layers (thickness or *z*-direction) or regions (*x,y*-directions, but mostly length) of the core. The superabsorbent polymer may be concentrated more toward the back-sheet, as in Figure 7.1b, or more toward the ends of the diaper (more distant from the crotch). These techniques concentrate the fluff near the point of initial wetting to

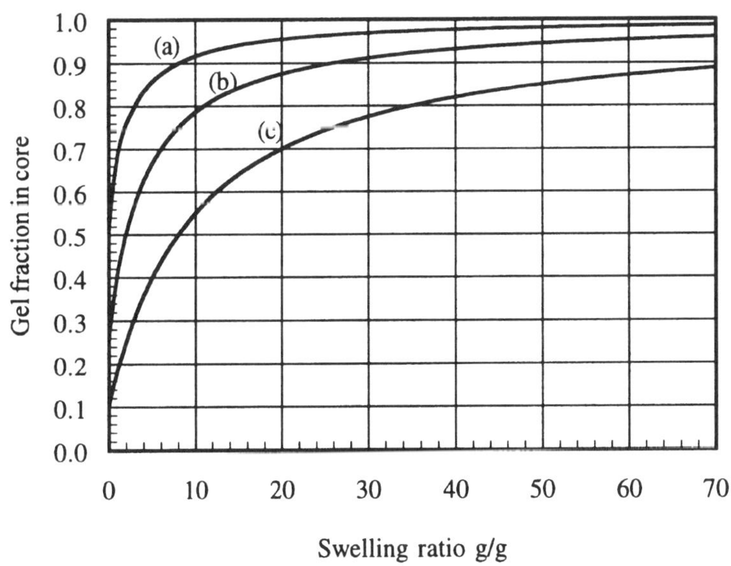

Figure 7.2. The fraction of gel in the absorbent core as a function of the swollen volume of the polymer for three different initial weight ratios of polymer to fluff: **(a)** $f_{SAP} = 0.5$, **(b)** $f_{SAP} = 0.25$, and **(c)** $f_{SAP} = 0.1$.

improve liquid flow before the superabsorbent polymer swells and during subsequent wettings of the same diaper. In practice, this is accomplished primarily by reducing the amount and density of the absorbent materials, particularly the superabsorbent polymer, at the urine insult point, as shown for example in Figure 7.1. This structure tends to draw urine away from the insult point as a result of higher density and hence higher capillary pressure of the distant regions and reduces the impact of gel blocking on initial containment of the liquid insult. In this way the design accommodates the superabsorbent polymer. Conversely, superabsorbent polymer design may be used to improve diaper performance.

If the superabsorbent polymer is more highly crosslinked, it is more rigid in the swollen state. Improving the rigidity of the particles enables the swelling particles actually to push aside the fiber component of the composite, thereby maintaining the porosity and permeability during subsequent contacts with liquid. Of course, if the swelling polymer opens the structure of the composite too much, the integrity of the composite may be destroyed, and leakage may increase. Therefore, the optimization of the polymer properties and location is required and is accomplished by a trial and error approach. A number of design challenges, and improvements, are evident from recent patent literature, which includes patents on sophisticated designs[11–17] and potential new attributes of the diaper, such as flushability.[18–20] In addition to the equilibrium properties of the superabsorbent polymer, the kinetic aspects of the swelling and absorption must be considered.

The absorption rate of the diaper must be optimized to the urination rate of the baby. If the absorbent core absorbs too slowly, the liquid will overflow the core and leak from the diaper. The absorption rate of the composite is influenced by, but not synonymous with, the absorption rate of the superabsorbent polymer. Fast swelling of the polymer may be desirable if it helps contain the liquid quickly. However, in some diaper designs, fast swelling may cause the diaper to leak if the porosity and permeability of the composite is reduced. As described in Chapter 5, the absorption rate of superabsorbent polymers is affected by the maximum absorption capacity of the polymer and its particle size and shape.[21] The placement of fast- or slow-absorbing polymers in the composite structure therefore has important implications for the effectiveness of the composite. Many different schemes for mixing fluff and superabsorbent polymer have been investigated in order to find optimum diaper performance. It should be expected that the particle size and shape of the superabsorbent polymer particles, and their change with swelling in relation to the size and shape of the fluff fibers, affect the porosity and permeability of the core. The relative amount of each component should also affect the optimum. Some effects of particle size and shape have been studied,[22,23] but much remains unknown. When the superabsorbent swelling is delayed in the wetting region of some diaper designs, there is more time to distribute urine through the diaper.[24,25] Improved distribution results in lower saturation of the core in the wetting region, so that subsequent insults may be more readily absorbed. The improved understanding of the impact of superabsorbent polymer on the absorbent core has resulted in some very thin diapers.

In the early 1990s, ultrathin diapers became possible. The amount of cellulose pulp fluff used in these diapers was reduced by half, yielding a thinner diaper with a higher concentration of superabsorbent polymer in the absorbent core. A separate layer of nonwoven fibers was added to improve urine distribution in the diaper. This distribution layer was placed between the composite absorbent core (cellulose fiber and superabsorbent) and the porous cover sheet. The distribution layer has lower absorbency than either the standard cellulose fluff or superabsorbent polymer and a low density that allows fast liquid distribution within the diaper. It has been made from either chemically crosslinked cellulose fibers or a nonabsorbent nonwoven material, such as polypropylene fiber, and it is sufficiently porous to allow liquid to pass through freely.

To maintain the total absorbent capacity of the ultrathin diapers, the amount of superabsorbent polymer was about doubled over earlier versions, replacing bulky cellulose fluff. This increase in concentration of superabsorbent required the polymer to have still higher resistance to gel blocking because there were fewer fibers to help maintain open pores between swollen superabsorbent particles. The most common approach to improving resistance to gel blocking was to increase gel modulus by creating a structured superabsorbent particle, as detailed in Chapter 3, section 3.2.8.1. These products have a crosslink density gradient, increasing from the center of the particle to its exterior. The higher density crosslinks near the particle surface produce a rigid gel early in the swelling process, which reduces the tendency to gel-block and improves the absorbency under load of the superabsorbent. The low crosslink density in the particle interior maintains the free swell capacity compared to a homogeneously crosslinked particle, with the potential to improve the absorption efficiency of the diaper.

The structured particle has been a significant advance in superabsorbent polymer technology for diapers, and superabsorbent polymers are likely to change further as the composite absorbent core structure becomes understood even better. Other applications of superabsorbent polymer have different requirements, however, and these requirements are also in need of study and optimization.

7.1.2. Adult Incontinence Products

Adult incontinence products also employ superabsorbent polymer to absorb urine. The market for these products is growing as the members of the post-World War II baby boom age. Because the products are used to absorb urine, generally the technology learned and used for baby diapers is applied directly to adult incontinence products (certainly most of the criteria are similar). However, the application is unique in several ways and could likely benefit from some research focused on finding the optimum superabsorbent polymer properties. Adults are more concerned about discreetness when using the products and desire very thin, easily disposed-of products. The thinness of these products has improved dramatically with the use of superabsorbent polymer. Urination volume and rate are both larger for adults than infants, and the products must also absorb under the larger pressures exerted by adults. Fast-swelling superabsorbent polymer has long been desired for this application and has been incorporated into multilayered absorbent pads.[26] Approaches to fast-swelling superabsorbent polymer involve increasing the specific area of the superabsorbent polymer, including using small particle sizes (higher surface area to volume ratio), using porous agglomerates of smaller particles,[27] and making highly wrinkled particles.[23] There is also more variability in the extent of incontinence in individuals, and, as a result, many different styles of products are available, including light- and heavy-duty personal pads, bed pads, and briefs.[28] Each of these is likely to need a slightly different superabsorbent polymer for optimum performance. Another application with many different product styles is feminine hygiene products.

7.1.3. Feminine Hygiene Products

Superabsorbent polymers are also used in some feminine hygiene products. In fact, the first commercial personal hygiene product with superabsorbent polymer was a feminine napkin in Japan. The reasons for using superabsorbent polymer in feminine care products are similar to those for diapers and incontinence products: leakage control and comfort. Good fitting, very thin, absorbent products are enabled by the properties of superabsorbent polymer. As in diapers, the superabsorbent must be able to absorb fluid under load, and therefore a high gel modulus product is useful. A challenge for the product developers, however, is the very significant difference between the fluids to be absorbed by feminine hygiene products and diapers. For diapers the fluid is a very low viscosity aqueous solution of (primarily) salt and urea. For feminine hygiene, the fluid is a complex, more viscous mixture of water, salts, and cells. The cells are far too large to diffuse into the network structure of the superabsorbent polymer, and may instead *adsorb* onto the surfaces of the superabsorbent polymer particles. The high osmotic pressure of partially swollen superabsorbent polymer can de-water the cells if they are in direct contact, and this can lead to a nearly impermeable surface layer of cells and protein and a severe reduction in the efficacy of the polymer. These factors suggest that the nature of the superabsorbent polymer for feminine hygiene should be different from the polymer used for urine absorption. In some respects the factors influencing the nature of the superabsorbent polymer for this application may be similar to those used to specify polymers for use in other blood applications. For example, polymers used in surgery and surgical appliances have been chemically modified to prevent undesired blood clotting.[29] Incorporating similar modifications into superabsorbent polymers may improve their performance in feminine hygiene products. Surface coating the superabsorbent polymer with surface-active agents improves blood dispersibility,[30] as does the addition of a variety of salts.[31] Potassium or lithium salt-neutralized superabsorbent polymers have also been disclosed for specific feminine hygiene needs.[32]

7.2. AGRICULTURAL AND HORTICULTURAL APPLICATIONS

Superabsorbents can also help conserve water in a variety of agricultural and horticultural applications. Mainly the polymers are used in the same way as mulch, to help the soil retain moisture. A comparison of the water-retention effectiveness of the traditional moisture retention aid, peat, and of superabsorbent polymer shows excellent retention by both materials and an expected decrease in water holding ability of the superabsorbents as salinity of the water increases.[33] As soil additives, the polymers are mixed into soil at a concentration of about 0.1 wt%. Little benefit is achieved below this concentration, and the soil can become too spongy and soft if much more is used and it becomes fully saturated with water. Soils that receive little rainfall, or that are too porous to hold moisture in the root zone, can be improved by adding superabsorbent polymers. The resulting soil mixture retains moisture longer, and plants live longer after germination. In clay soils, superabsorbent polymer helps improve the air content of the soil as the polymer particles expand and contract during the cycle of moisture absorption and evaporation, at least when conditions are not too wet. Under wet and soggy soil conditions, which are often not conducive to plant growth due to low oxygen content in the soil, the use of absorbent polymers may actually prolong the unfavorable soil conditions. An excellent review of how the properties of the superabsorbents influence the behavior of soils is by Kazanskii and Dubrovskii.[34]

Superabsorbent polymers have found limited utility as soil additives because results from studies on the efficacy of superabsorbents in improving plant health or yield have been mixed. The conflicting results are a consequence of the varied soil conditions present in the experiments.[35,36] Most of the positive results occurred when small ratios of superabsorbent polymer to soil were used and the soil was normally drier than average. The superabsorbent polymer absorbs water more effectively and controls its release into the soil as conditions become drier. The polymers have shown positive results, for example, in cabbage seedling germination and survival[37] and under conditions of water economy in growing corn.[38] In one experiment, the yields of Chinese cabbage increased about 10% compared to the control plot without superabsorbent polymer, using the same amount of water.[39] Seed pellets have been disclosed that contain seed, a superabsorbent polymer, fertilizers, binders, and an inorganic filler such as vermiculite.[40] These pellets can be spread in remote areas and help the seed germinate and grow with no further attention.

The nursery/floriculture market consumes a large percentage of the superabsorbent volume used in the agricultural market. The primary function of the absorbent polymer in this case is to reduce the volume and frequency of watering for ornamentals such as flowering plants, trees, shrubs, and hanging baskets. In the professional landscaping business, superabsorbents also are used to reduce transplant failures resulting from improper watering. For the nursery market, light-weight, sterile, soil-less mixtures are often used to prevent soil-borne diseases from affecting the plants. These mixtures can benefit from the addition of relatively lightweight superabsorbent polymers as an efficient water-retention agent.[41]

Superabsorbent polymer may also be used as a soil additive for turf grasses on high value land such as golf courses, sod production, and athletic fields, where the health and vigor of the vegetation is vital, and maintenance is expensive. In addition to reducing maintenance costs, another advantage is improved durability for soccer, football, and other athletic fields. The superabsorbents are applied by broadcasting, by injecting the polymer under the turf after the turf has been sliced and lifted, or by a liquid spray application wherein small, swollen, deformable gel particles can pass through the spray nozzles with relative ease.[42] The goal is to evenly distribute the polymer and any additives such as fertilizers or pesticides.

The type of superabsorbent used for these markets is typically based on polyacrylamide chemistry. Superabsorbents containing ionic functional groups, such as the polyacrylates used for infant diapers, show a reduction in swelling capacity when exposed to multivalent cations present in the soil or in the irrigation water. These cations form irreversible crosslinks between ionic functional groups, reducing the ability of the superabsorbent polymer to swell after numerous absorption cycles. Nonionic polyacrylamides maintain a larger percentage of the original absorbent capacity than polyacrylates when exposed to the polyvalent cations present in ground and irrigation waters.[39]

7.3. CONTROLLED RELEASE

Superabsorbent polymer may also be used to control the release of substances that are initially dissolved or trapped within the polymer phase, such as pesticides, fertilizers, and pharmaceuticals. Just as the absorption of water by superabsorbent polymer is caused by unequal activities of water inside the gel and in the external liquid phase, substances that are initially at a higher activity in the polymer will diffuse out of the particle and into the surroundings. The initially dry polymer system is activated for release by swelling, and the control depends on the change in diffusibility of the releasable substance as the polymer phase changes its concentration. Numerous water-absorbent polymers share this essential feature for controlled release by swelling,[43] but superabsorbent polyelectrolytes possess additional useful features. The swelling of poly(acrylic acid) and other acidic or basic polymers depends on the pH and ionic strength of the swelling medium, and changes in pH or ionic strength may serve as switches for controlled release.[44] The crosslink density also affects the extent of swelling and is used to control the release characteristics of the polymers. To make the controlled-release particles, usually the polymer is prepared first and then the active substance is added to the polymer by swelling the polymer in a solution of the active substance, followed by drying of the impregnated gel. Adding the active substance to the polymerization may affect the course of the polymerization or alter the active substance, either of which is undesirable.

As an example, Levy et al.[45] describe using superabsorbent polymer as a reservoir for insecticide for the control of mosquitoes. Polyelectrolyte superabsorbent polymers were impregnated with insecticidal, film-forming compounds such as tridecyl alcohol or with insecticides such as malathion, or insect growth regulators. The formulations were stable when dry but swelled in water to release the active ingredients. The release was relatively slow, enabling the formulation to be active over long periods of time and in situations where the polymer cycled through periods of wetness and dryness. In a dry period, the water in the superabsorbent polymer gel evaporates. The remaining active ingredient is trapped and can be released again when the polymer becomes wet and swollen. The formulation with superabsorbent polymers also allows for a combination of active ingredients that is effective at different stages of the mosquito's life-cycle.

Antimicrobial compounds have also been impregnated into superabsorbent polymer particles. These materials are useful in diapers or in sheets for absorbing food drips and help control odors caused by microbial action.[46] Antimicrobial compounds, such as organobromides, benzimidazoles, and silanes, are incorporated by dissolving the active ingredient in a volatile organic solvent, mixing the solution with dry superabsorbent polymer, and then evaporating the solvent, leaving a coating of the antimicrobial substance on the polymer particles.

Superabsorbent polymers have also been used in formulations of sustained-release fertilizers. The partially water-swollen, gel intermediate from the polymerization process may be blended with a fertilizer mixture (e.g., urea and potassium nitrate) to provide the formulation.[47] Poly(acrylic acid)–based superabsorbents have a high tolerance for added simple salts within the network structure, so fertilizer loading can be high. The soluble salts of the fertilizer dissolve and penetrate into the gel, which is subsequently dried and milled to yield the granular polymer product.[48] The fertilizer salts are then leached from the polymer in the field gradually, as the polymer swells and dries with the cycles of rainfall. Because the release rates vary for different substances, depending on whether the substance is nonionic (urea) or a salt (potassium nitrate), a major challenge in this application is to modify the superabsorbent polymer to yield the optimum release rate for the specific nutrient.[49]

Superabsorbent polymers also have been studied as delivery systems for pharmaceuticals. Because pharmaceuticals differ greatly in their solubility characteristics in water and in polymer gels, a variety of methods have been developed to load the drug into the polymer and to control its release. For example, hydrophobic drugs are difficult to incorporate into hydrophilic, water-swellable polyelectrolytes. However, this difficulty may be overcome by means of suitable polymer–surfactant complexes.[50] First, a more hydrophobic polymer system is prepared by mixing a polycarboxylic acid and an ethoxylated surfactant, such as ethoxylated nonylphenol. The surfactant is held within the network structure by hydrogen bonds between the ether groups of the surfactant and the carboxylic acid groups of the polymer. Then, the hydrophobic drug is absorbed into the association complex. The surfactant increases the solubility of hydrophobic substances in the polymer network, by forming hydrophobic micelles within the gel. Upon swelling, the hydrophobe-laden polymer complex releases the active agent. The release may be enhanced if the pH is also raised during the swelling process, because the hydrogen bonds between surfactant and polymer are destroyed as the carboxylic acid groups become neutralized.

The pH dependence of swelling is useful for specific targeting of pharmaceuticals. For example, a therapeutic agent such as insulin may be delivered specifically to the lower gastrointestinal tract by delivering it inside of a poly(ethylene-co-maleic acid) hydrogel.[51] The carboxylic acid–functional polymer network swells very little in the highly acidic environment of the stomach, and as a result the drug release in inhibited. In the intestines, where the pH is elevated, the polymer swells more and the drug is released, as shown schematically in Figure 7.3. The choice of polymer depends to some degree on the chemical nature and solubility of the therapeutic agent. A cationic or amine-containing drug may be trapped within the polyelectrolyte gel at low pH and released at high pH as a result of a change in the acid–base interaction.

Drug release from polyelectrolyte gels may also be controlled by means of electrical stimulation of the gel substance. Application of an electrical potential to a polyelectrolyte gel causes small changes in its extent of ionization, which affects the swelling and shrinking of the gel and the permeation rate of substances through the gel.[52] In addition, the charge characteristics of the permeating substance will determine if the substance diffuses toward the anode or cathode of the system.[53] These features may permit the design and application of sophisticated pharmaceutical delivery systems.

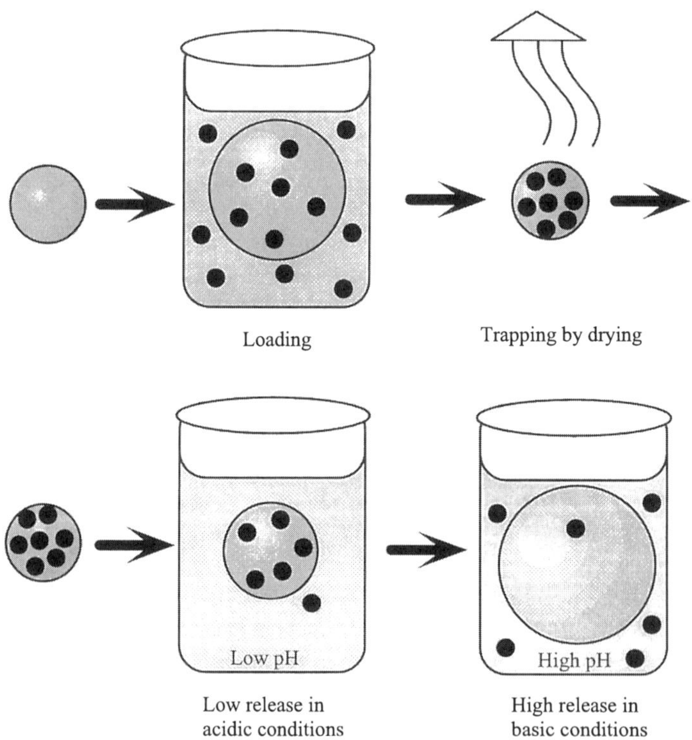

Figure 7.3. A schematic representation of a pH-triggered, controlled-release system. The active ingredient is absorbed into the crosslinked, carboxylic acid–functional polymer, and trapped in the matrix by drying. Release of the active ingredient is controlled by the swelling of the polymer, which occurs mainly at high pH.

7.4. WATER-ABSORBING CONSTRUCTION MATERIALS

Superabsorbent polymers have a number of uses in the construction industry where the control of water is important. One advantage to using superabsorbents in construction is to use the volume increase of the gel to form a barrier to further water flow. Sealing composites made by blending a superabsorbent into a rubber such as chloroprene[54] or a thermoplastic elastomer such as poly(ethylene-co-vinyl acetate)[55] have been developed for sealing around the joints of various building materials. Because of the incompatibility of rubber or elastomer and the superabsorbent polymer, appropriate modification of the interface of the polymer phases is critical to the success of the composite material.[56] Modified superabsorbent polymers, for example, cationic polyelectrolytes or poly(acrylic acid-co-vinyl acetate), may be more compatible with some elastomers. Alternatively, the rubber composite may be prepared by blending the finely divided superabsorbent polymer with isocyanate prepolymers and polyols and then molding the composite rubber into the final desired shape.[57] The sealing composite may be used like mortar between the concrete blocks that make up the walls of the structure. If any gaps were left during construction or created by shifting after construction, the superabsorbent swells in any leaking water to fill any gaps and prevents water from leaking through the joint.[39] Similar formulations are used as swellable gaskets for pipes, where the limited swelling results in a tight seal at the pipe joint. Such a composite was used in construction of the Channel Tunnel between France and England.

A water-blocking construction filler that is composed of cement, water-absorbing polymer, and an asphalt emulsion has also been developed.[58] The components may be mixed at the construction site to quickly form a gelled solid that serves as a backfill material, for example, in tunnel construction to fill the gap between the tunnel liner and the walls of the boring. The principal benefit of the fill material is its higher ductility than conventional construction backfills such as gravel or sand. The developers suggest that this may be beneficial in earthquake zones as a cushioning backfill that would be less susceptible to fracturing during ground movement. The novel filler can be delivered at a lower viscosity, which aids in the complete filling of the space. In addition, the viscosity increases at a lower rate than with conventional tunnel backfills, helping to eliminate voids and the resulting higher propensity of the construction to seep water from the surrounding rock. The solidifying filler also expands somewhat during its curing time, again aiding in the complete filling of the space.

Superabsorbent polymer may also be used as a concrete additive in construction applications. The polymer may be mixed with cement and other materials to form a composite soil stabilizer for wet soils.[59] These composites may be added directly to the soil to absorb and gel any water present and at the same time form a rigid surface upon which to place the foundation footings of a building. The addition of finely divided superabsorbent polymer to concrete formulations also improves the strength and durability of the resulting hardened concrete.[60] Furthermore, the consistency of the concrete slurry can be adjusted by adding superabsorbent polymer. The fluidity of the concrete slurry must be controlled differently depending on the conditions at the construction site, and the addition of superabsorbents to increase the viscosity of the slurry can be easily accomplished on-site.[61] As much as 2 kg superabsorbent polymer per cubic meter of concrete is used, depending on the desired consistency. In another application, concrete patching formulations also may contain superabsorbent polymer to help control the water necessary for proper curing of the patch and optimum bonding of the patch to the old concrete.[62] On a similar theme, concrete surfaces are especially prone to damage caused by poor, water-starved curing conditions. A high evaporation rate of water from the concrete depletes the water needed for optimum curing and strength and can be prevented by applying a water-laden gel sheet to the concrete surface during the cure. The gel provides water to the curing surface only as required for the curing reaction and prevents evaporation of the surface water. The gel layer is strengthened and protected from evaporation by applying a latex rubber coating on top of the gel.[63] Decorative raised-stone or pebbled surfaces on pre-cast concrete panels have also been made with superabsorbent polymer as an aid for easy preparation of the surface.[64] A layer of stone or pebbles intermixed with superabsorbent polymer is placed in a form and overlain with wet concrete. The superabsorbent polymer swells into the spaces between the stones and prevents the concrete from hardening there. When the optimum amount of superabsorbent polymer is used, the stones are only partially embedded in the concrete. After the concrete cures, the gel layer can be easily washed from the surface of the panel, revealing the decorative stone finish. A concrete-protection system has also been disclosed that uses superabsorbent polymer in a sealing layer to protect concrete and the steel reinforcing bars from water damage.[65]

7.5. ELECTRONICS AND CABLING

The swelling property of superabsorbents is also applied to protect communication cables from water damage,[66] using principles similar to those used in the construction sealer applications.[67] Leaking water degrades the performance of fiberoptic communication cables and power transmission cables. Water-blocking tapes prevent intrusion of water into the cables.[68] The flexible, water-blocking tape may be made by mixing a superabsorbent polymer and a polymeric binder and then spreading the mixture onto a nonwoven fabric.[69] Alternatively the superabsorbent is mixed with a rubber such as butyl rubber and solvents and then coated onto a polyester tape.[70] The tape is wrapped around the cable, beneath the plastic covering, as in Figure 7.4. Any water that penetrates the covering of the cable encounters the layer of water-absorbent tape and swells the superabsorbent polymer to seal the crack or perforation in the covering material.[71] Because cables may be exposed to sea water or ground water containing divalent cations, superabsorbents containing sulfonate functional groups, rather than carboxylic acid groups, are preferred for this application.[72]

During assembly, a superabsorbent-containing resin may be dusted into the cable structure to surround the wire or optical fiber bundles inside the protective covering of the cable. With either this approach or the method using a reinforced wrapping tape, water entering the cable as a result of cracks, splits, or cuts to the outer cover-

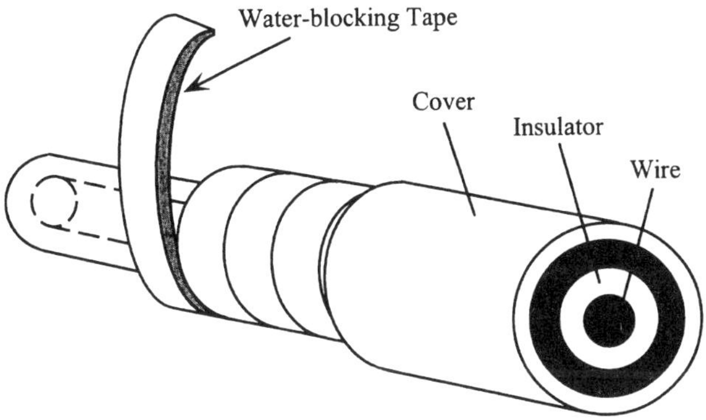

Figure 7.4. Wire protected with a water-blocking tape containing superabsorbent polymer.

ing will cause the superabsorbent to swell, sealing the damaged area and slowing water penetration further into the cable.

7.6. FOOD PACKAGING

Another market for superabsorbent polymer is as a liquid-absorber in food packaging systems. In these systems, the superabsorbent polymer absorbs juice or water from fresh foods such as raw chicken, shellfish, and other meats or from frozen foods as they thaw.[73] Chilled superabsorbent polymer gels can also be used as a dry-cooling medium. The water-swollen gel, contained in a durable plastic bag, is frozen and used to keep perishable foods cold. As the chilled gel thaws, the gel retains the water and its shape, which keeps the cooling medium in place.[74] Composite cooling-sheets also can be made by impregnating a polyester nonwoven sheet with a solution of partially neutralized acrylic acid and crosslinker and polymerizing the absorbent polymer directly on the fiber web.[75] If a temperature-sensitive dye is added to the gel, the temperature change of the composite sheet can be monitored visually. This would be useful in either food storage or in the cold storage and transport of medical tissues and specimens.

In addition to its *liquid*-water absorption characteristic, superabsorbent polymers absorb water from the *vapor* state and therefore can be used to control humidity. The water vapor absorption from humid air is shown in Figure 7.5 for a superabsorbent polymer composed of 68 mol% neutralized poly(acrylic acid) (T.C. Johnson, The Dow Chemical Co., unpublished result). In fact, superabsorbents absorb and release more moisture from the air per mass of absorbent than does silica gel.[39] This property can be used, for example, to prevent damage due to moisture condensing on walls and ceilings in humid buildings. Because powdered superabsorbents are difficult to use directly in these applications, superabsorbent sheets or superabsorbent fibers have been developed. For some uses, such as absorbent pads in meat and poultry packages, the polymer is laminated between two layers of tis-

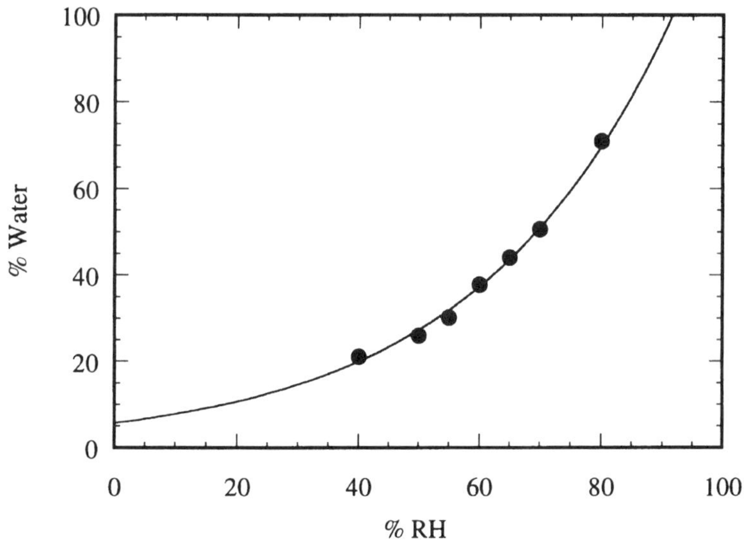

Figure 7.5. Water vapor uptake of superabsorbent polymer at different relative humidities (RH).

sue. Superabsorbent sheets can also be made by polymerizing the monomers directly onto nonwoven fabrics, as with the composite cooling sheets,[76] or by coating a polyester film or nonwoven polyester sheet with the finely milled polymer[77] and adding an adhesive to allow the sheet to stick to any surface. These products help maintain a constant humidity in vegetable and fruit storage buildings and prevent spotting of the produce caused by water dripping from the surfaces of the structure.[39]

7.7. RECREATIONAL ACTIVITIES

There has been some interest in artificial snow for indoor skiing arenas. The fake snow can be made by mixing a superabsorbent polymer with up to 100 times its mass of water and then aerating the gel while freezing it.[78] The artificial snow is maintained with a refrigeration system in indoor ski arenas in Japan, such as an indoor ski center near Tokyo. The frozen gel layer is groomed to yield snow with a realistic feel, similar to "powder" snow.[79] By using superabsorbent polymer in this way, the air temperature in the building can be at least 10°C higher than with snow made from only water, and it is more comfortable to the skiers.

The particle size of the frozen gel and the ratio of water to polymer in the gel determine the quality of the skiing conditions. Other factors affecting the quality of the snow-like material are the particle shape and the rheological properties of the surface of the swollen gel. Swollen superabsorbent polymer spheres coated with fluorocarbon or silicone oils to reduce the coefficient of friction have been disclosed as improved snow-making materials.[80] Presumably the spherical particles impart higher slip as a result of a ball-bearing-like action, and the oils reduce interparticle adhesion of the gel particles. Mixtures of real snow and the gel-snow have better strength properties than does the real snow alone.[81] In addition to the ratio of water to polymer, the crosslink density of the superabsorbent polymer affects the rigidity of the gel particle and the stickiness of the surface. In this regard, the surface-crosslinked superabsorbents described in previous chapters might be an improvement because the particle surfaces would be expected to be much less sticky in this application than gels with the same average crosslink density but with a homogeneous distribution of crosslinks.

7.8. SENSORS

The swellability of superabsorbent polymer gels, their mechanical modulus and rubbery character, and their sensitivity to changes in water content, pH, and ionic strength make them suitable for use in various sensing systems. Sawahata et al.[82] have developed a pressure-sensitive switch based on polyelectrolyte hydrogel. An electrical potential of a few millivolts may be induced in a soft hydrogel by applying mechanical stress to one part of the gel. The potential develops between the stressed and unstressed parts of the gel. The hydrogel used was a co-polymer of acrylic acid and acrylamide, crosslinked with N,N′-methylenebisacrylamide. Small changes in the pH of the gel may be induced by deforming the gel. These pH changes cause an electrical potential to develop as a result of a change in the ionization of the functional groups of the gel. By attaching wires to the gel, the potential difference between deformed gel and undeformed gel generates a signal (subsequently amplified) that lights a photo diode. The intensity of the light emitted by the diode array depends on the magnitude of the mechanical deformation.

In the reverse process, application of a voltage across a polyelectrolyte gel causes a change in volume, which may be applied to perform work on attached objects. This may be used to stimulate hydrogels in applications such as robotics, where soft "fingers" need to be used for gentle grasping.[83] The principles involved in these applications were shown in systems called *artificial muscle* by Kuhn et al.[84] in 1950. Networks formed by reacting poly(acrylic acid) with glycerol or poly(vinyl alcohol) changed shape depending on the degree of ionization, and this behavior could be modified by orientation of the network during the curing reaction. Upon swelling in water the oriented networks contracted along the "stretch" direction and swelled strongly in the perpendicular direction. This behavior also could be applied to perform work on an attached mass, as a muscle contracts to perform work.

Water-sensing devices have been constructed that use the high conductivity of swollen polyelectrolyte gels as the detection switch. The dry polymer does not conduct electricity and corresponds to an open switch, but the swollen polymer conducts electricity and completes the electrical circuit. The magnitude of the resulting gel conductivity indicates the degree of water absorption, for example in a diaper.[85] Another water-sensing system relies on the color change of a substance such as cobalt chloride that can be trapped in the polyelectrolyte gel.[86] As the gel absorbs water, the changing hydration state of cobalt chloride causes a change in color from blue to pink, indicating the degree of saturation of the gel. This product was used to remove water from liquid aromatic hydrocarbons. A third system senses liquid water in frozen foods and thus serves as a temperature indicator for the package.[87] Upon thawing, the water formed is drawn into a laminate that contains both a water-swellable polymer and printed graphics made of water-soluble inks. The water in the indicating laminate causes the graphic to change as a result of the solubility of the ink, thereby alerting the consumer to the thawed state of the product. Simple water-sensing valves have incorporated polyelectrolyte gels to act as valve activators in the presence of water. For example, a valve could operate in a phase-separation process for water and an organic liquid. If the water phase begins to flow through the valve, the polymer swells and blocks the valve toward further flow of water. By proper selection of the crosslink density and extent of ionization of the polymer in such a valve, the valve could activate as the ionic strength of a process brine stream dropped, preventing flow of an aqueous solution with a low electrolyte concentration.

7.9. AQUEOUS WASTE MANAGEMENT

Aqueous waste is a ubiquitous result of life and human endeavor. It may become a difficult problem if the volume is too large, if the contaminants are too hazardous, or if the liquid form itself adds to concerns about spills and possible contamination of groundwater. Absorption and solidification of aqueous wastes can improve their handling characteristics and reduce the potential of environmental contamination as a result of leaks or spills during storage, transport, and final disposal. Aqueous waste mixtures may also be dewatered by means of treatment with a superabsorbent polymer. Various kinds of waste can be treated in this manner, including sewage plant sludge, dredged lake and river bottom sludge,[88] and liquid radioactive wastes.[89] The goal is to make the liquids more handleable by gelling them into a rubbery, solid phase. Containment of spilled medical waste or gelling of the biological liquids to prevent a liquid spill is also possible.[90]

Devices containing superabsorbent polymers that work in concert with surgical suctioning equipment have been developed. In one technique the superabsorbent polymer is contained in a water-soluble bag, which dissolves in the aqueous medical waste to release the polymer.[91] The liquid is then rapidly absorbed, and the waste is solidified.

For industrial waste applications, usually the superabsorbent polymer is part of a formulation that can also contain other viscous thickeners, clays, cements, and the like. The formulation is either added to containers of the waste, or the waste stream and superabsorbent may be mixed and then pumped or poured into a container or mold to make the final shape. Although the principal desirable property of a superabsorbent for this application would seem to be the maximum swelling capacity of the polymer, the goal is frequently to make a rubbery solid that has adequate rigidity to immobilize the waste. This requires a higher crosslink density to ensure that the modulus of the gelled waste is high.

7.10. MISCELLANEOUS APPLICATIONS

Superabsorbent polymers have been used in several cosmetic formulations. A skin-cream emulsion that was prepared with a crosslinked superabsorbent in addition to Carbopol 934 (B.F. Goodrich), a commonly used, branched poly(acrylic acid) thickener for cosmetics, exhibited a longer term moisturizing effect than did the emulsion without the superabsorbent polymer.[92] Finely powdered superabsorbent polymer was also incorporated into a face powder with talc. The powder absorbed perspiration and adhered well to the skin, and hence was long-lasting in application.[93] Superabsorbents have also been used as a component in a gel-form cosmetic face mask.[94] A dental filler has also been disclosed that contains a small amount of superabsorbent polymer. The filler has improved sealing properties when it is used to fill cavities or in root canal therapy.[95] A polyelectrolyte gel formed from the polysaccharide alginic acid has been used as a delivery device for biological cells.[96] The gel was prepared by crosslinking sodium alginate with divalent metal ions in the presence of bovine osteoblasts (bone-forming cells). The novel cell-gel was injected into mice, where it served as a scaffold for early-stage bone formation. The technique shows promise as a way to induce new bone growth at fractures or other sites. While these applications of superabsorbent polymer are certainly in the early stages of development, and may not amount to much volume usage, they give an idea of the broad range of uses for the materials and give an indication of the kinds of developmental effort that may be needed to optimize superabsorbent polymers for specific uses.

REFERENCES

1. B.J. Obenski, *Nonwovens Industry* (November 1994).
2. K. Nishizawa, T. Shirase, H. Mizutani, *U.S. Patent 4,306,559* (1981).
3. R.C. Duncan, N.L. Baker, *U.S. Patent 3,180,335* (1961).
4. B. Harper, R. Bashaw, B. Atkins, *U.S. Patent 3,669,103* (1972).
5. S.R. Kellenberger, *U.S. Patent 5,147,343* (1992).
6. S.R. Kellenberger, W.-H. Shih-Schroeder, A.J. Wisneski, *U.S. Patent 5,149,335* (1992).
7. M.J. Iskra, *U.S. Patent 5,021,050* (1991).
8. R.A. Young, In *Absorbency,* P.K Chatterjee, ed., Elsevier Science Publishers B.V., Amsterdam, p. 231 (1985).
9. H. Nagorski, In *Superabsorbent Polymers, Science and Technology,* Symposium Series 573, F.L. Buchholz, N.A. Peppas, eds., American Chemical Society, Washington, D.C., pp. 99–100 (1994).
10. M. Plischke, M. Schmidt, C.P. Bewick-Sonntag, *Eur. Patent Appl. 640,330* (1995).

11. M.C. Wanek et al., *U.S. Patent 5,294,478* (1994).
12. D.M. Jackson, B.J. Matthews, *U.S. Patent 5,350,370* (1994).
13. M.J. Faulks, D.R.Schlinz, *U.S. Patent 5,356,403* (1994).
14. G.A. Young, G.D. LaVon, G.W. Taylor, *U.S. Patent 5,318,554* (1994).
15. J.R. Noel, N.A. Ahr, *U.S. Patent 5,439,458* (1995).
16. D.V. Phan, P.D. Trokhan, *U.S. Patent 5,451,452* (1995).
17. M.K. Melius et al., *Eur. Patent Appl. 615,736* (1994).
18. D.D. Pratt, *U.S. Patent 5,026,363* (1991).
19. R.S. Yeo, *Br. Patent Appl. 2,284,820* (1995).
20. G.R. Evers, *U.S. Patent 5,300,358* (1994).
21. F.L. Buchholz, In *Absorbent Polymer Technology,* L. Brannon-Peppas, R.S. Harland, eds., Elsevier Science Publishers, New York, pp. 233–247 (1990).
22. S. Goldman, J. Horney, H. Retzsch, *U.S. Patent 5,061,259* (1995).
23. F. Stanley, J. Lamphere, L. Wilson, *WIPO Int. Appl. 92-16,565* (1992).
24. M. Plischke, M. Schmidt, C.P. Bewick-Sonntag, *Eur. Patent Appl. 631,768* (1995).
25. C. Steger et al., *WIPO Int. Appl. 95-00183* (1995).
26. J.R. Noel, A. Ahr, *U.S. Patent 5,439,458* (1995).
27. E.F. Tai, *WIPO Int. Appl. 90-8,789* (1990).
28. C.B. Gartley, *Managing Incontinence,* Jameson Books, Inc., Ottawa, IL (1985).
29. E.W. Merrill, R.W. Pekala, N.A. Mahmud, In *Hydrogels in Medicine and Pharmacy, Vol. 3: Properties and Applications,* N.A. Peppas, ed., CRC Press, Boca Raton, FL pp. 1–16 (1987).
30. J.A. Bosley, V. Farmer, A.A. McKinnon, *Eur. Patent Appl. 9,977* (1980).
31. M. Chmelir et al., *U.S. Patent 4,693,713* (1987).
32. R.S. Harland, R.T. Shet, S.M. Yarbrough, *U.S. Patent 5,241,009* (1994).
33. T. Yamaguchi, S. Tsukakoshi, *Chem. Express,* **7,** 165–168 (1992).
34. K.S. Kazanskii, S.A. Dubrovskii, *Adv. Polym. Sci.,* **104,** 97–133 (1992).
35. T. Furuta, R. Autio, *California Agriculture,* **42,** 19 (1988).
36. J.M. Woodhouse, M.S. Johnson, *J. Arid Environments,* **20,** 375–380 (1991).
37. P.M. Perkins-Veazie et al., *Scientia Horticulturae,* **45,** 183–190 (1991).
38. S. El-Amir et al., *Soil Technol.,* **4,** 177–181 (1991).
39. T. Shimomura, T. Namba, In *Superabsorbent Polymers, Science and Technology,* Symposium Series 573, F.L. Buchholz, N.A. Peppas, eds., American Chemical Society, Washington, D.C., pp. 112–115 (1994).
40. C. Schneider, *Eur. Patent Appl. 380,448* (1990).
41. A. Oguri, T. Sakaki, *Jpn. Kokai Tokkyo Koho 01-287,193* (1989).
42. S.R. Siemer, L.L. Wood, G.J. Calton, *U.S. Patent 5,185,024* (1993).
43. N.A. Peppas, *Hydrogels in Medicine and Pharmacy,* Vols. 1–3, CRC Press, Inc., Boca Raton, FL (1994).
44. H. Brondsted, J. Kopecek, In *Polyelectrolyte Gels, Properties, Preparation and Applications,* R.S. Harland, R.K. Prud'homme, eds., American Chemical Society, Washington, D.C., pp. 285–304 (1992).
45. R. Levy, M.A. Nichols, T.W. Miller Jr., In *Polymeric Delivery Systems,* M.A. El-Nokaly,

D.M. Piatt, B.A. Charpentier, eds., American Chemical Society, Washington, D.C., pp. 202–212 (1993).

46. Sekisui Plastics Co. Ltd., *Jpn. Kokai Tokkyo Koho 03-059,075* (1991).
47. F.H. Otey et al., *J. Agric. Food Chem.*, **32,** 1095–1098 (1984).
48. M. Chmelir, *U.S. Patent 5,264,471* (1993).
49. R.L. Mikkelsen, *Fertilizer Res.*, **38,** 53–59 (1994).
50. M.S. Goldenberg et al., *Eur. Patent Appl. 302,824* (1989).
51. R.Z. Greenley et al., In *Polymeric Drugs and Drug Delivery Systems,* R.L. Dunn, R.M. Ottenbrite, eds., American Chemical Society, Washington, D.C., pp. 213–236 (1991).
52. K. Sawahata et al., *J. Controlled Release,* **14,** 253–62 (1990).
53. T. Yamauchi, E. Kkufuta, Y. Osada, *Polym. Gels Networks,* **1,** 247–55 (1993).
54. T. Tsubakimoto, T. Shimomura, H. Kobayashi, *Jpn. Kokai Tokkyo Koho 62-149,335* (1987).
55. Suetsugu, *Jpn. Kokai Tokkyo Koho 06-157,839* (1994).
56. T. Tsubakimoto, T. Shimomura, H. Kobayashi, *Jpn. Kokai Tokkyo Koho 62-149,335* (1987).
57. M. Mueller, *Ger. Pat. 4,211,302* (1993).
58. A. Moriyoshi, I. Fukai, M. Takeguchi, *Nature,* **344,** 230–232 (1990).
59. Nippon Synthetic Chemicals Ltd., *Jpn. Kokai Tokkyo Koho 06-287,556* (1994).
60. Sekisui House KK, *Jpn. Kokai Tokkyo Koho 63-11,558* (1988).
61. E. Yooda, F. Itoh, *U.S. Patent 5,332,434* (1994).
62. D.I. Haynes, R. E. Hefner, *U.S. Patent 4,611,015* (1986).
63. Onoda Cement Company Ltd., *Jpn. Kokai Tokkyo Koho 62-56,382* (1987).
64. H. Osada, *U.S. Patent 5,171,497* (1992).
65. I. Yamaguchi et al., *U.S. Patent 4,634,626* (1987).
66. K.E. Bow, *Eur. Patent Appl. 24,631* (1981).
67. K. Hogari, F. Ashiya, In *Superabsorbent Polymers: Science and Technology,* F.L. Buchholz, N.A. Peppas, eds., The American Chemical Society, Washington, D.C., pp. 128–140 (1994).
68. H. Polle, *Eur. Patent Appl. 375,685* (1990).
69. H. Kobayashi et al., *WIPO Int. Appl. 88-10,001* (1988).
70. C. Hayashi et al., *Jpn. Kokai Tokkyo Koho 05-314,826* (1993).
71. H. Polle, *WIPO Int. Appl. 88-10,000* (1988).
72. J.J. Sheu, *U.S. Patent 5,163,115* (1992).
73. A.M. DeGouw, J. Prins, J. Dingermans, *Eur. Patent Appl. 68,530* (1983).
74. P. Clerk, *Aust. Patent 8,777,327* (1988).
75. K. Kato, K. Fujitani, T. Fujimoto, *Jpn. Kokai Tokkyo Koho 88-25,455* (1988).
76. P.H. Ericksen et al., *Eur. Patent Appl. 40,087* (1981)
77. Nippon Shokubai Co. Ltd., *Jpn. Kokai Tokkyo Koho 04-140,135* (1992).
78. A. Bucceri, *U.S. Patent 4,742,958* (1988).
79. K. Morioka, S. Nakahigashi, *Refrigeration,* **67,** 28 (1992).
80. T. Kambayashi et al., *Eur. Patent Appl. 620,260* (1994).

81. Y. Miura et al., *Eur. Patent Appl. 440,256* (1991).
82. K. Sawahata, J. P. Gong, Y. Osada, *Macromol. Rapid Commun.*, **16,** 713–716 (1995).
83. T. Kurauchi et al., *Polymer Gels,* D. DeRossi et al., eds., Plenum Press, New York, pp.237–246 (1991).
84. W. Kuhn, B. Hargitay, A. Katchalsky, H. Eisenberg, *Nature,* **165,** 514–516 (1950).
85. Nippondenso Co. Ltd., *Jpn. Kokai Tokkyo Koho 06-300,724* (1994).
86. S. Takemori et al., *Jpn. Kokai Tokkyo Koho 01-201,364* (1989).
87. M. Shigemitsu et al., *Jpn. Kokai Tokkyo Koho 04-313,033* (1992).
88. M. Kawai, *Jpn. Kokai Tokkyo Koho 06-339,698* (1994).
89. T. Seto, R. Yoshikawa, A. Komatsu, *Jpn. Kokai Tokkyo Koho 06-238,261* (1994).
90. M. Karakelle et al., *Eur. Patent Appl. 440,962* (1991).
91. Y. Fujiura, K. Mita, K. Shinoda, *Eur. Patent Appl. 494,599* (1992).
92. J.M. Kendall, D.H. Maes, R. Figueroa Jr., *Eur. Patent Appl. 281,395* (1988).
93. T. Sano, Y. Suzuki, *Jpn. Kokai Tokkyo Koho 63-190,811* (1988).
94. L. Ayache, *Eur. Patent Appl. 309,309* (1989).
95. N. Hatsutori, O. Iwamoto, *Jpn. Kokai Tokkyo Koho 06-135,813* (1994).
96. L. Griffith-Cima et al., *WIPO Int. Appl. 94-25,080* (1994).

INDEX